Aqueous Organometallic Chemistry and Catalysis

NATO ASI Series

Advanced Science Institutes Series

A Series presenting the results of activities sponsored by the NATO Science Committee, which aims at the dissemination of advanced scientific and technological knowledge, with a view to strengthening links between scientific communities.

The Series is published by an international board of publishers in conjunction with the NATO Scientific Affairs Division

A Life Sciences **B Physics**	Plenum Publishing Corporation London and New York
C Mathematical and Physical Sciences **D Behavioural and Social Sciences** **E Applied Sciences**	Kluwer Academic Publishers Dordrecht, Boston and London
F Computer and Systems Sciences **G Ecological Sciences** **H Cell Biology** **I Global Environmental Change**	Springer-Verlag Berlin, Heidelberg, New York, London, Paris and Tokyo

PARTNERSHIP SUB-SERIES

1. Disarmament Technologies	Kluwer Academic Publishers
2. Environment	Springer-Verlag / Kluwer Academic Publishers
3. High Technology	Kluwer Academic Publishers
4. Science and Technology Policy	Kluwer Academic Publishers
5. Computer Networking	Kluwer Academic Publishers

The Partnership Sub-Series incorporates activities undertaken in collaboration with NATO's Cooperation Partners, the countries of the CIS and Central and Eastern Europe, in Priority Areas of concern to those countries.

NATO-PCO-DATA BASE

The electronic index to the NATO ASI Series provides full bibliographical references (with keywords and/or abstracts) to more than 50000 contributions from international scientists published in all sections of the NATO ASI Series.
Access to the NATO-PCO-DATA BASE is possible in two ways:

– via online FILE 128 (NATO-PCO-DATA BASE) hosted by ESRIN, Via Galileo Galilei, I-00044 Frascati, Italy.

– via CD-ROM "NATO-PCO-DATA BASE" with user-friendly retrieval software in English, French and German (© WTV GmbH and DATAWARE Technologies Inc. 1989).

The CD-ROM can be ordered through any member of the Board of Publishers or through NATO-PCO, Overijse, Belgium.

Series 3: High Technology – Vol. 5

Aqueous Organometallic Chemistry and Catalysis

edited by

István T. Horváth

Corporate Research Laboratories,
Exxon Research and Engineering Company,
Annandale, New Jersey, U.S.A.

and

Ferenc Joó

Institute of Physical Chemistry,
Lajos Kossuth University,
Debrecen, Hungary

Kluwer Academic Publishers

Dordrecht / Boston / London

Published in cooperation with NATO Scientific Affairs Division

Proceedings of the NATO Advanced Research Workshop on
Aqueous Organometallic Chemistry and Catalysis
Debrecen, Hungary
August 29–September 1, 1994

A C.I.P. Catalogue record for this book is available from the Library of Congress.

ISBN 0-7923-3703-4

Published by Kluwer Academic Publishers,
P.O. Box 17, 3300 AA Dordrecht, The Netherlands.

Kluwer Academic Publishers incorporates the publishing programmes of
D. Reidel, Martinus Nijhoff, Dr W. Junk and MTP Press.

Sold and distributed in the U.S.A. and Canada
by Kluwer Academic Publishers,
101 Philip Drive, Norwell, MA 02061, U.S.A.

In all other countries, sold and distributed
by Kluwer Academic Publishers Group,
P.O. Box 322, 3300 AH Dordrecht, The Netherlands.

Printed on acid-free paper

Printed in the Netherlands

TABLE OF CONTENTS

PREFACE

The development of *high-tech catalytic processes* based on organometallic catalysts requires an in-depth understanding of how to control structure and reactivity in the catalytic materials on the molecular level. Rapid advances in organometallic chemistry over the last twenty years have allowed us to design and prepare very sophisticated molecules, many of which are catalysts of important processes. However, the separation of such homogeneous catalysts from products remains a limiting factor in large scale application. It is now generally accepted that the use of aqueous catalysts systems is a logical approach that could overcome of the separation problems by allowing the phase separation of the products from the catalyst phase as an inherent feature of the medium.

The design and synthesis of aqueous transition-metal catalysts requires broad knowledge of organometallic chemistry in water as well as of physical and chemical properties of water itself. Water is now regarded as a unique solvent for certain stoichiometric and catalytic reactions involving organometallic compounds. Because of its highly polar, protic nature water strongly influences the acid-base behavior of solutes, the formation and disruption of ion-pairs and hydrogen bonds, and the extent of hydrophobic interactions. These effects often lead to unexpected reaction rates and selectivities, but the phenomena are still not fully understood.

A very important practical and environmentally beneficial aspect of the use of a separate aqueous phase in catalytic reactions producing water-insoluble products is the potential for relatively easy and complete recovery of water soluble catalysts. This could lead to the elimination of further steps for removal of traces of heavy metals from the product and to considerable savings.

When we envisioned this workshop, we felt that a specialized meeting was timely, and should aid in summarizing previous findings and facilitate further progress. Many of the major contributors in this field have already individually established personal contacts. However, bringing together most of the eminent scientists in such a diverse field would further strengthen these contacts and should permit the establishment of new relationships and new patterns of thinking.

István T. Horváth

Ferenc Joó

ACKNOWLEDGMENTS

We thank the Scientific and Environmental Affairs Division of NATO for the sponsorship of the workshop.

We are especially grateful to the following companies and institutions for their financial contributions:

Exxon Research and Engineering Company

F. Hoffmann-LaRoche Ltd.

Kossuth Lajos Tudományegyetem

Országos Müszaki Fejlesztési Bizottság

Környezetvédelmi és Területfejlesztési Minisztérium

Magyar Tudományos Akadémia

Kersekedelmi Bank Rt., Universitas Alapitvány

Vegyész Alapitvány

We thank Professor Jean-Mary Basset, Ecole Supérieure de Chimie Industrielle de Lyon, France, Professor Robert H. Grubbs, California Institute of Technology, Pasadena, USA, and Professor Wolfgang A. Herrmann, Technische Universität München, Germany, members of the organizing committee, for their encouragement and support.

Finally, this workshop would not have been possible without the assistance and encouragement of colleagues at Lajos Kossuth University and Exxon Research and Engineering Company too numerous to mention, and we are grateful to each and every one of them.

INTRODUCTORY REMARKS: NATO ADVANCED RESEARCH WORKSHOP ON AQUEOUS ORGANOMETALLIC CHEMISTRY AND CATALYSIS IN DEBRECEN, HUNGARY

ISTVÁN T. HORVÁTH
Corporate Research Laboratories
Exxon Research and Engineering Company
Annandale, NJ 08801, USA

The dramatic political changes in Central and Eastern Europe in 1989 and 1990 have captured my heart and attention, as I grew up in Buda, the hilly part of Budapest, Hungary. Like many other Hungarian scientists living and working in the west at those exciting times, I felt that I should be doing something to accelerate the change and helping to find the right directions in the sometimes messy democratic processes.

The idea of organizing a NATO Advanced Research Workshop came up when Professor Ferenc Joó (Lajos Kossuth University, Debrecen, Hungary) visited me at Exxon in the spring of 1991. Besides our long time friendship and common interest in organometallic chemistry and homogeneous catalysis, we were both studying aqueous homogeneous catalyst systems at the time. Scientists in Debrecen such as Professors Mihály T. Beck, Ferenc Joó, Zoltán Tóth, and their colleagues were among the pioneers of aqueous organometallic catalysis [1], and some recollections of the early research in this field can be found in the next few pages. I became interested in the mechanisms of aqueous organometallic catalysis at the 1987 Königstein Conference organized by Professor Wolfgang A. Herrmann (Technische Universität der München, Germany) [2]. Professor Boy Cornils (Hoechst AG, Frankfurt, Germany) delivered there a stimulating lecture on the commercialization of the water soluble phosphine modified rhodium hydroformylation catalyst system, originally developed by Dr. Emil Kuntz [3] at Rhône Poulenc. In the same year, I started to work on the application of high pressure NMR in organometallic chemistry and catalysis, and did the first high pressure NMR investigation of the mechanisms of the Rhône-Poulenc/Rurhchemie process [4].

I hope that bringing the NATO Advanced Research Workshop on Aqueous Organometallic Chemistry and Catalysis to Debrecen is seen as a clear recognition of the leadership role of the scientists of the Institute of Physical Chemistry of the Lajos Kossuth University. The fact that so many eminent scientists gathered here to present and discuss their latest results will undoubtedly stimulate further research, lead to scientific collaborations and last, but not least, will help in giving more recognition of quality science and in strengthening the positions of Hungarian research in the international scientific arena. By making the organization of this meeting (one of the very first ARW-s in Hungary) possible, NATO and the other foreign and domestic sponsors contribute a significant piece both to the advancement of top science on the international level and to the long awaited changes of the Hungarian scientific scene.

References

1. Joó, F. and Tóth, Z. (1980) Catalysis by water-soluble phosphine complexes of transition metals in aqueous and two-phase media, J. Mol. Catal., 8, 369-383.

2. Haggin, J. (1987) Chemical & Engineering News, October 5, 31-44.

3. Kuntz, E.G. (1987) Homogeneous catalysis ... in water, CHEMTECH, 17, 570-575.

4. Horváth, I. T., Kastrup, R. V., Oswald, A. A. Mozeleski, E. J. (1989) High-pressure NMR studies of the water-soluble rhodium hydroformylation system, Catal. Lett., 2, 85-90.

INTRODUCTORY REMARKS: THE CATALYTIC PROPERTIES OF WATER SOLUBLE PHOSPHINE COMPLEXES

M.T. BECK
Institute of Physical Chemistry, Lajos Kossuth University
Debrecen 10, P.O. Box 7, H-4010 Hungary

I am greatly honored to have the opportunity to say a few words about the very beginning of a work which lead to the formation of a new and flourishing field of fundamental research and its practical applications. The best evidence for the importance of this rapidly developing field is just this workshop.

In 1960/61 I spent one semester in the laboratory of Professor Jannik Bjerrum at the University of Copenhagen where I mainly dealt with the determination of stability constants of mixed ligand complexes of nickel(II). That time this lab was the Mecca, or at least one of the few Meccas, of chemists dealing with the study of coordination compounds. During my stay Dr. Graham Wright of New Zealand studied the stepwise formation of bismuth(III) complexes with diphenylphosphinobenzene-m-sulphonate ligand [1]. This water soluble phosphine (TPPMS) was prepared a few years earlier by Chatt and coworkers who pioneered the thermodynamic studies of the complexes of this intriguing ligand with *class b* metal ions [2]. It occurred to me that time or a little later that it would be very interesting to study the catalytic properties of complexes of transition metal ions with TPPMS. In 1968 I was appointed to the Lajos Kossuth University as professor of physical chemistry which greatly expanded my research possibilities. I was particularly interested in the activation of small molecules by coordination. The now classical papers of Vaska and Wilkinson on the catalytic properties of transition metal complexes with triphenylphosphine were published in the early sixties. Since one of the aims of those studies was to find models for metallo-enzymes, the insolubility of PPh_3 complexes in aqueous media meant a serious limitation. It was quite plausible for me to achieve the solubility of these complexes in water by using TPPMS instead of PPh_3. For several reasons, however, I did not start these studies until 1971 when a brilliant student joined my team to prepare his M.S. thesis.

It was found soon that some of the Ru-complexes showed remarkable hydrogenation activity in aqueous solutions. We were sure that the Thesis and the jointly prepared paper will be greeted by the referees. Indeed, our preliminary communication was published in 1973 in the Magyar Kémiai Folyóirat (Hungarian Journal of Chemistry) [3], abstracted by the Chemical Abstracts. Before sending a manuscript to the Chemical Communications, I asked the opinion of Professor Chatt whom I already knew quite well and whose contributions in coordination chemistry I highly appreciated. Although, he found our results on the catalytic effect of the complexes interesting, he did not encourage us to publish the results at that stage and suggested to delay the publication until obtaining well defined and spectroscopically well characterized complexes [4]. Since we needed considerable time to achieve these aims, we sent the manuscript of a short paper to the Reaction Kinetics and Catalysis Letters where it was published [5]. We were glad to find that later this paper was found interesting enough to be referred to in the Heart Cut column of CHEMTECH [6]. It may be added that Professor Chatt himself studied the catalytic effects of transition metal phosphine complexes in water, but for solubility reasons they used $P(CH_2OH)_3$. He was kind enough to send us the preprint of this paper which was then soon published [7]. Unfortunately for Professor Chatt, in aqueous solutions their well defined and well soluble complexes "showed no obviously useful catalytic properties" [7].

I believe that most of the scientists regard their research and the publication of their results as a kind of adventure, and our work contributing to the initiation of this exciting field was quite adventurous, indeed.

References

1. Wright, G. and Bjerrum, J. (1962) Studies on metal ion coordination in solution II. The complex formation of diphenylphosphinobenzene-m-sulphonate with bismuth, *Acta Chem. Scand.* **16**, 1262-1270
2. This ligand was first abbreviated as *Dpm*; Ahrland, S., Chatt, J., Davies, N.R. and Williams, A.A. (1958) The relative affinities of co-ordinating atoms for silver ion. Part II. Nitrogen, phosphorus and arsenic, *J. Chem. Soc.*, 276-288
3. Joó, F. and Beck, M. (1973) Redox reactions of phosphine complexes in aqueous solution - new possibilities in homogeneous solution catalysis (*in Hungarian*), *Magy. Kém. Folyóirat* **79**, 189-191; *C.A.* (1973) **73**, 167966b
4. Chatt, J., *personal communication* (letter of March 2, 1973)
5. Joó, F. and Beck, M.T. (1975) Formation and catalytic properties of water-soluble phosphine complexes, *React. Kin. Catal. Let.* **2**, 257-263
6. Hughes, R. (1978) Heart Cut, CHEMTECH **8**, 43
7. Chatt, J., Leigh, G.J. and Slade, M. (1973) Rhodium(I), rhodium(III), palladium(II), and platinum (II) complexes containing ligands of the type PR_nQ_{3-n} (n=0, 1, or 2; R= Me, Et, Bu^t, or Ph; Q=CH_2OCOMe or CH_2OH), *J.C.S. Dalton Trans.*, 2021-2028

EARLY WORK ON CATALYSIS BY WATER SOLUBLE TRANSITION METAL PHOSPHINE COMPLEXES

F. JOÓ
Institute of Physical Chemistry, Lajos Kossuth University
Debrecen 10, P.O. Box 7, H-4010 Hungary

In 1970, as a 3rd year undergraduate I was asked by Professor Mihály T. Beck (then head of the Institute of Physical Chemistry at the Debrecen University, later the supervisor of my M.S. thesis work) to try the synthesis of analogs of some better known transition metal triphenylphosphine compounds with *mono*-sulfonated triphenylphosphine in the hope of obtaining complexes with useful catalytic properties in aqueous solutions. By that time the superb hydrogenation ability of $RhCl(PPh_3)_3$ *(Wilkinson's catalyst)* and $HRuCl(PPh_3)_3$ was already widely known as was the oxygen carrying property of *trans*-$IrCl(CO)(PPh_3)_2$ *(Vaska's complex)*. In that flourishing era of bioinorganic chemistry we were much interested in catalytic fixation of biologically important small molecules (e.g. N_2 and CO_2) and realized that a proper synthetic model should be based on aqueous chemistry.

Since in our Institute we had only limited possibilities for synthesis and catalytic studies under anaerobic conditions our first experiments were rather simple and were designed to demonstrate *qualitatively* whether the *mono*-sulfonated triphenylphosphine (abbreviated nowadays as TPPMS) was able to stabilize lower oxidation states of transition metal ions in aqueous solutions. Catalytic hydrogenation of olefinic and 2-oxo-carboxylic acids with an *in situ* Ru/TPPMS catalyst system was also studied. Later, with the help of Professor László Markó and his colleagues at the Veszprém University, improved synthesis of $RhCl(TPPMS)_3$, $RuCl_2(TPPMS)_2$ and a few other complexes was also completed. I have summarized the early results in my M.S. Thesis [1] submitted at the Lajos Kossuth University, Debrecen in 1972. A paragraph of its summary reads:

"In the systems where we studied the catalytic activity of these complexes only the activation of molecular hydrogen was investigated. It is, however, probable, that these compounds will also catalyze other reactions (such as oxidation, hydroformylation, hydration and isomerization). The transition metal-sulfonated triphenylphosphine complexes can be catalysts for the fixation of dinitrogen, too. Furthermore, in the future asymmetric syntheses might be made possible by using catalysts containing asymmetric, water soluble sulfonated phosphine ligands."

The referee of the thesis was not completely convinced, writing: "The theories derived therein are no more than mere hypotheses. No experimental evidence exists to support the suggestions made in the Thesis summary with regard to the possible applications of the complexes under investigation. Realistically, these ideas are nothing more than the fruits of a fertile imagination, belonging to a very enthusiastic young man. It is unfortunate that the lack of necessary and sufficient evidence did not contain this otherwise essential imagination during the analysis of the results."

To be fair, the referee was correct in most of his factual remarks. On the other hand this quote characteristically shows the general reception of the idea of working with organometallic catalysts in aqueous solutions. The attitude changed only later, but then dramatically, with the successful introduction of biphasic hydroformylation [2,3] into the industrial scene. Nowadays one can find an aqueous version of almost all the reactions which previously had been found important in organometallic catalysis [4]. I feel personally indebted to all those who participate in this workshop, present their latest results and work together on pointing out further directions of development in this fascinating field of chemistry.

This is another story already. However, I still encourage enthusiastic young men and women to stretch their imagination to the maximum extent when exploring an exciting new field of chemistry such as aqueous organometallic catalysis. But look for the necessary and sufficient evidence!

References

1. Joó, F. (1972) Investigation of transition metal complexes containing sulfonated triphenylphosphine. Attempts of carbon dioxide fixation in aqueous solutions under mild conditions, *M.S. Thesis*, L. Kossuth University, Debrecen, Hungary
2. Kuntz, E.G. (1987) Homogeneous catalysis... *in water*, *CHEMTECH* **17**, 570-575
3. Bach, H., Gick, W., Wiebus, E. and Cornils, B. (1984) Continous hydroformylation with water-soluble rhodium catalysts, *Proc. Int. Congress Catal.* (Berlin, 1984) V-417-427
4. Herrmann, W.A. and Kohlpaintner, C.W. (1993) Water-soluble ligands, metal complexes and catalysts: synergism of homogeneous and heterogeneous catalysis, *Angew. Chem. Int. Ed. Engl.* **32**, 1524-1544

SUMMARY OF THE ROUND TABLE DISCUSSION OF THE STATE OF THE ART AND FUTURE DIRECTIONS OF AQUEOUS ORGANOMETALLIC CHEMISTRY AND CATALYSIS

ISTVÁN T. HORVÁTH
Corporate Research Laboratories
Exxon Research and Engineering Company
Annandale, NJ 08801, USA

FERENC JOÓ
Institute of Physical Chemistry
Lajos Kossuth University
P.O.Box 7, H-4010 Debrecen, Hungary

1. Introduction

The state of the art and the future directions of aqueous organometallic chemistry and catalysis was critically evaluated and summarized during a round table discussion of the last day of the NATO Advanced Research Workshop on Aqueous Organometallic Chemistry and Catalysis, August 29 - September 1, 1994, Lajos Kossuth University, Debrecen, Hungary. Several small groups have been assembled to address various aspects of the subject matter of the workshop. Members of the groups worked together during the workshop. Each group elected a spoke-person who communicated orally their evaluation and suggestions to the rest of the workshop. This summary is based on these reports and the minutes taken during the discussion.

2. Organic synthesis in water

There is an increasing interest in using water as a solvent for various organic reactions [1,2]. One of the important findings that generated such an interest is the observed acceleration and modified selectivities in certain Diels-Alder reactions [3]. These effects are probably due to hydrophobic interactions which, to some extent, also govern the catalytic activity of metallo-enzymes in aqueous environments. In some cases the use of water eliminates the need of protection and deprotection of polar functions of the reactants. However, the primary role of water as a solvent in organic synthesis is the separation of reactants, catalysts and products into

I. T. Horváth and F. Joó (eds.), Aqueous Organometallic Chemistry and Catalysis, 1–13.

immiscible phases. In many cases the main advantage of such a phase separation is the easy isolation of a reactive product or recovery of an expensive catalyst [4]. In addition, there are examples where such a phase separation eliminates side reactions or substrate inhibition [5]. More importantly, there are also examples when the catalyst is formed exclusively in water. Aqueous organic synthesis can be carried out either by using water soluble catalysts [6] or in the presence of surfactants [7] and phase transfer agents [8]. Although the majority of investigations is still focused on activation of small molecules (hydrogenation, hydroformylation, carbonylation, hydrocyanation, oxygenation etc.), other types of reactions (isomerization, alkylation [9], coupling, oligomerization and polymerization [10] etc.) are gaining wider interest. Homochiral synthesis is facilitated by the preparation and use of water soluble chiral ligands [11], primarily tertiary phosphines, and catalysts as exemplified by asymmetric hydrogenations [12], epoxide hydrogenolysis [13], and epoxidation and dihydroxylation of alkenes in homogeneous aqueous solutions or in two phase systems.

In addition to devising successful and ingenious methods of catalyst recovery future research is needed in this area to explore the unique role of water in determining catalyst activity and selectivity. Extensive fundamental, mostly physicochemical studies are required to relate changes in selectivity to intrinsic properties of water (ionizing power, H-bonding ability, triggering of hydrophobic interactions) which may influence properties of both the catalyst and reactants. In some cases the functionalization of the ligands (e.g., sulfonation of tertiary phosphines) bestows upon them properties (e.g., surfactant activity [14]) which are completely unknown in organic media, but simple introduction of large substituents may change important ligand properties, as exemplified by the comparison of the Tolman cone angle of PPh_3 and TPPTS [15]. The quest for new and properly modified ligands and catalysts will continue in order to devise synthetic procedures best suited to water; oxidations with aqueous H_2O_2 and polymerization can be such processes. Environmentally benign synthetic processes may also favor water as solvent [16] though this question is not completely free of ambiguity. Further efforts will be made to explore the use of supercritical and superheated water as solvent for organometallic catalysis [17,18].

3. Synthesis of aqueous organometallic catalysts

The most frequently used approach to obtain water soluble organometallic catalysts has been the attachment of hydrophilic polar (ionic) groups to known ligands [19,20]. A variety of water soluble tertiary phosphines, bearing sulfonate, carboxylate, phosphonate or ammonium groups have been prepared. To a lesser extent ligands with polyoxyethylene chains of various length were also investigated. Although, direct sulfonation of tertiary arylphosphines is a one-pot procedure, it often leads to a complex product mixture especially in the case of chelating diphosphines. However, observing the sulfonation by NMR [21] or HPLC [22] techniques allows the preparation of well characterized compounds. In these reactions the sulfonate groups are introduced at the meta-position with respect to the phosphorus atom. It should be noted that the sulfonate groups may interfere with the coordination chemistry of the phosphine itself. Synthetic procedures have been developed for tertiary phosphines with ionic substituents in para-position in order to minimize such interference [12,23,24]. Omega-sulfoalkylphosphines have been prepared [25]. In certain cases non-ionic ligands are also soluble in water (e.g. 1,3,5-triaza-7-phosphaadamantane [26], $P(CH_2OH)_3$ [27] and phosphines with polyethylene glycol ether chains [28]). Transition metal complexes with highly basic ligands may dissolve well in aqueous solvents, usually due to the formation of ionic metal complex species [29]. Metals with simple ligands, i.e. aqua complexes like $[Ru(H_2O)_6]^{2+}$ are useful not only as precursors to more complex catalysts [30] but as catalytically active species themselves [10].

Further synthesis based on these approaches will undoubtedly increase the number of water soluble ligands. However, more attention should be paid to the deliberate positioning of polar substituents on the parent ligand structure. In order to mimic known non-aqueous catalysts the polar groups should be the furthest away from the metal center. On the other hand new reactivity may be explored by bringing polar groups into the vicinity of the metal center. The development of such systems may require total synthesis of novel ligands. Non-ionic derivatives may gain more importance and oxidation resistant ligands are clearly desirable to extend the scope of aqueous organometallic catalysis. Catalyst characterization is important in order to understand the mechanisms of catalytic transformations. In this respect FTIR spectroscopy of aqueous solutions using CIRCLE cells or reactors [31] is a valuable addition to multinuclear, often high pressure NMR [32]. It should be emphasized, that acid-base behavior of both the catalyst and reactants may strongly influence the outcome of a catalytic reaction and therefore should be thoroughly studied. Finally, the concept of supported aqueous phase catalysts (SAPC-s) [33,34] will be worked out in further detail and deposition of water soluble catalysts on hydrophilic gels, dextrans, etc. is anticipated.

4. Biphasic and phase transfer catalysis

An intrinsic limitation of almost all biphasic reactions is in the usually low rates. The reaction takes place either at the interface of the two phases, which is normally rather small, or inside one of the bulk phases, but then solubility problems arise. In many cases mass transfer between the phases can be facilitated by appropriate phase transfer agents (catalysts) and this approach will be further pursued especially in reactions of hydrophobic substrates. Use of micelle forming agents allows investigation of microheterogeneous systems and pronounced effects on oligomerization, polymerization reactions, as well as on enantioselection in asymmetric hydrogenation have been reported [35]. Selective molecular hosts (e.g., cyclodextrins [36]) were already used for bringing substrates from the organic phase to the aqueous one. In some cases an appropriately substituted phosphine, like the one bearing a crown-ether substituent [37], can itself act as a phase transfer catalyst. Liposomes, as micro-heterogeneous biphasic systems, are capable of imposing selectivity in hydroformylation [38] of structurally closely related substrates and this phenomenon deserves further investigation.

Future research may center around the increased use of physical techniques for agitation and separation. An unusual approach for catalyst recovery in the aqueous phase can be the phase separation by freezing the aqueous phase. Ultrasound irradiation is useful for enlarging the interface between the phases and inherent mixing. In continuous systems membranes may be applied for holding the aqueous solution of the catalyst inside the reactor while permitting the organics to pass through. Reactions within micelles will be undoubtedly further investigated, including photocatalytic reactions and polymerizations. Micelles and lyotropic liquid crystalline phases can be useful for directing substrates towards the catalytically active metal center via organized structures.

The idea of biphasic catalytic processes preceeded the advent of aqueous organometallic catalysis [39]. Recent results also stimulate the critical evaluation of alternatives to water as a second liquid phase. A very important generalization of biphasic catalysis is the fluorous biphase concept [40]: transition metal catalysts with appropriately fluorinated ligands are well soluble in perfluoroalkanes which do not mix with solutions of substrates in common organic solvents at room temperature. In this field ligand synthesis is of key importance to widespread application. Moreover, since perfluoroalkanes have many unique properties as solvents, fundamental characterization of the behavior of catalysts and other solutes in such media is necessary to devise useful catalytic systems. Data and discoveries from research on artificial blood (a fluorous mixture itself) should be systematically evaluated and transferred to fluorous organometallic catalysis where applicable.

5. Asymmetric organometallic chemistry and catalysis

Catalytic asymmetric synthesis in aqueous media has been fundamentally advanced by successful modification of known chiral ligands (BINAP, CHIRAPHOS, BDPP, etc.) via functionalization by sulfonate [41] or quaternary ammonium groups. Even the water soluble derivatives of acid sensitive molecules like DIOP could be prepared by introducing quaternary ammonium substituents [42]. The hydrophilic analogs of well known enantioselective rhodium and ruthenium catalysts were mainly used in the reduction of dehydroaminoacid derivatives [11,41], imines [43] and hydrogenolysis of epoxides [13]. In several cases very high enantioselectivities, close to the ones observed in the analogous non-aqueous systems, have been achieved [11]. However, as a general rule, enantioselectivity was found to be lower in aqueous systems than in organic solutions [44]. Addition of micelle forming agents to aqueous systems increase the rate and enantioselectivity of hydrogenation of dehydroaminoacids [35]. An important feature of these biphasic reactions is that the catalysts could be recycled in the aqueous phase with no or marginal loss in enantioselecitvity. A few examples of enantioselective reductions by transfer hydrogenation are also known. Although ionic ligands may be easily anchored to ion exchange resins, this method of preparing supported chiral catalysts was only scarcely studied [42, 45].

Further work is needed in the preparation of new chiral ligands, both via the modification of known water-insoluble molecules and by making use of water soluble, chiral natural substances as starting materials. Functional groups others than $-SO_3H$ and $-NR_3$ should also be considered. A phosphonate group, far from the actual donor atom in the ligand would be useful for monitoring the general environment of the catalyst by ^{31}P NMR independent of the NMR effects accompanying the changes in the metal coordination sphere. Further progress is required in the preparation of water-soluble analogs of BINAP and of ligands with C1 symmetry. Many of the water soluble ligands are surface active and their micelle forming ability should be studied as well as the effect of micellar agents on reaction rates and enantioselectivities. Water soluble dendrimers can be vehicles for immobilization of chiral catalysts. As in non-aqueous systems, extending the scope of asymmetric syntheses and moving from enantioselective hydrogenation towards enantioselective hydroformylation, C-H activation, C-C coupling, kinetic resolution and others will be emphasized. Selective cleavage of aryl-chloride bonds can be a method for C-C bond formation. Asymmetric hydroformylation in water may be hampered by problems of aldehyde racemization and the use of supported catalysts, like SAPC-s, can be useful in preventing such problems. Until now no attention has been paid to enantioselective isomerizations in water. Extension of the fluorous biphase concept to homochiral synthesis will require the preparation of fluorous-soluble ligands and catalysts. In general, since enantioselection relies on thermodynamics of ligand and substrate binding and kinetics of the successive steps of the reaction, basic research into the kinetic behavior of chiral catalysts and complete kinetic characterization of enantioselective processes will remain of paramount importance.

6. Biochemical applications

A traditional field of organometallic catalysis is the activation of biologically important small molecules (O_2, CO_2, N_2 etc). Recently more attempts has been made to perform such reactions in aqueous media [46,47] under biologically acceptable conditions to construct working analogs of enzymatic systems. A closely related area is the study of organometallic catalysis taking place in enzyme catalyzed reactions (hydrogenase reactions [48], biological methylation [49], methane formation, etc.). Biomimetic organometallic catalysis relies heavily on molecular recognition in aqueous solutions the mechanism of which is widely different of that observed in organic solvents. Catalytic hydrogenation of unsaturated lipids in biomembranes in model systems and in living cells is of far reaching importance [50,51]. Deuteration of membrane lipids either using D_2 or D_2O offers a unique way of characterization of specific membrane domains [52]. Obviously, such modifications take place in aqueous media and a variety of investigations was carried out both with soluble and with supported catalysts. Due to the enormously complex architecture of living systems very high degree of selectivity is required including spatial (topological) selectivity [53]. Traditionally, the use of metal complex compounds as anticancer agents or as other remedies was studied by bioinorganic and biocoordination chemistry. Much is still left to be done in the area of bioorganometallic chemistry as exemplified by the binding of DNA to surfaces using (eta5-pentamethylcyclopentadienyl)rhodium complexes [54] and by preparation of haptens for monoclonal antibody generation of catalysts.

Future research will focus on further aspects of all these topics. In modification of living cells selectivity will remain the central and crucial question and will require means of targeting and following the catalyst inside the cells; fluorescent and spin labelled catalysts will serve as useful probes for topological distribution. In order to carry out local membrane modifications in higher organisms, reactions such as hydrogen transfer from biocompatible donor molecules (e.g. ascorbic acid) have to be developed. Reactions of membrane lipids other than hydrogenation should also be considered. Effects of modification of biomembranes in living cells will be followed more extensively by techniques of molecular biology to establish the connection between the physical state of cell membranes and physiological functions of the cell. Mechanistic study of the pharmaceutical effects of anticancer, antirheumatic, antiarthritic etc. compounds will gain much from investigations on model organometallic systems. Research into fundamental questions of enzymatic processes and construction of more elaborate enzyme models will remain of prime importance within aqueous organometallic catalysis. Reactions of biomolecules (such as DNA) with organometallic compounds will be further studied in order to devise probes and catalysts for binding, fluorescent, radioimmune or other kinds of assay.

7. Mechanistic implications of water as a solvent

Several phenomena draw attention to the effects of water on reaction mechanisms. The consequences of such mechanistic implications are seen in all the preceding paragraphs. Pronounced salt effects have been observed in hydrogenation of aldehydes in aqueous and aldehyde biphasic systems [55]. The dehydrohalogenation of metal hydride complexes favors formation of monohydrides vs. dihydrides [e.g. $HRh(TPPMS)_3$ vs. $H_2RhCl(TPPMS)_3$] in aqueous solutions has also been demonstrated [56]. Water facilitates reactions which are of marginal importance in aprotic solvents; examples are the reactions of TPPMS with activated olefins [57,58] and aldehydes [59] even if the phosphine is originally bound in a metal complex. Though the mechanism of some important reactions was investigated in detail [60] in general there is still a lack of fundamental mechanistic studies of aqueous organometallic reactions. A special question is that of the solubility of reactants (e.g. gases) which may result in changes of mechanism. It is most important therefore to establish the region where the reaction actually takes place (one or both of the bulk phases or the interface). Again, the surfactant behavior of functionalized ligands (such as of sulfonated tertiary phosphines) may influence the course of reactions by causing aggregation of catalyst molecules and solubilization of lipophilic substrates.

Future research will undoubtedly bring more fundamental investigations of reaction parameters: determination of binding constants of ligands and reactants, rate and activation parameters of catalyzed reactions. Kinetic parameters for some of the fundamental organometallic reactions such as CO-insertions will be determined for aqueous systems as well. Exchange of ligands by and self-exchange of H_2O in the coordination sphere of catalytic metal complexes will be studied mainly by multinuclear NMR techniques. Water is an ideal solvent for study of radical reactions [49,61] and such investigations, either based on or independent of photochemical activation will continue. Redox chemistry is facilitated by water and water activation in the coordination sphere of low valent metal complexes should be studied in more details. Kinetic measurements using homogeneous solutions made with water-miscible organic solvents as a function of solvent composition will shed light on the solvation effects implied in the mechanism of aqueous organic solvents. The effect of salts and micelle forming agents should further be investigated as well as the possible enantioselection induced by chiral salts and surfactants. Kinetic characterization of organometallic reactions in superheated and in supercritical water is highly desirable; in this respect microwave activation should also be considered. In the emerging field of fluorous biphase catalysis mechanistic research is indispensable for further progress. In general, the wealth of information contained in the earlier literature on physical organic chemistry should be thoroughly exploited.

8. Industrial applications

The first large scale industrial process utilizing a water soluble organometallic catalyst was developed by Rhône-Poulenc and commercialized by Rurhchemie in 1984 [4]. The aqueous biphasic hydroformylation of propylene is catalyzed by the water soluble trisulfonated triphenylphosphine (TPPTS) modified rhodium system and produces 300,00 tonnes/year of n-butanal (with about 4% *iso*-butanal). Another major application of biphasic, but not aqueous transition-metal catalyst, has been Shell's SHOP process for the oligomerization of ethylene to higher olefins [62]. The reaction is carried out in butanediol in the presence of a diol soluble phosphinocarboxylic acid modified nickel catalyst. In addition, several smaller scale biphasic processes have been developed such as Kuraray's hydrodimerization of butadiene [63] and Rhône-Poulenc's manufacture of geranylaceton by the addition of methyl acetoacetate to myrcene [64]. In all of these processes the biphasic approach, a unique immobilization technique, offers several advantages over other schemes of anchoring ligands to solid carriers. Most biphasic catalyst systems are mobile and cause only small change in reaction mechanism, they allow counter current extraction, facile separation of products, and easy catalyst reprocessing. In contrasts, most heterogenized catalysts are leaching the anchored metals to the reaction media, more expensive and difficult to synthesize, cause significant change in reaction mechanism, and finally the presence of "hot spots" could lead to rapid catalyst deactivation. It should be noted, that the two limiting factors for aqueous catalyst systems could be the low solubility of certain organic substrates and potential side reactions with water. Non-aqueous biphasic systems could overcome of these limitations, however [40].

There are several fundamental chemistry issues crucial to the development and commercialization of novel aqueous processes. The solubility of reactants and the mass transfer between the organic and aqueous phase should be investigated in the very early stage of research. The effects of surfactants on the fundamental physico-chemical data should be also studied and the resulting information should be used for the design and synthesis of tailored "surfactants". Long term stability of catalysts and recovery should be also carefully investigated. The biphasic aqueous catalyst approach could be extended to reactions like carbonylation, oxidation, and reactions where water is also a reagent.

Finally, the future of aqueous organometallic chemistry and catalysis requires a better communication between scientists and engineers in academia, national laboratories and industry. Better interaction with government regulators, environmentalist and the general public should also be developed.

9. References

1. Lubineau, A., Auge, J. and Queneau, Y. (1994) Water-promoted organic reactions, Synthesis, 741-760.

2. Li, C.J. (1993) Organic reactions in aqueous media - with focus on carbon-carbon bond formation, Chem. Rev. 93, 2023-2035.

3. Cativiela, C., Garcia, J.I., Mayoral, J.A. and Salvatella, L. (1994) Solvent effects on endo/exo- and regio-selectivities of Diels-Alder reactions of carbonyl-containing dienophiles, J. Chem. Soc. Perkin trans. 2, 847-851.

4. Kuntz, E.G. (1987) Homogeneous catalysis ... in water, CHEMTECH, 17, 570-575; see also This Volume, pp. 177-181.

5. Bényei, A. and Joó, F. (1990) Organometallic catalysis in aqueous solutions: The biphasic transfer hydrogenation of aldehydes catalyzed by water soluble phosphine complexes of ruthenium, rhodium and iridium, J. Mol. Catal., 58, 151-163.

6. Herrmann, W.A. and Kohlpaintner, C.W. (1993) Water-soluble ligands, metal complexes and catalysts: Synergism of homogeneous and heterogeneous catalysis, Angew. Chem. Int. Ed. Engl., 32, 1524-1544.

7. Russel, M.J.H.(1988) Water soluble rhodium catalysts, Platinum Metals Rev., 32, 179-186.

8. Goldberg, Yu. (1992) Phase Transfer Catalysis. Selected Problems and Applications. Gordon and Breach, Yverdon, Switzerland; see also Grushin, V. and Alper, H., This Volume, pp. 81-96.

9. Grotjahn, D.B. (1994) Effect of aqueous solvent media on Heck reactions, This Volume, pp. 123-125.

10. Novak, B.M. and Grubbs, R.H. (1988) Catalytic organometallic chemistry in water: The aqueous ring-opening polymerization of 7-oxanorbornene derivatives, J. Am. Chem. Soc., 110, 7542-7543; see also This Volume, pp. 15-22.

11. Wan, K. and Davis, M.E. (1993) Asymmetric hydrogenation in water by a rhodium complex of sulfonated 2,2`-bis(diphenylphosphino)-1,1`-binaphtyl (binap), J. Chem. Soc. Chem. Commun, 1262-1264.

12. Broger, E.A., Schmid, R., Crameri, Y., Lalonde, M., Meili, A. and Schoettel, G. (1994) Synthesis of a water-soluble atropisomeric diphosphine ligand and its evaluation in asymmetric hydrogenations, Abstracts of NATO ARW "Aqueous Organometallic Chemistry and Catalysis", Debrecen, Hungary, 1994, p. 10.

13. Bakos, J., Orosz, A., Sinou, D. (1990) Rhodium(I) sulphonated phosphine complexes as catalysts for the asymmetric hydrogenolysis of epoxides, Proc. ISHC-7 (Lyon), p. 43; see also This Volume, pp. 231-244.

14. Hanson, B.E., Ding, H., Bartik, T. and Bartik, B. (1994), This Volume, pp. 149-158.

15. Darensbourg, D.J. and Bischoff, C.J. (1993) Water-soluble organometallic compounds. 3. Kinetic investigations of dissociative phosphine substitution processes involving water-soluble group 6 metal derivatives in miscible aqueous organic/media, Inorg. Chem. 32, 47-53.

16. Sheldon, R.A. (1994) Consider the environmental quotient, CHEMTECH, 24, 38-47.

17. Kuhlmann, B., Arnett, E.M. and Siskin, M. (1994) Classical organic reactions in pure superheated water, J. Org. Chem., 59, 3098-3101.

18. Crittendon, R.C. and Parsons, E.J. (1994) Transformations of cyclohexane derivatives in supercritical water, Organometallics, 13, 2587-2591.

19. Kalck, P. and Monteil, F. (1992) Use of water soluble ligands in homogeneous catalysis, Adv. Organometal. Chem., 34, 219-284.

20. Joó, F. and Tóth, Z. (1980) Catalysis by water-soluble phosphine complexes of transition metals in aqueous and two-phase media, J. Mol. Catal., 8, 369-383.

21. Bartik, T., Bartik, B., Hanson, B.E., Glass, T. and Bebout, W. (1992) Comments on the synthesis of trisulfonated triphenylphosphine: Reaction monitoring by NMR spectroscopy, Inorg. Chem., 31, 2667-2670.

22. Lecomte, L., Triolet, J., Sinou, D., Bakos, J. and Heil, B. (1987) High-performance liquid chromatographic separation of sulphonated phosphines, J. Chromatography, 408, 416-419.

23. Herd, O., Healer, A., Langhans, K.P., Stelzer, O., Sheldrick, W.S. and Weferling, N. (1994) Wasserlösliche Phosphane. II. Ein neuer Syntheseweg für wasserlösliche sekundäre und tertiäre Phosphane mit sulfonierten aromatischen Resten - Kristall-struktur von $P(p\text{-}C_6H_4\text{-}SO_3K)_3 \cdot KCl \cdot 0.5H_2O$, J. Organometal. Chem. 475, 99-111.

24. Ravindar, V., Hemling, H., Schumann, H. and Blum, J. (1992) A new synthesis of hydrophilic carboxylated arylphosphines, Synth. Commun., 22, 841-851.

25. Paetzold, E., Kinting, A. and Oehme, G. (1987) Synthesis of phosphino alkane sulfonates and their corresponding sulfonic acids by reaction of alkalimetal-phosphides with sultones, J. prakt. Chem., 329, 725-731.

26. Darensbourg, D.J., Joó, F., Kannisto, M., Kathó, Á. and Reibenspies, H. (1992) Water-soluble organometallic compounds. 2. Catalytic hydrogenation of aldehydes and olefins by new water-soluble 1,3,5-triaza-7-phosphaadamantane complexes of ruthenium and rhodium, Organometallics, 11, 1990-1993.

27. Ellis, J.W., Harrison, K.N., Hoye, P.A.T., Orpen, A.G., Pringle, P.G. and Smith, M.B. (1992) Water-soluble tris(hydroxymethyl)phosphine complexes with nickel, palladium, and platinum. Crystal structure of $[Pd\{P(CH_2OH)_3\}_4].CH_3OH$, Inorg. Chem., 31, 3026-3033; see also This Volume, pp. 111-122.

28. Nuzzo, R.G., Haynie, S.L., Wilson, M.E. and Whitesides, G.M. (1981) Synthesis of functional chelating diphosphines containing the bis[2-(diphenylphosphino)-ethyl]amino moiety and the use of these materials in the preparation of water-soluble diphosphine complexes of transition metals, J. Org. Chem., 46, 2861-2867.

29. Le, T.X. and Merola, J.S. (1993) Synthesis and reaction chemistry of water-soluble mer-$(Me_3P)_3Ir(H)(H)Cl$: Activation by water of alkyne insertion into an Ir-H bond, Organometallics, 12, 3798-3799; see also This Volume, pp. 33-45.

30. Bényei, A.C. (1994) Hexaaquaruthenate(II) as catalyst precursor in aqueous organometallic catalysis, This Volume, pp. 159-171.

31. Darensbourg, D. J. and Gibson, G. (1987) New methods for acquiring IR spectral data in organometallic chemistry and catalysis, in Experimental Organometallic Chemistry (ACS Symp. Ser. 357), A.L. Wayda and M.Y. Darensbourg, eds., American Chemical Society, Washington, pp. 230-249.

32. Horváth, I.T. and Millar, J.M. (1991) NMR under high gas pressure, Chem. Rev., 91, 1339-1351.

33. Arhancet, J.P., Davis, M.E., Merola, J.S. and Hanson, B.E. (1989) Hydroformylation by supported aqueous-phase catalysis: a new class of heterogeneous catalysts, Nature, 339, 454-455.

34. Horváth, I.T. (1990) Hydroformylation of olefins with the water-soluble $HRh(CO)[P(m\text{-}C_6H_4SO_3Na)_3]_3$ in supported aqueous phase. Is it really aqueous?, Catal. Lett., 6, 43-48.

35. Kumar, A., Oehme, G., Roque, J.P., Schwarze, M. and Selke R. (1994) Increase in the enantioselectivity of asymmetric hydrogenation in water influenced by surfactants or polymerized micelles, Angew. Chem. Int. Ed. Engl., 33, 2197-2199; see also This Volume, pp. 245-257.

36. Anderson, J.R., Campi, E.M. and Jackson, R.W. (1991) Hydroformylation of olefins with water-soluble rhodium catalysts in the presence of α-cyclodextrin, Catal. Lett., 9, 55-58; see also This Volume, pp. 187-194.

37. Okano, T., Iwahara, M., Konishi, H. and Kiji, J. (1988) Synthesis and properties of crown ether-modified phosphines and their use as ligands in transition metal catalysts, J. Organometal. Chem., 346, 267-275; see also This Volume, pp. 97-110.

38. Quinn, P.J. and Taylor, C.E. (1981) Selective homogeneous catalysis in heterogeneous reaction systems: phospholipid bilayers in water, J. Mol. Catal., 13, 389-396.

39. Durocher, A., Keim, W. and Voncken, P. (1976) Telomerisation von Olefinen in Zwei-Phasen-Systemen, COMPENDIUM, Erdöl und Kohle, 75/76, 347-354.

40. Horváth, I.T. and Rábai, J. (1994) Facile catalyst separation without water: fluorous biphase hydroformylation of olefins, Science, 266, 72-75.

41. Amrani, Y., Lecomte, L., Sinou, D., Bakos, J., Tóth, I. and Heil, B. (1989) Chiral sulphonated phosphines. Syntheses and use as ligands in asymmetric hydrogenations using an aqueous-organic two-phase solvent system, Organometallics, 8, 542-547.

42. Tóth, I., Hanson, B.E. and Davis, M.E. (1990) Novel chiral water soluble phosphines II. Applications in catalytic asymmetric hydrogenation, Tetrahedron: Asymmetry, 1, 913-930.

43. Lensink, C. and de Vries, J.G. (1992) Improving enantioselectivity by using a mono-sulphonated diphosphine as ligand for homogeneous imine hydrogenation, Tetrahedron: Asymmetry, 3, 235-238.

44. Lecomte, L., Sinou, D., Bakos, J., Tóth, I. and Heil, B. (1989) Chiral sulphonated phosphines II. Influence of water on the enantioselectivity in the reduction of dehydro-aminoacids, J. Organometal. Chem., 370, 277-284.

45. Flach, H.N., Grassert, I. and Oehme, G. (1994) Polymeric surfactant systems in the asymmetric hydrogenation of amino acid precursors with a rhodium complex, Macromol. Chem. Phys., 195, 3289-3301; see also This Volume, pp. xx-yy.

46. Graf, E. and Leitner, W. (1992) Direct Formation of Formic Acid from Carbon Dioxide and Dihydrogen using the $[\{Rh(cod)Cl\}_2]$ - $Ph_2P(CH_2)_4PPh_2$ Catalyst System J. Chem. Soc. Chem. Commun., 623-624.

47. Willner, I., Maidan, R. and Shapira, M. (1990) Thermal and photochemical regeneration of nicotinamide cofactors and a nicotinamide model compound using a water soluble rhodium phosphine catalyst, J. Chem. Soc. Perkin. Trans. 2., 559-564.

48. Chaloner, P.A., Esteruelas, M.A., Joó, F. and Oro, L.A. (1994) Homogeneous Hydrogenation, Kluwer, Dordrecht, Chapter 7, pp. 255-270.

49. Kräutler, B. (1994) On Nature`s aqueous organometallic and free radical chemistry with B12-derivatives, This Volume, pp. 277-279.

50. Vígh, L., Los, D.A., Horváth, I., and Murata, N. (1993) The primary signal of the biological perception of temperature, Proc. Natn. Acad. Sci. USA, 90, 9090-9094; see also This Volume, pp. 281-287.

51. Maresca, B. and Cossins, R. (1993) Fatty feedback and fluidity, Nature, 365, 606-607.

52. Török, Z., Szalontai, B., Joó, F. Wistrom, C.A. and Vígh (1993) Homogeneous catalytic deuteration of fatty acyl chains as a tool to detect lipid phase transitions in specific membrane domains: a Fourier transform infrared spectroscopic study, Biochem. Biophys. Res. Commun., 192, 518-524.

53. Joó, F., Chevy, F., Colard, O. and Wolf, C. (1993) The activation of rat platelets increases the exposure of polyunsaturated fatty acid enriched phospholipids on the external leaflet of the plasma membrane, Biochim. Biophys. Acta, 1149, 231-240.

54. Fish, R.H., Smith, D.P., Chen, H., Maestre, M.F., Olmstead, M.M., Eisen, M.S. and Haskel, A. (1994) Bioorganometallic Chemistry: The Reaction of a (η^5-Pentamethylcyclo-pentadienyl)rhodium Aqua Complex with Nucleobases, Nucleosides, Nucleotides, and Oligonucleotides, This Volume, pp. 259-276.

55. Fache, E., Santini, C., Senocq, F. and Basset, J.M. (1992) Homogeneous catalysis in water Part III. The catalytic hydrogenation of propionaldehyde with $(RuCl_2L_2)_2$, $RuHClL_3$, $RuH(OAc)L_3$, RuH_2L_4, $RuHIL_3$, $RuCl_2CO)_2L_2$ and $[Ru(OAc)(CO)_2L]_2$, $(L{=}P(C_6H_4\text{-}mSO_3Na)_3.3H_2O$: a kinetic investigation of the salt effect in water, J. Mol. Catal., 72, 337-350.

56. Joó, F., Csiba, P. and Bényei (1993) Effect of water on the mechanism of hydrogenations catalyzed by rhodium phosphine complexes, J. Chem. Soc. Chem. Commun., 1602-1604; see Also This Volume, pp. 23-32.

57. Larpent, C. and Patin, H. (1988) Organic chemistry in water. Nucleophilic addition of water-soluble phosphines on activated olefins, Tetrahedron, 44, 6107-6118.

58. Bényei, A., Stafford, J.N.W., Kathó, Á., Darensbourg, D.J. and Joó, F. (1993) The effect of phosphonium salt formation on the kinetics of homogeneous hydrogenations in water utilizing a rhodium-meta-sulfonatophenyl-diphenylphosphine complex, J. Mol. Catal., 84, 157-163.

59. Darensbourg, D.J., Joó, F., Kathó, Á., Stafford, J.N.W., Bényei, A. and Reibenspies, J. (1994) Nucleophilic addition of a water-soluble phosphine to aldehydes. Isolation of (1- hydroxyalkyl)phosphonium salts and the crystal structure of the (1-methoxy-1-benzyl)-(m-sulfonatophenyl)-diphenylphosphonium salt, Inorg. Chem., 33, 175-177.

60. Horváth, I.T., Kastrup, R.V., Oswald, A.A. and Mozeleski, E.J. (1989) High-pressure NMR studies of the water-soluble rhodium hydroformylation system, Catal. Lett., 2, 85-90.

61. Avey, A., Schut, D.M., Weakley, T.J.R. and Tyler, D.R. (1993) A new water-soluble phosphine for use in aqueous organometallic systems. Products from the reactions of 2,3- bis(diphenylphosphino)maleic anhydride with water and oxygen, Inorg. Chem., 32, 233-236; see also This Volume, pp. 47-60.

62. Moulijn, J. A., Sheldon, R. A., van Bekkum, H.l, van Leeuwen, P. W. N. M. (1993) Catalytic processes in industry, in Moulijn, J. A., van Leeuwen, P. W. N. M., van Santen, R. A. (eds), Catalysis, Elsevier, Amsterdam, pp. 45-48.

63. Yoshimura, N. and Tamura, M. (1982) Process for preparing n-octanedienol, US Patent 4,356,333 to Kuraray Co. Ltd.

64. Mercier, C. and Chabardes (1994) Organometallic chemistry in vitamin A and vitamin E synthesis, Pure & Appl. Chem. 66, 1509-1518.

TRANSITION METAL CATALYZED REACTIONS OF OLEFINS IN WATER: OLEFIN METATHESIS AND ISOMERIZATION

R. H. Grubbs
Division of Chemistry and Chemical Engineering
California Institute of Technology
Pasadena, California 91125 USA

1. Aqueous Catalysis

Water can play a number of roles in organometallic chemistry related to catalysis. Water can either be a critical ligand on the active catalytic species or can serve only as a convenient reaction medium. In all cases the catalyst must be able to function in the presence of water. This water stability makes possible a variety of new processes and materials. Our research into the development of new metathesis catalysts provides examples of many of these applications.

2. Water as Part of a Catalyst System

In some cases water plays a non-innocent role in both catalyst formation and catalyst activity. Two cases from our research in olefin metathesis demonstrate the role of water in the initiation of catalysis.

2.1 OLEFIN METATHESIS

In an attempt to prepare polymers from 7-oxonorbornenes it was found that the best catalyst for the ring opening metathesis polymerization (ROMP) of these monomers was ruthenium chloride in a mixture of ethanol and benzene. It was found that drying the solvents decreased the activity of the catalyst. In fact it was found that water was essential for catalyst initiation and the system worked best in water [1].

I. T. Horváth and F. Joó (eds.), Aqueous Organometallic Chemistry and Catalysis, 15–22.

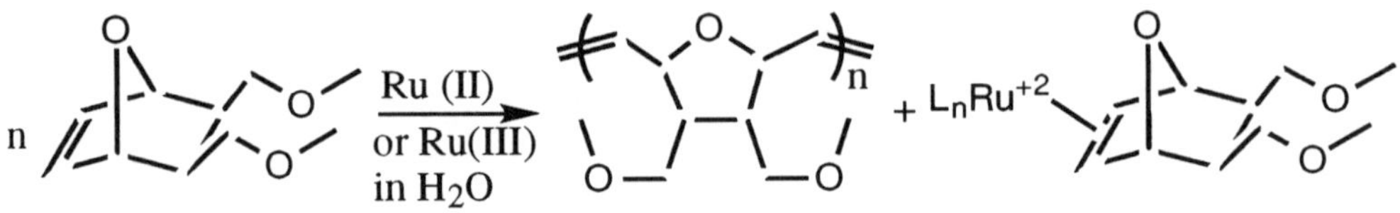

For the ROMP of many monomers, the best catalyst precursor was the prereduced aqueous complex, $Ru(H_2O)_6(OTs)_2$ (**1**) [2]. At the end of the polymerization reaction an aquoRuthenium (II) olefin complex of the monomer was isolated. This olefin complex was more active than the original complex in the polymerization of additional monomer [3]. It is assumed that the olefin complex is a precursor to the active catalyst and rearranges by one of a number of possible mechanisms to a ruthenium carbene complex.

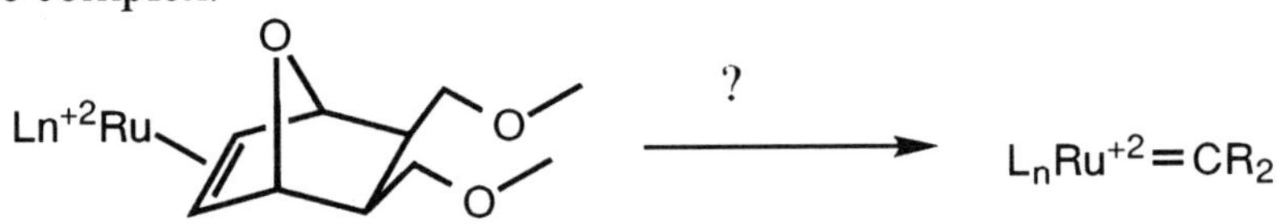

2.2 OLEFIN ISOMERIZATION

In an attempt to understand the possible reactions of coordinated olefins, a variety of olefins were reacted with (**1**), hexaaquoruthenium (II). Although many olefins gave stable complexes [4], allylalcohols and ethers rearranged to the corresponding aldehyde or ketone.

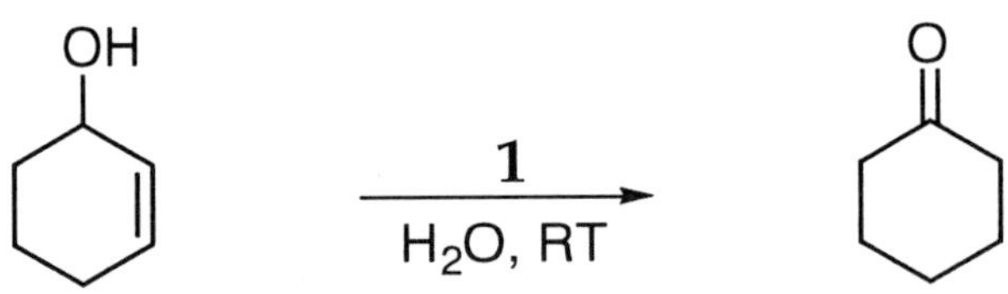

For example, allylmethylether rearranged to propanaldehyde and methanol. In order to elucidate the mechanism of this reaction, the following series of labeling studies were carried out to define the source of the protons in the products [5].

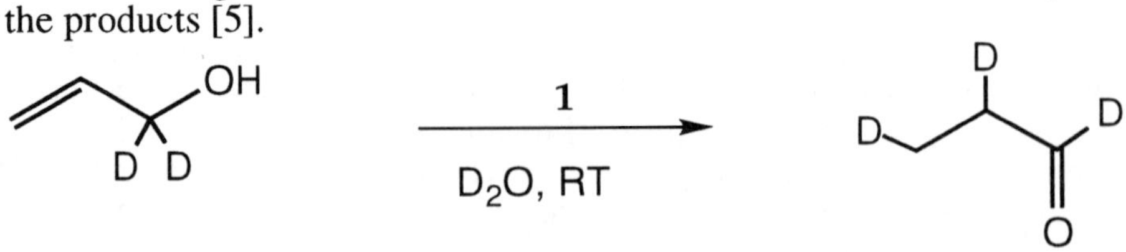

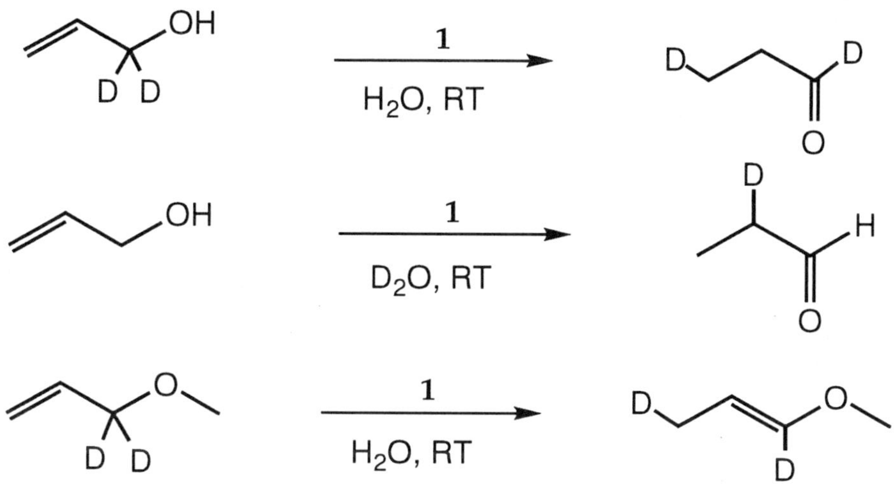

From this series of labeling studies the source and path of all the protons in the system can be defined.

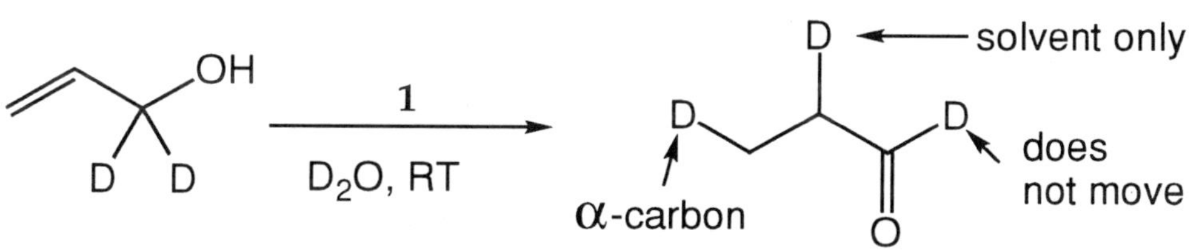

Since the labeling patterns suggested that a clean 1,3-migration was involved, a double labeling study was performed to determine if the hydrogen transfer occurred by an intra- or inter-molecular mechanismß.

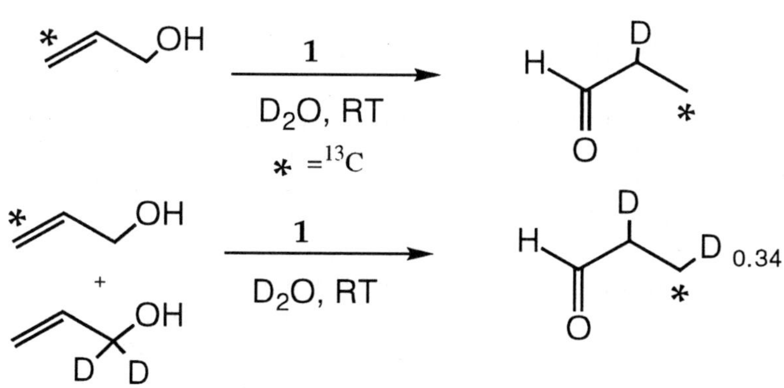

Based on the fact that there is cross over between molecules, the following mechanism is proposed.

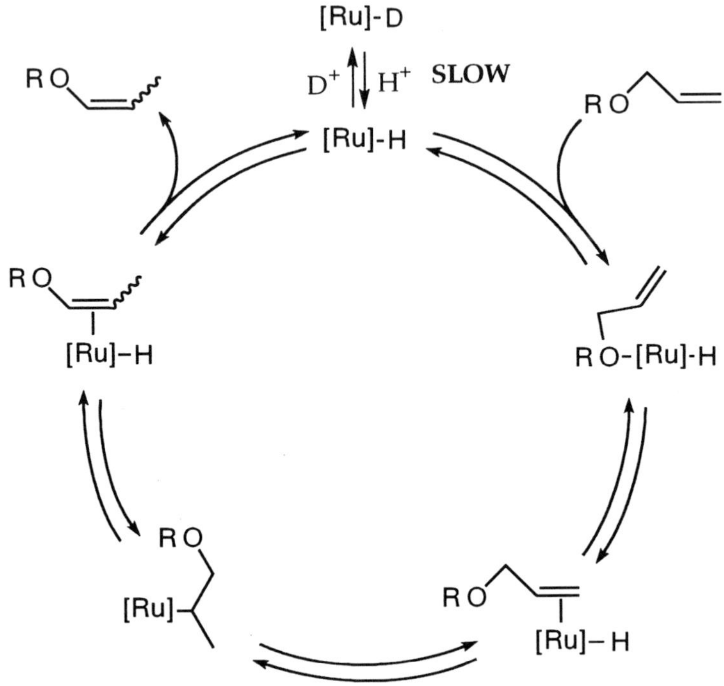

In this case the catalyst is a metal hydride that is formed in situ. The lack of exchange requires that exchange of the metal hydride with water is slow. In this case water is essential for the catalytic process and it is surprising that the intermediate metal hydrides do not exchange with the protons of the solvent.

3. Water Stable Metathesis Catalysts

Since it was observed that an extremely active catalyst could be prepared in extremely low yield from Ru(II) and a strained olefin and that this catalyst was tolerant of a variety of functionality as well as water, routes to a well-defined ruthenium carbene complex were explored since it was anticipated that such a complex would be particularly tolerant of functionality and processing conditions. It was found that the reaction of a simple ruthenium carbene complex, tetrakistriphenylphosphinerutheniumdichloride, with diphenylcyclopropene (one of the most strained olefins available) resulted in the generation of a stable carbene complex [6].

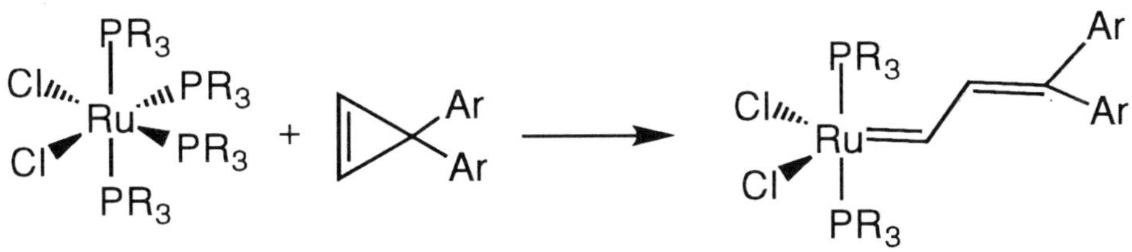

As anticipated from the work with simple ruthenium salts in water, this catalyst was stable to water and a variety of organic functional groups and was an active olefin metathesis catalyst.

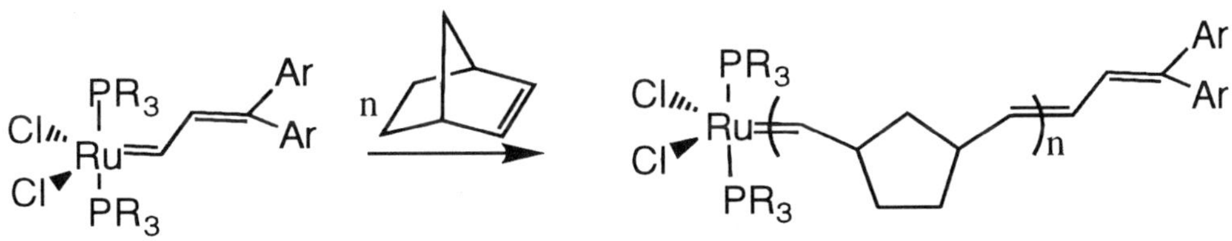

As an added bonus, the catalyst gave "living" polymers and allowed the synthesis of a variety of block polymers [7]. With a well defined complex available, the activity and selectivity of the catalyst was tuned by usual organometallic ligand substitution reactions [8].

3.1 WATER SOLUBLE LIGANDS

Although this catalyst was not soluble in water , it was stable in the presence of water and other protic solvents. This characteristic opened a large number of possibilities. For example the catalyst could be made soluble in water by using a water soluble ligand of the type that has been used in other catalytic systems [9].

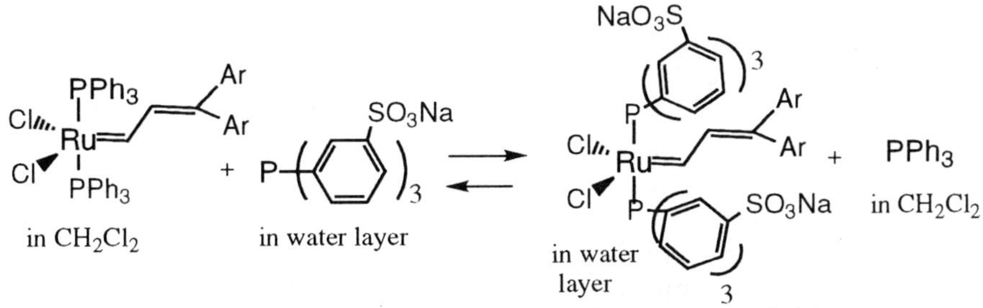

These catalysts could be used to polymerize water-soluble monomers.

3.2 WATER STABLE CATALYSTS

A catalyst stable to water has several processing advantages over traditional organic based systems. The most obvious advantage is the ability to use monomers and substrates without the rigorous purification and drying that is nomrally required when early transition metal catalysts are used. However, the most promising application result from the ability to carry out metal catalyzed emulsion polymerizations [10]. There are a number of advantages to emulsion systems.

Small Size of Polymer Particles (0.05 - 0.5 μm)

- Easy Penetration into the Smallest Crevices of Substrates
- Maximum Efficiency in Deposition of Thin Films
- Optimum Optical Control of Films

Use As Latex

- Rubber Goods
- Construction Applications
- Bonding Applications
  - Adhesives
  - Fiber Bonding
  - Paints and Industrial Coatings
  - Paper Coatings

It has recently been demonstrated that norbornene and cyclooctadiene can be polymerized in water using the preformed carbene complex **2** as the catalyst and with dodecyltrimethylammonium bromide (DTAB) as the emulsifying agent [10].

4. Choice of Metal for Aqueous Organometallic Catalysts

Perhaps the most improtant lesson learned from the series of metathesis catalysts that has been prepared over the past few years is the increased stability of late transition metal catalysts relative to those based on early metal systems [11]. The following table shows the relative reactivity of olefins relative to other functional groups.

It is apparent that the reactivity of olefins relative to other organic functional groups increases as the central metal in the metal carbene complex is chosen from later transition metals. As the central metal of the catalyst is chosen from further right in the periodic table it becomes softer, and contains more d-electrons. This trend favors the complexation and reaction with soft base, π-acidic olefins over the harder oxygen containing functional groups. This may be a lesson that will apply to other olefin polymerization reactions.

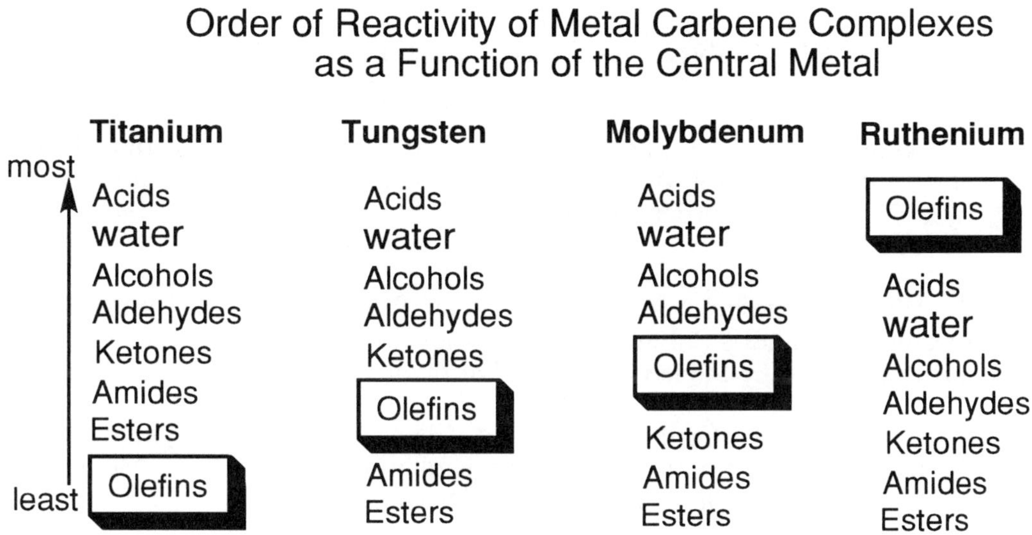

5. References

1. Novak, B. M. and Grubbs, R. H. (1988) The Ring Opening Metathesis Polymerization of 7-Oxabicylo[2.2.1]hex-5-ene Derivatives: A New Acyclic Polymeric Ionophore, *J. Am. Chem. Soc.* **110**, 960-961.

2. Novak, B. M. and Grubbs, R. H. (1988) Catalytic Organometallic Chemistry in Water: The Aqueous Ring Opening Metathesis Polymerization of 7- Oxanorbornene Derivatives, *J. Am. Chem. Soc.* **110**, 7542-7543.

3. McGrath, D. V., Novak, B. M., Grubbs, R. H. (1989) Aqueous Ring-Opening Metathesis Polymerizations of 7-Oxanorbornene Derivatives Using Ruthenium Catalysts, *Proceedings of the NATO ASI* in Akcay, Turkey, September, 1989. McGrath, D. V., Novak, B. M., Grubbs, R. H. (1990) Aqueous Ring-Opening Metathesis Polymerizations of 7-Oxanorbornene Derivatives Using Ruthenium Catalysts, in Y. Imamoglu (Ed.), *Olefin Metathesis and Polymerization Catalysts*, Kluwer Academic Publishers: the Netherlands, pp. 525-536.

4. McGrath, D. V., Grubbs, R. H., and Ziller, J. W. (1991) Aqueous Ruthenium(II) Complexes of Functionalized Olefins: The X-Ray Structure

of Ru(H_2O)$_2$(h^2-(*O*),h^2-(*C*,*C'*)-OCOCH_2CH=CHCH_3)$_2$, *J. Am. Chem. Soc.* **113,** 3611-3613.

5. McGrath, D. V. and Grubbs, R. H. (1994) The Mechanism of Aqueous Ruthenium(II)-Catalyzed Olefin Isomerization, *Organometallics* **13**, 224-235.

6. Nguyen, S. T., Johnson, L. K., Grubbs, R. H. and Ziller, J. W. (1992) Ring-Opening Metathesis Polymerization (ROMP) of Norbornene by a Group VIII Carbene Complex in Protic Media, *J. Am. Chem. Soc.* **114**, 3974-3975.

7. Wu, Z., Benedicto, A. D., and Grubbs, R. H. (1993) The Living Ring-Opening Metathesis Polymerization of Bicyclo[3.2.0]-heptene Catalyzed by a Ruthenium Alkylidene Complex, *Macromolecules* **26**, 4975-4977.

8. Nguyen, S. T., Grubbs, R. H. and Ziller, J. W. (1993) Syntheses and Activities of New Single-Component, Ruthenium-based Olefin Metathesis Catalysts, *J. Am. Chem. Soc.* **115**, 9858-9859.

9. Nguyen, S. T. and Grubbs, R. H, unpublished results.

10. Kanaoka, S. and Grubbs, R. H., unpublished results.

11. Grubbs, R. H. (1994) The Development of Functional Group Tolerant ROMP Catalysts, *Pure Appl. Chem.* **A31(11)**, 1829-1833.

AQUEOUS ORGANOMETALLIC CHEMISTRY. EFFECTS OF A REACTIVE SOLVENT

F. JOÓ, L. NÁDASDI, A. BÉNYEI, P. CSIBA and Á. KATHÓ
Institute of Physical Chemistry, Lajos Kossuth University
Debrecen 10, P.O. Box 7, H-4010 Hungary

1. Introduction

In aqueous organometallic chemistry water is often considered solely as a solvent. However, there are several examples when such an assumption is not justified [1]. Water is a strongly coordinating, polar solvent, capable of assisting reactions yielding polar or charged (ionic) products. H_2O itself can oxidatively add to a low valent transition metal complex with formation of hydrido-hydroxometal complex compounds. Hydroxometal species are also formed by replacement of anionic ligands (*e.g.* halides) by OH^- (metal ion hydrolysis equilibria). Dissolved salts strongly influence reaction mechanisms involving polar or charged transition states and intermediates. In addition acid/base catalytic reactions of the substrates or products effect rates and selectivities. All these reactions are influenced by the pH which should be regarded of paramount importance in aqueous organometallic catalysis. Presence of a separate water phase in aqueous/organic biphasic reactions may have important consequences on the reaction mechanism due to the separation of reactants, products and intermediates (physical or solubility effects). This paper focusses on a few selected examples of such effects from our research on reactions of or catalyzed by platinum metal complexes with the water soluble phosphines: *m*-sulfonatophenyldiphenylphosphine (TPPMS) and 1,3,5-triaza-7-phosphaadamantane (PTA).

2. Effect of water on the formation of metal hydrides

Detailed kinetic studies have shown that $RhCl(TPPMS)_3$ is an active catalyst for hydrogenation of water soluble olefinic substrates like unsaturated

I. T. Horváth and F. Joó (eds.), Aqueous Organometallic Chemistry and Catalysis, 23–32.

carboxylic acids in aqueous solutions, at 60°C, under 1 bar total pressure [2]. In most respects the kinetics of such catalyzed hydrogenations resembled that of olefin reduction in organic solvents catalyzed by $RhCl(PPh_3)_3$. However, in aqueous solutions maleic acid (*i.e.* a *cis*-olefin) was hydrogenated much slower than fumaric acid (its *trans*-isomer). This and other peculiar features of the reactions made it necessary to investigate the effect of water on catalysis by the water soluble analog of Wilkinson's compound [3].

After screening several solvents we choose diglyme (diethylene glycol dimethyl ether), since it dissolves well both maleic and fumaric acids and is freely miscible with water. $RhCl(TPPMS)_3$ generally contained the sodium salt of the *mono*-sulfonated triphenylphosphine and was soluble in diglyme/water mixtures up to 90% diglyme concentration. However, for studies in anhydrous diglyme Na^+ had to be replaced by $^nBu_4N^+$ to achieve good solubility. $RhCl(PPh_3)_3$, too, is sufficiently soluble in diglyme.

A direct comparison of the catalytic activity of $RhCl(PPh_3)_3$ and $RhCl(TPPMS/^nBu_4N)_3$ in anhydrous diglyme under strictly identical conditions showed that *both* catalysts were more active for the reduction of maleic acid (Table 1).

TABLE 1. Hydrogenation of maleic and fumaric acid in anhydrous diglyme with $RhClP_3$ catalysts

Catalyst	Substrate	Rate (ml/min)
$RhCl(PPh_3)_3$	Maleic acid	0.49
	Fumaric acid	0.14
$RhCl(TPPMS/^nBu_4N)_3$	Maleic acid	1.64
	Fumaric acid	0.13

$[Rh]=10^{-3}$ M, $[MA]=[FA]=5.10^{-2}$ M, T=60°C, $P_{total}=1$ bar

It should be concluded that the differences in the catalytic behavior of Rh-TPPMS complexes in aqueous solutions relative to their Rh-PPh_3 counterparts in purely organic solvents *do not* arise as a result of the sulfonation of the ligand. This is the first strict comparison of the behavior of a water soluble transition metal phosphine complex catalyst in aqueous and anhydrous solutions, furthermore the comparison of the catalytic properties of the same catalyst with PPh_3 or TPPMS ligands.

A continuous variation of the solvent composition showed that the rate of maleic acid hydrogenation somewhat slowed down (initial rate is 0.6 ml/min in water under conditions of Table 1). On the other hand, the rate of fumaric acid hydrogenation increased steadily with increasing water proportion and reached 7.9 ml/min in purely aqueous solution. *Water has a dramatic effect on the rate and selectivity!*

During the hydrogenation it was observed that while diglyme solutions remained clear yellow the aqueous solutions turned dark brown after an initial phase of the reaction especially in the absence of the substrate. It was also noticed that the reaction of $RhCl(TPPMS)_3$ and H_2 in water is accompanied by *acidification* of the solution. Therefore a quantitative study of proton production was undertaken.

In a thermostatted cell the solution of $RhCl(TPPMS)_3$ was placed under argon and its pH was measured by a combined electrode connected to a Radiometer PHM 93 pH meter and a Radiometer ABU 91 autoburette. This setup allows addition of μl amounts of NaOH solution either to set the pH to a desired value or to keep it constant during a reaction (pH-static mode). When the $RhCl(TPPMS)_3$ solution was bubbled with H_2 (together with proper stirring) the reaction was accompanied by the formation of protons which were continuously neutralized to keep the pH constant. Measuring the amount of the base consumed it was established that at pH 7.00 approximately 1 mole H^+ is liberated during the reaction:

$$RhCl(TPPMS)_3 + H_2 = HRh(TPPMS)_3 + H^+ + Cl^- \qquad (1)$$

Independent gas-volumetric measurements have shown that at this pH (phosphate buffered solution) 1 mole H_2 is consumed per 1 mole Rh and the rate of hydrogen uptake is identical to the rate of proton production. (Conditions: [Rh]=0.02 M, 60°C, 1 bar total pressure). It is very likely therefore that contrary to the well known reaction of Wilkinson's catalyst with hydrogen in an organic solvent:

$$RhCl(PPh_3)_3 + H_2 = H_2RhCl(PPh_3)_3 \qquad (2)$$

in aqueous system the formation of a *monohydridorhodium(I)* complex takes place. The driving force of the reaction comes from the very large solvation energy of the ionic products, especially H^+ in water.

Rather disappointingly, despite all our efforts we could not detect this monohydridorhodium(I) species by ^1H NMR. However, its formation is further substantiated by the following observations.

The rate of proton production and the amount of H^+ formed in a reasonable time (*e.g.* 1 hr) depends on the pH of the solution (note that this is kept constant throughout the reaction). While at pH 9.00 the reaction is over in a few minutes at pH 4.00 only 65% conversion is reached in 1 hr. When about all the rhodium reacted neutral and basic solutions turn brown; in acidic solutions this does not happen. We can envisage the formation of $HRh(TPPMS)_3$ in a reversible dehydrochlorination reaction:

$$H_2RhCl(TPPMS)_3 = HRh(TPPMS)_3 + H^+ + Cl^- \quad (3)$$

where the relative concentrations of $H_2RhCl(TPPMS)_3$ and $HRh(TPPMS)_3$ depend on the pH of the solution. Such dehydrochlorination reactions in organic solutions can be effected by the addition of appropriate bases [4].

The same features ("reversed" *cis/trans* selectivity, proton production in the reaction of the catalyst and H_2) were observed with the analog 1,3,5-triaza-7-phosphaadamantane complex, $RhCl(PTA)_3$. In this case a pH-variation study showed that the rate of hydrogenation of both crotonic acid (CA) and allyl alcohol (AA) was sharply dependent on solution acidity (Table 2).

TABLE 2. Effect of pH on the hydrogenation rate (turnover/h) of crotonic acid and allyl alcohol catalyzed by $RhCl(PTA)_3$

pH	Crotonic acid	Allyl alcohol
4.1	242.6	111.5
4.6	275.4	137.7
4.7	**289.0**	**172.3**
4.9	260.3	140,9
5.2	252.8	138.5

[Rh]=4.9.10^{-4}M, [CA]=0.017 M, [AA]=0.045 M, T=37°C, P_{total}=1 bar

This behavior can be rationalized by assuming that the monohydride, $HRh(PTA)_3$ gives an intermediate alkylrhodium species and the product of crotonic acid (CA) hydrogenation (butyric acid, BA) is formed by the *protolysis* of this intermediate:

$$(HCA)Rh(PTA)_3 + H^+ = [Rh(PTA)_3]^+ + BA \quad (4)$$

Accordingly, the pH has a dual effect on the overall hydrogenation rate. The formation of $HRhP_3$ is favored by high pH, however its further reaction is disfavored; the opposite is true at low pH of the reaction mixture.

Deuteration studies give further support to the assumption of the role of monohydridorhodium(I) species in the above hydrogenations. When unsaturated carboxylic acids were hydrogenated with H_2 in D_2O substantial incorporation of deuterium into the products was observed similar to the findings of Laghmari and Sinou in a related system [5]. *E.g.* α-acetamidocinnamic acid was deuterated *fully* and *exclusively* on the α-carbon atom when treated with H_2 in a D_2O solution under $RhCl(PTA)_3$ catalysis. In case of an H/D exchange on Rh such a selective deuteration is highly improbable, therefore labelling of the product implies a protolysis (deuterolysis) step in the reaction mechanism, similar to reaction (4).

It can be concluded therefore, that water plays an intimate role in the formation of transition metal hydrides in aqueous solutions. It can react as a base and favor dehydrohalogenation reactions of halogenodihydrides, like $H_2RhCl(PPh_3)_3$ which often form in oxidative addition reactions of dihydrogen. The same reaction is facilitated by strong solvation of both H^+ and the halide.

There can be important consequences of this change of mechanism from a dihydridic to a monohydridic one. It is well known that in many cases Rh-phosphine catalyzed asymmetric hydrogenations proceed with lower enantioselectivity if carried out in a fully or partially aqueous system using a phosphine solubilized *e.g.* by sulfonation. A solvent variation study established a direct connection between the enantioselectivity of enamide hydrogenation catalyzed by various Rh-phosphine complexes and the solvofobicity parameter of the solvent [6]. Also, it was found that enantioselectivity in the hydrogenation of (*Z*)-methyl α-acetamidocinnamate catalyzed by cationic rhodium(I) phosphine complexes decreased considerably when carried out in water/methanol mixtures. Addition of surfactants dramatically restored both activity and selectivity [7]. In our opinion these effects can be simply rationalized: in the less hydrophilic environment of non-polar organic solvents or of the inner core of micelles the dihydridic mechanism becomes overwhelming with an obvious increase in the probability of enantioselection.

3. Formation of phosphonium salts

3.1 REACTION OF TERTIARY PHOSPHINES WITH ACTIVATED OLEFINS

In the course of our studies of maleic acid hydrogenation catalyzed by $RhCl(TPPMS)_3$ it was established that the rate was independent of the TPPMS in excess [2], a rather unusual phenomenon taking the classical mechanism of hydrogenations catalyzed by Wilkinson's catalyst [1]. On the other hand, crotonic acid hydrogenations were strongly inhibited by excess TPPMS in complete accordance with the expectations.

This unusual phenomenon was resolved when it was found [8] that in aqueous (protic) solution TPPMS and TPPTS [TPPTS = *tris*(*m*-sulfonatophenyl)phosphine] readily react with *activated* olefins giving phosphonium salts ($Ar=C_6H_4$-*3*-SO_3Na, R=COOH, CN, R'=R or Ph):

$$ArPh_2P + R\text{-}CH{=}CH\text{-}R' + H^+ = ArPh_2P^+\text{-}CH(R)\text{-}CH_2R' \qquad (5)$$

The driving force of this reaction comes from the *protonation* of the carbanionic intermediate, the primary product of the addition of the phosphine across the olefinic double bond. Further reactions of such phosphonium salts, especially in basic solutions, yield phosphine oxides.

Using ^{31}P NMR spectroscopy it was established that in aqueous solution at room temperature TPPMS reacted with maleic acid instantaneously and quantitatively, while the reaction of crotonic acid was slow and incomplete [9].

During hydrogenation of unsaturated acids the substrate is in a fairly large excess, usually 50-100 times more than the catalyst. Therefore, when TPPMS was added in a 5-10 times excess relative to Rh it immediately reacted *with maleic acid.* The rate of hydrogenation did not change since the reaction proceeded with a saturation rate both before and after the addition of the phosphine. On the other hand, with crotonic acid as substrate, no phosphonium salt formation occured, and the added phosphine could exert its inhibitory effect on the hydrogenation.

There is another influence of phoshonium salt formation on hydrogenations catalyzed by water soluble phosphine complexes. We have found that maleic acid reacted even with the ligand of the $RhCl(TPPTS)_3$ catalyst: in a solution containing maleic acid and $RhCl(TPPMS)_3$ in a 5:1 ratio under H_2, 18.5% of all phosphorus introduced with the catalyst was in form of the corresponding phosphonium salt (room temperature, ^{31}P NMR) [9]. Under catalytic conditions, when a large excess a phosphonium salt forming substrate is present, this reaction can effectively remove phosphine ligands from the coordination sphere of rhodium, facilitating its entering into the catalytic cycle.

Two other points deserve mentioning. First, in anhydrous diglyme no formation of phosphonium salts was detected and maleic acid hydrogenation *was* inhibited by the TPPMS added in excess to the $RhCl(TPPMS)_3$ catalyst (in form of $^nBu_4N^+$ salt). That is, the above peculiar kinetic behaviour was brought about on the effect of water, a protic, reactive solvent. Second, in contrast to the sulfonated triphenylphosphine derivatives, PTA *does not* form alkylphosphonium salts with activated olefins. This may be a beneficial feature when used for the preparation of water soluble transition metal phosphine complexes as catalysts for reactions of olefins.

3.2 REACTION OF TERTIARY PHOSPHINES WITH ALDEHYDES

According to the literature basic tertiary phosphines react with aldehydes in *strictly unhydrous* solvents [10]. We have found, however, that such a reaction takes place in aqueous acids between TPPMS and aliphatic as well as aromatic aldehydes yielding (1-hydroxyalkyl)phosphonium salts ($Ar{=}C_6H_4$-*3*-SO_3Na) [11]:

$$ArPh_2P + RCHO + H^+ = ArPh_2P^+\text{-}CH(OH)\text{-}R \qquad (6)$$

These products could be prepared in batch quantitites by reacting TPPMS with the aldehydes (*inter alia* pentanal, hexanal, benzaldehyde, cinnamaldehyde) in ethanol in the presence of 1 eq aqueous HCl. After filtration of the solution and evaporation to dryness the (1-hydroxyalkyl)phosphonium salts were obtained as white, hygroscopic solids characterized in aqueous solution by a ^{1}P NMR signal in the 20.5-21.2 ppm range. Good quality crystals of the benzaldehyde adduct were grown by recrystallization of the product from methanol and the structure was determined by X-ray diffraction [11]. Again, we *did not* observe such an interaction between the aldehydes and 1,3,5-triaza-7-phosphaadamantane.

The significance of these findings is in that in principle such a reaction may influence all reactions of aldehydes (hydroformylation, hydrogenation, etc.) catalyzed by water soluble phosphine complexes. It may account for the severe phosphine losses encountered in many hydroformylation systems [12]. A study of the effect of the pH on aqueous hydroformylation catalysis with phosphine complexes is therefore a must in each case.

4. Extraction phenomena

Some years ago we prepared a series of phosphinoacetamides through the dicyclohexylcarbodiimide condensation reaction of diphenylphosphinoacetic acid and amines or amino acid esters [13]:

$$Ph_2PCH_2COOH + R^1\text{-}\underset{\substack{|\\ NH_2}}{CH}\text{-}R^2 \rightarrow Ph_2PCH_2\underset{\substack{\| \\ O}}{C}NH\underset{\substack{|\\ R^1}}{CH}\text{-}R^2 \qquad (7)$$

These chiral phosphines were tested as ligands in the hydrogenation of acetophenone and styrene catalyzed by rhodium(I) complexes prepared *in situ* from $[Rh(COD)Cl]_2$ and the ligands. The most active though only moderately enantioselective catalysts were obtained using the phosphines derived from L-α-phenylethylamine (FEA) and L-proline tertiary butyl ester (PT). Since these phosphines and their complexes are insoluble in water hydrogenations were carried out either in the neat substrate or in benzene solutions. In all cases an external base (usually triethylamine) was added to achieve measurable rates.

It was established that a small amount of water had no appreciable effect on the rate. However, in the presence of a *separate aqueous phase* (*e.g.* 20 v/v%) the rate increased very considerably (Table 3). The Rh-FEA catalyst which in the absence of an aqueous phase hardly showed any activity in styrene hydrogenation showed approximately the same activity in the presence of 20 v/v% water as the catalyst obtained in the reaction of $[Rh(COD)Cl]_2$ and PPh_3.

TABLE 3. Effect of water on the hydrogenation of acetophenone and styrene catalyzed by rhodium(I) phosphine complexes in biphasic systems

Substrate	Ligand	H_2O (v/v%)	Initial rate (turnover/h)
Acetophenone	PPh_3	-	2.5^a
Acetophenone	PPh_3	20	19.0
Acetophenone	FEA	-	-
Acetophenone	FEA	20	3.5
Acetophenone	PT	-	0.7^b
Acetophenone	PT	20	6.0^b
Styrene	PPh_3	-	35.0^c
Styrene	PPh_3	20	37.5^c
Styrene	FEA	-	0.5^c
Styrene	FEA	20	38.0^c

Conditions: T=60°C, P_{total}=1 bar, [Ligand]/[Rh]=3, $[Et_3N]/[Rh]=2$ [acetophenone]/[Rh]=435 (neat acetophenone), [styrene]/[Rh]=100 (benzene solution), ^{a} 1 v/v% water does not influence the rate
^{b} $[Et_3N]/[Rh]=1$ ^{c} $[Et_3N]/[Rh]=4$

The effect of bases on the rate of rhodium(I) phosphine catalyzed ketone hydrogenations is well understood [1] and we suggest that in the above cases, too, a dehydrochlorination of the initially formed H_2RhClP_3 takes place

($P=$ PPh_3, FEA or PT):

$$H_2RhClP_3 + Et_3N \rightleftharpoons HRhP_3 + Et_3N^+ + Cl^- \tag{8}$$

In nonpolar solvents, this equilibrium is governed by the reactivity of the dihydridochlororhodium(III) species. However, in the presence of a separate aqueous phase of sufficient volume the product triethylammonium chloride is extracted to the aqueous phase driving the equilibrium close to completion. Indeed, up to 85% of the theoretically required chloride could be determined in the aqueous phase by titration with $Hg(NO_3)_2$ (depending on the conditions and the phosphine used).

As in the case of homogeneous aqueous solutions (**Section 2**) in the above biphasic reactions, too, water influences the relative concentration of possible metal hydride intermediates. A literature example shows that such effects may also be exhibited in the change of selectivity: while in benzene solutions 3,6-octadienoic acid was hydrogenated by $RhCl(PPh_3)_3$ to yield 18% 6-octenoic and 70% 3-octenoic acid, in a biphasic mixture of benzene:water 1:1 the product contained 72% 6-octenoic acid and 15% 3-octenoic acid [14].

5. Conclusions

The examples discussed in the previous sections show conclusively that the presence of water both in homogeneous and in biphasic reaction mixtures may cause important changes in the mechanism of reactions catalyzed by water soluble tertiary phosphine complexes. Such mechanistic changes may manifest themselves in dramatic changes in reaction rates or in regio- and enantioselectivity. Care should be taken in controlling the acidity/basicity of such reaction mixtures since many of the equilibria involving the catalytically active metal complex intermediates is governed by the pH.

6. Acknowledgement

Part of this work was carried out in a joint program with Professor D.J. Darensbourg (Texas A&M University) and the authors are grateful for the fruitful collaboration. Financial support of this program by the National Science Foundation (Grants CHE91-19737 and INT90-08227) and the Hungarian Academy of Sciences (30.008/134/90) is greatly appreciated. F. Joó and Á. Kathó are grateful for financial support provided by the Hungarian National Research Foundation (OTKA 1699, T7527 and T4022) as well as Á.

Kathó for a travel grant from the Széchenyi Foundation, Budapest, Hungary.

7. References

1. Chaloner, P.A., Esteruelas, M.A., Joó, F. and Oro, L. (1994) *Homogeneous Hydrogenation (Catalysis by Metal Complexes)*, Kluwer Academic Publishers, Dordrecht, The Netherlands
2. Joó, F., Somsák, L. and Beck, M.T. (1984) Peculiar kinetics of hydrogenations catalyzed by chlorotris(sulphonated triphenylphosphine)rhodium(I) in aqueous solutions, *J. Mol. Catal.* **24**, 71-75
3. Joó, F., Csiba, P. and Bényei, A. (1993) Effect of water on the mechanism of hydrogenations catalyzed by rhodium phosphine complexes, *J.C.S., Chem. Commun.*, 1602-1604
4. Grushin, V.V. (1993) Reductive elimination of hydrogen chloride from chloro hydrido transition metal complexes. An efficient and simple method for generation of electron-rich, coordinatively unsaturated, reactive intermediates, *Acc. Chem. Res.* **26**, 279-286
5. Laghmari, M. and Sinou, D. (1991) Chiral sulfonated phosphines Part VI. Reduction in a two-phase system: is water only a solvent? *J. Mol. Catal.* **66**, L15-L18
6. Lecomte, L., Sinou, D., Bakos, J., Tóth, I. and Heil, B. (1989) Chiral sulphonated phosphines II. Influence of water on the enantioselectivity in the reduction of dehydro-aminoacids, *J. Organometal. Chem.* **370**, 277-284
7. Oehme, G., Paetzold, E., Selke, R. (1992) Increase in activity and enantioselectivity in asymmetric hydrogenation reactions catalysed by chiral rhodium(I) complexes as a consequence of the action of amphiphiles, *J. Mol. Catal.* **71**, L1-L5
8. Larpent, C. and Patin, H. (1988) Organic chemistry in water (part II). Nucleophilic addition of water-soluble phosphines on activated olefins, *Tetrahedron* **44**, 6107-6118
9. Bényei, A., Stafford, J.N.W., Kathó, Á., Darensbourg, D.J and Joó, F. (1993) The effect of phosphonium salt formation on the kinetics of homogeneous hydrogenations in water utilizing a rhodium *meta*-sulfonatophenyl-diphenylphosphine complex, *J. Mol. Catal.* **84** 157-163
10. Lee, S.W. and Trogler, W.C. (1990) Nucleophilic addition of phosphines to carbonyl groups. Isolation of 1-hydroxy phosphonium and 1-(trimethylsiloxy)phosphonium salts and the crystal structure of (1-hydroxy-1-methylethyl)-triphenylphosphonium bromide, *J. Org. Chem.* **55**, 2644-2648
11. Darensbourg, D.J., Joó, F., Kathó, Á., Stafford, J.N.W., Bényei, A. and Reibenspies, J.H. (1994) Nucleophilic addition of a water-soluble phosphine to aldehydes. Isolation of (1-hydroxyalkyl)phosphonium salts and the crystal structure of the (1-methoxy-1-benzyl)(*m*-sulfonatophenyl)-diphenylphosphonium salt, *Inorg. Chem.* **33**, 175-177
12. Abatjoglou, A.G. and Bryant, D.R. (1985) Mechanisms of catalyst activity/selectivity changes during the rhodium-triphenylphosphine-catalyzed low-pressure hydroformylation of allyl alcohol, *Arabian J. Sci. Eng.*, **10**, 428-436
13. Joó, F. and Trócsányi, E. (1982) Asymmetric hydrogenation of acetophenone with rhodium(I) complexes of new chiral phosphines derived from amino acids. An unusual modification of the catalyst system, *J. Organometal. Chem.* **231**, 63-70
14. Okano, T., Kaji, M., Isotani, S. and Kiji, J. (1992) Novel regioselective hydrogenation of alkadienoic acids caused by the addition of water, *Tetrahedron L.* **38**, 5547-5550

AQUEOUS CHEMISTRY AND CATALYTIC ACTIVITY OF ORGANOMETALLIC IRIDIUM COMPLEXES

J.S. MEROLA, T.L. HUSEBO, K. E. MATTHEWS, M.A. FRANKS, R. PAFFORD AND P. CHIRIK
Department of Chemistry
Virginia Polytechnic Institute and State University
Blacksburg, VA 24061-0212
USA

1.Introduction

Moving catalytic processes from organic solvents into water may have significant advantages: not only is water more environmentally benign, but product separation may be accomplished more readily. The bulk of the activity directed toward developing water-soluble catalysts centers around modifying currently used ligands, such as triphenylphosphine, to make them water soluble[1]. In turn, attachment of these now water-soluble ligands imparts water solubility to the metal complex catalyst.

Systems in which the aqueous solubility of an organometallic catalyst is not dependent on the attachment of a water-soluble ligand are not as well known[2]. In this paper, we report on the aqueous chemistry and catalytic activity of some organometallic iridium complexes. The water solubility of these compounds derives, not from the water solubility of a functionalized ligand, but rather from specific water-metal interactions.

2. The Chemistry of $[Ir(COD)(PMe_3)_3]Cl$

$[Ir(COD)(PMe_3)_3]Cl$, **1**, (COD = 1,5-cyclooctadiene) is readily synthesized by the reaction between $[Ir(COD)Cl]_2$ and trimethylphosphine (equation 1). The complex

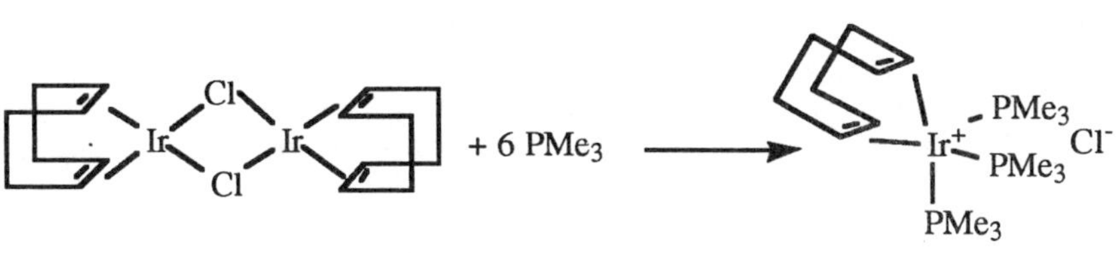 (1)

1

precipitates in nearly quantitative yield when the reaction is carried out in toluene[3]. The structure of **1** is square pyramidal in the solid state, although it is fluxional in solution. **1** displays significant water solubility: 1 g of **1** will dissolve in 1 mL of H_2O yielding a solution which is 1.8 M in iridium.

1 is also quite basic and is readily protonated on iridium. Simple aqueous titration with HCl yields a pK_a value for the conjugate acid of approximately 5.0 which indicates that **1** is approximately as basic as pyridine. Reaction between **1** and pyridinium chloride in water results in the equilibrium shown in equation 2. An equilibrium

I. T. Horváth and F. Joó (eds.), Aqueous Organometallic Chemistry and Catalysis, 33–45.

constant of approximately 0.8 is found for this reaction which again supports the notion that **1** is only slightly less basic than pyridine.

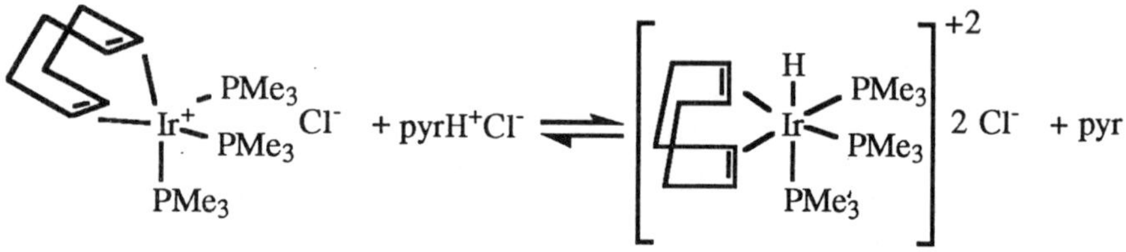

(2)

In aprotic solvents, **1** will react with anhydrous HCl to also form a protonated iridium species, **2**, but with two extra equivalents of HCl in the complex (equation 3).

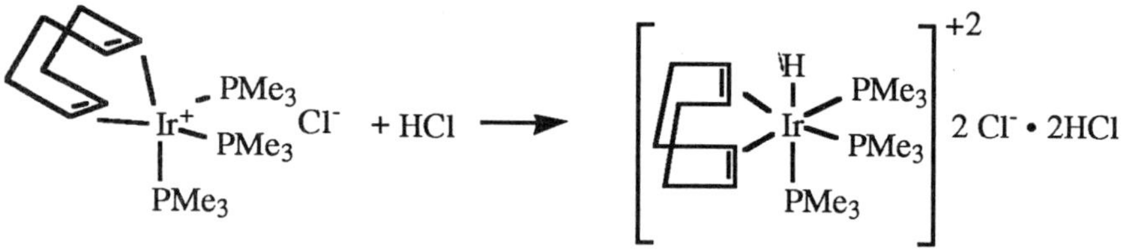

(3)

2

These extra equivalents of HCl are very tightly held and we have evidence that they are hydrogen-bonded to the chloride ions in the solid lattice. **2** is insoluble in essentially all common organic solvents, but is quite soluble in water. On standing at room temperature, aqueous solutions of **2** will deposit clear colorless crystals of fac-$IrCl_3(PMe_3)$ (equation 4). The fate of the COD ligand in this transformation is

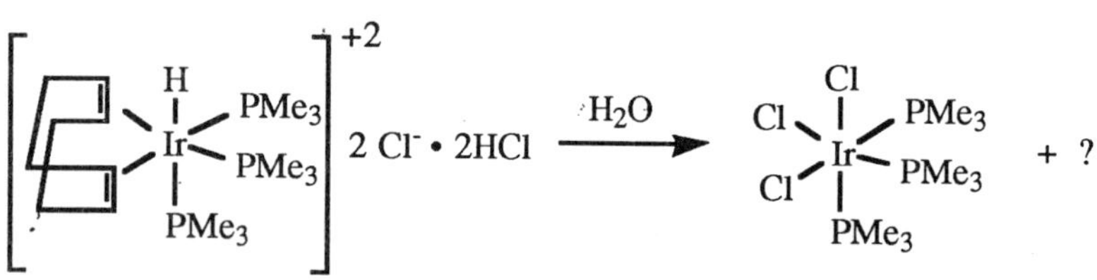

(4)

unknown. Apparently, the extra equivalents of HCl are responsible for this reaction.

Protonation of **1** produces a protonated complex, **3**, which contains one chloride ion, one BF_4 ion and one equivalent of HBF_4 (equation 5). Again, we believe that the extra

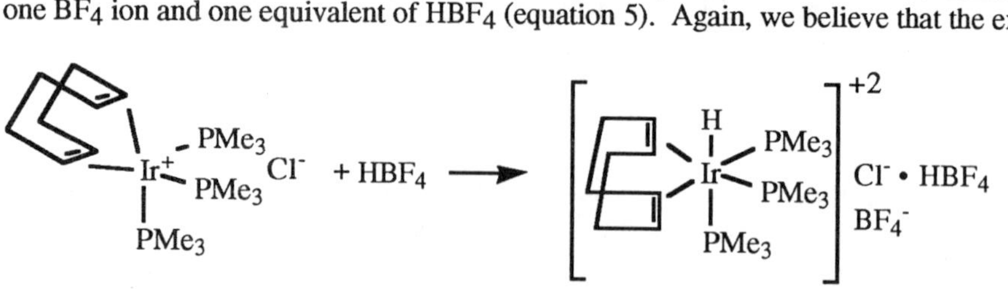

(5)

3

equivalent of acid is hydrogen bonded to the chloride ion in the solid lattice. Upon standing at room temperature in water, **3** transforms into a chloro-bridge dinuclear complex, **4** (equation 6 and figure 1).

The aqueous chemistry of chloroiridium compounds is largely unexplored and we plan to continue these investigations.

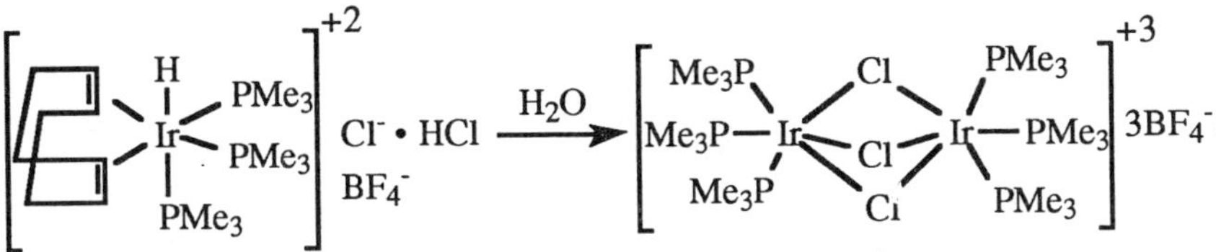

2   4   (6)

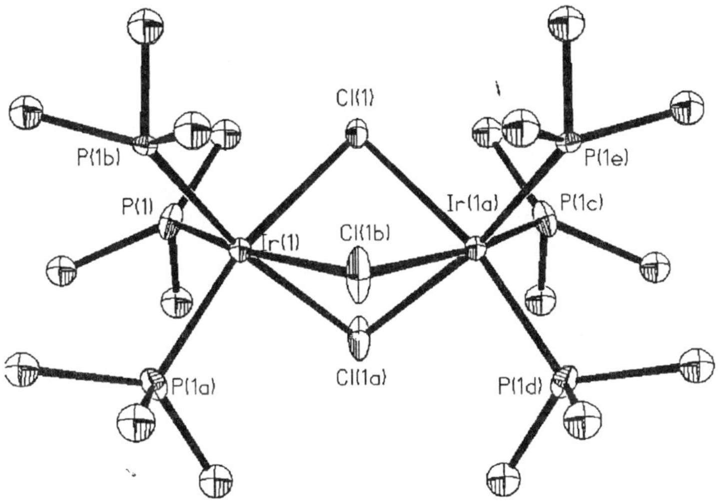

Figure 1. Crystal Structure of **4**

3. The Chemistry of cis,mer-$(Me_3P)_3IrH_2Cl$

The iridium(I) center in **1** is very electron rich and oxidatively adds H-H[4], B-H[5], C-H[6], N-H[7] and O-H[8] bonds. cis,mer-$(Me_3P)_3IrH_2Cl$, **5**, is the product from the oxidative addition

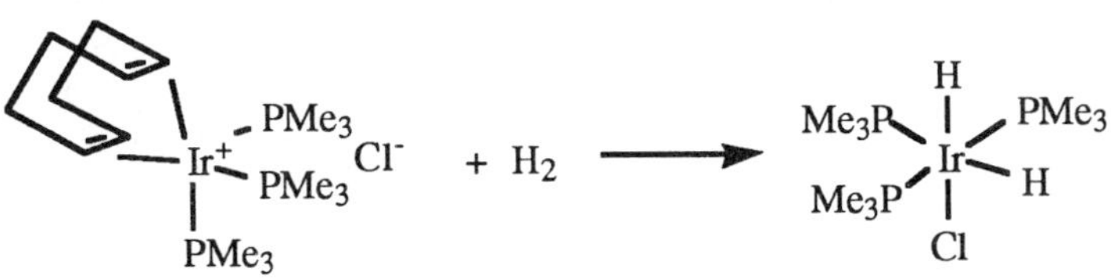

1   5   (7)

of H_2 to **1** (equation 7). This deceptively simple compound is water soluble and has a rich aqueous chemistry.

The spectral characterization of **5** is quite definitive and bears mentioning at this point. The ^{31}P NMR spectrum of **5** in CH_2Cl_2 displays a doublet for the mutually trans PMe_3 phosphorus atoms and a triplet for the center PMe_3 phosphorus atom. The ^1H NMR spectrum of **5** displays a resonance at δ -20 ppm (a quartet) attributable to the hydride ligand cis to all three PMe_3 groups and a doublet of triplets at δ -10 ppm arising from the hydride ligand trans to PMe_3. (There is additional fine structure to these resonances due to H-H coupling).

In probing the nature of the water solubility of **5**, we examined NMR spectra of this complex dissolved in D_2O. We were surprised to discover *two* sets of resonances with exactly the same features in both the ^1H and ^{31}P NMR spectra. So, on dissolution in water, two separate species are formed. (Removing the water and re-dissolving in CH_2Cl_2 regenerates the spectra described in the previous paragraph, so two different species exist only in water.] The hydride ligands on **5** will slowly exchange with the D in D_2O, so further investigations were carried out using ^{31}P NMR spectroscopy. Fresh solutions of **5** in D_2O display two triplets and two doublets in the ^{31}P NMR spectrum. After a period of time, the two triplets appear as a seven-line pattern as the hydride trans to the PMe_3 group responsible for that resonance becomes a trans deuteride and couples with that phosphorus.

That both species showed identical features in both ^1H and ^{31}P NMR spectra means that they both must have a meridional arrangement of PMe_3 ligands and cis dihydrides. This leaves only the sixth coordination site on the octahedron to differ between the two species. It seemed reasonable that one of the species would be the cationic aquo complex [mer,cis-$(Me_3P)_3IrH_2(OH_2)]^+$, **6**, which would result from dissociation of chloride from **5**. An attempt was made to isolate this (or whatever species exists in solution) by adding KPF_6 to an aqueous solution of **5**. A crystalline solid immediately precipitated and was identified as the unusual mono-chloro bridged diiridium complex $[(Me_3P)_3H_2Ir$-μ-Cl-$IrH_2(PMe_3)_3]^+$, **7** (equation 8 and figure 2). This is a rare example of a single

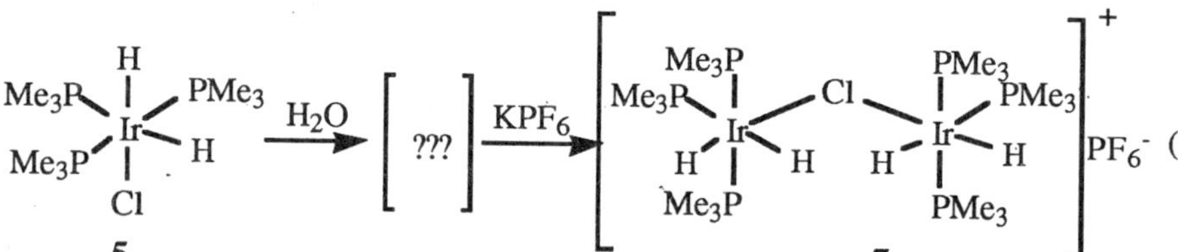

5 **7** (8)

unsupported halide bridge between two metal centers and is unprecedented for iridium.

With this structure in hand, we hypothesized that the two species in solution were the cationic aquo compound, **6**, and the diiridium cation, **7**. Further studies showed that the two species in solution were in equilibrium and were related by the loss or gain of chloride. Although **6** and **7** fit that criterion, a detailed investigation into the equilibrium showed that a dinuclear compound could *not* be part of the equilibrium. The only equilibrium expression that was consistent with the data is one involving the neutral cis,mer-$(Me_3P)_3IrH_2Cl$, **5**, and the cationic aquo complex, [cis,mer-$(Me_3P)_3IrH_2(OH_2)]^+$, **7** (equation 9). Apparently, the diiridium compound, **6**, is a very

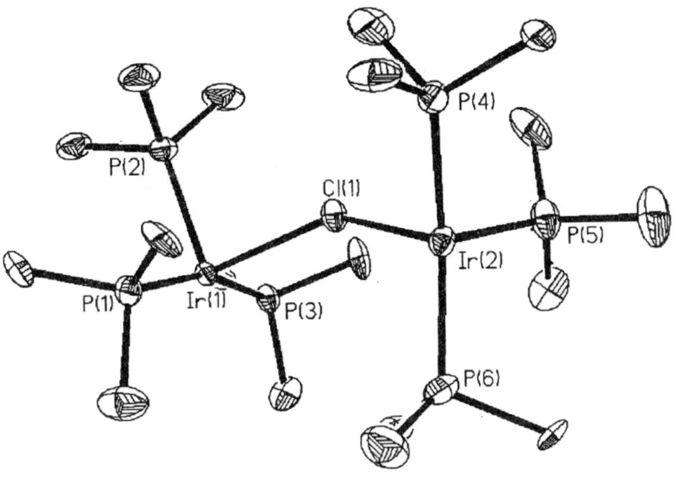

Figure 2. Crystal Structure of **7**.

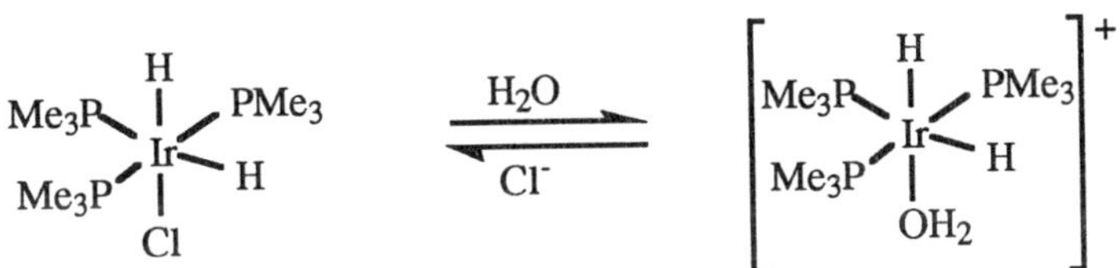

5 **6** (9)

minor component in the solution equilibrium, but precipitation drives the equilibrium to it.

The equilibrium between **5** and **7** was investigated as a function of temperature to extract some thermodynamic parameters. The van'tHoff plot of the data is shown in figure 3. ΔH was calculated to be 30.8 kJ/mol while ΔS is 56.0 J/K-mol. These

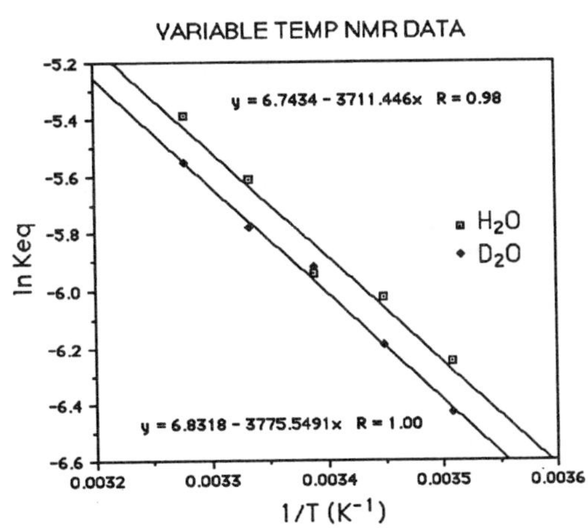

Figure 3. van't Hoff Plot from the Equilibrium between **5** and **7**.

values are consistent with the equilibrium shown in equation 9. Based on simple bonding concepts, the Ir-Cl bond is stronger than the Ir-O bond. Thus, on going from left to right (**5** to **7**), an Ir-Cl bond is broken and an Ir-O bond is made; so, the process should be endothermic. That the equilibrium proceeds to the right at all is due to an entropic driving force, the dissociation of chloride to form two separate ions.

Following this reasoning, we concluded that an iridium complex analogous to **5** but with and oxygen donor ligand in place of the chloride should more readily form the aquo species **7** since the bond energy differences would be negligible: The benzoato iridium complex, cis,mer-$(Me_3P)_3IrH_2(OC(O)Ph)$, **8**, was prepared via the series of reactions shown in equations 10-12.

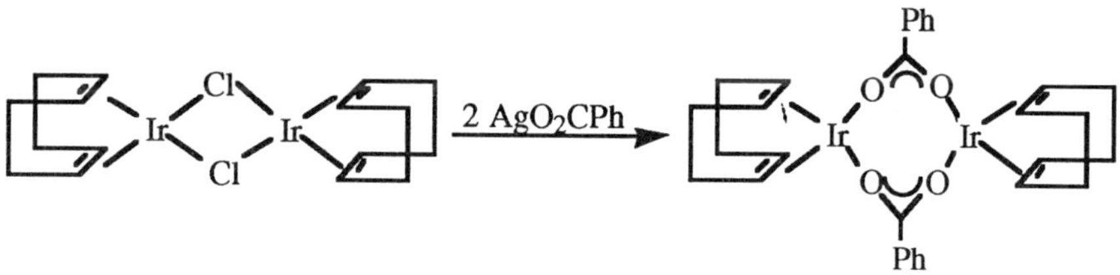

(10)

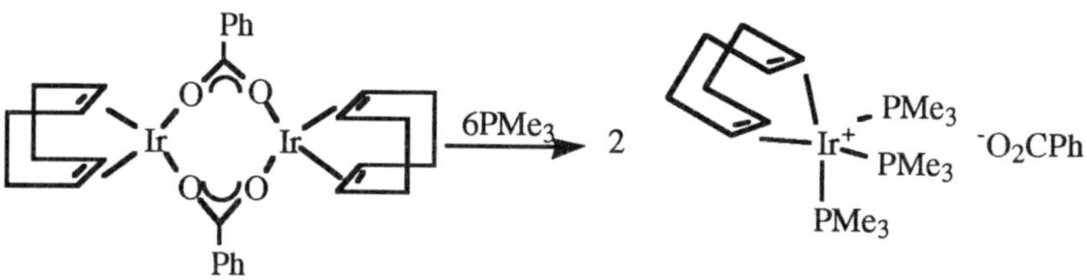

(11)

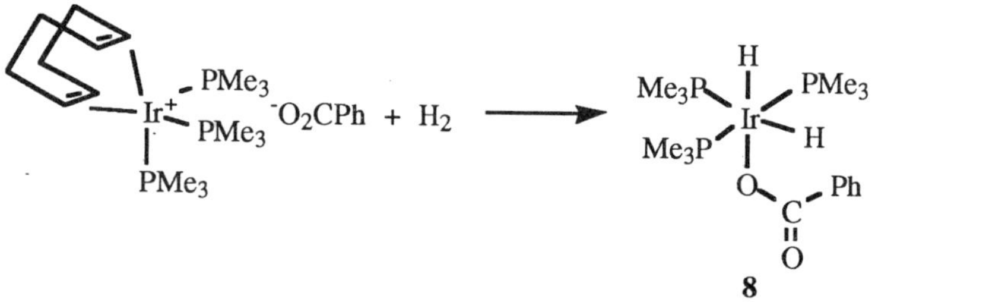

(12)

8

Complex **8** dissolves in water to form a single species whose spectroscopic data is identical to the aquo complex **7** (equation 13). Addition of one equivalent of chloride (as NaCl) yields the same equilibrium concentration of species as we obtained when dissolving **5** in water. This finding is consistent with the thermodynamic arguments stated above: ΔH for this reaction should be essentially 0, and so now the entropic term dominates and the benzoate completely dissociates.

One nagging question remains: why is the neutral complex **5** water soluble? Our only answer at this point is that significant hydrogen bonding occurs between the water and the hydride ligands and between the water and the bound chloride to bring that neutral species into aqueous solution. Conversely, replacing the chloride in **5** with a larger, softer ligand should make it more difficult to replace the ligand with water. An iodo analog of **5**, cis,mer-$(Me_3P)_3IrH_2I$ was synthesized and has negligible water solubility.

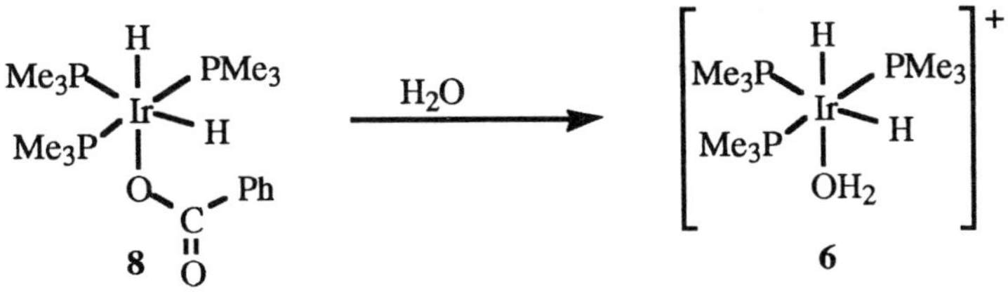

(13)

4. **Reactivity of cis,mer-$(Me_3P)_3IrH_2Cl$ and cis,mer-$(Me_3P)_3IrH_2(OC(O)Ph)$**

Cis,mer-$(Me_3P)_3IrH_2Cl$, **5**, is not reactive in common organic solvents, including polar ones such as DMSO. However, in water **5** will react rapidly with alkynes and alkenes. For example, **5** will react with both terminal and internal alkynes to form vinyl iridium hydride compounds, **9** (equation 14) and the structure of one member of this family of

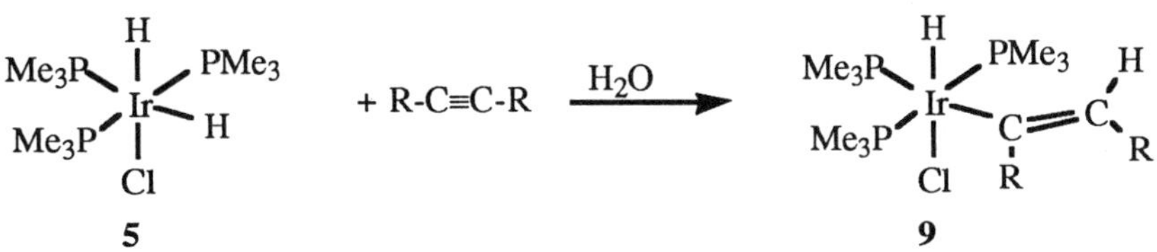

(14)

compounds is depicted in figure 4.

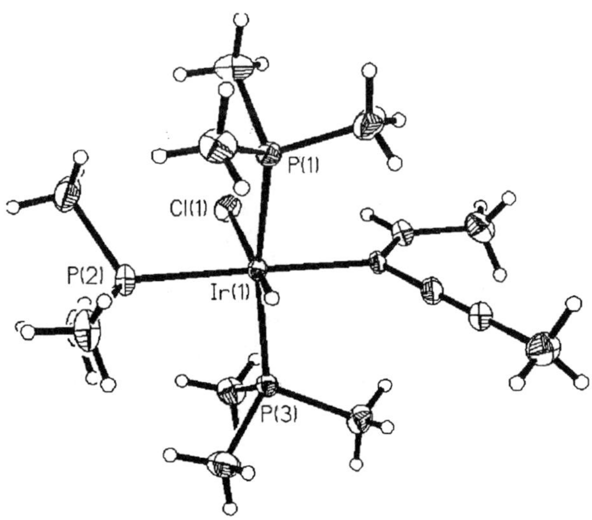

Figure 4. Crystal Structure of Vinyl Iridium Complex **9** Derived from Reaction Between **5** and 2,4-Hexadiyne

Complex **8** is even more reactive than the chloride, also reacting with alkynes in water to form vinyl iridium hydride complexes, **10** (equation 15). An example of one such compound is shown in figure 5.

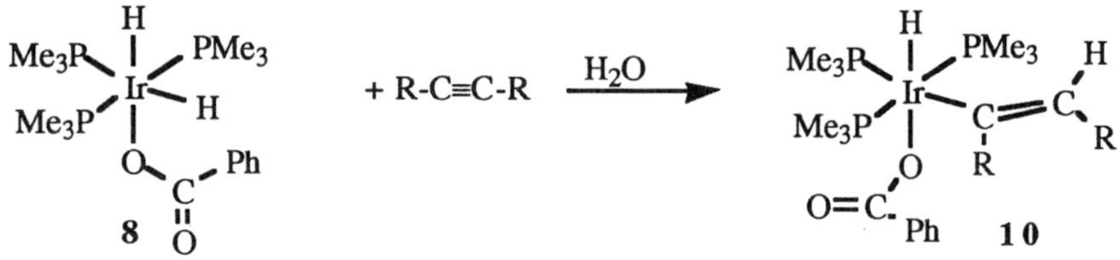

(15)

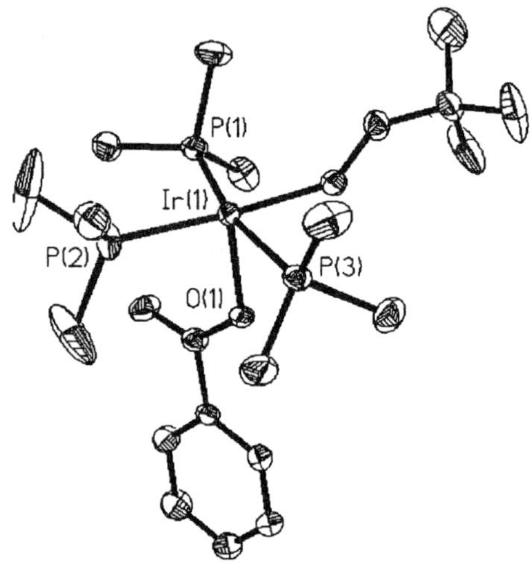

Figure 5. Crystal Structure of Vinyl Iridium Complex **10** Derived from Reaction Between **8** and t-Butylacetylene

5. Catalysis with cis,mer-$(Me_3P)_3IrH_2Cl$ and cis,mer-$(Me_3P)_3IrH_2(OC(O)Ph)$

Given that the initial insertion of organic unsaturates into the Ir-H bond is rapid in water to form a vinyl iridium hydride compound, it seemed reasonable that if reductive elimination of the vinyl and the hydride could be induced, that compounds **5** and **8** could act as hydrogenation catalysts in the presence of hydrogen. That is indeed the case, both compounds hydrogenating alkynes and alkenes to alkanes under relatively mild conditions (60° C, 6-8 atm H_2). We propose the mechanism depicted in figure 6 for the hydrogenation catalysis. This catalysis acts in a step-wise fashion, first hydrogenating an alkyne to an alkene and then hydrogenating the alkene to the alkane. We have discovered that this catalyst will also hydrogenate ketones to alcohols. These catalysts are active *only in water*. Attempts to carry out hydrogenation in a variety of organic solvents, both polar and non-polar, resulted in no reduction of alkynes or alkenes.

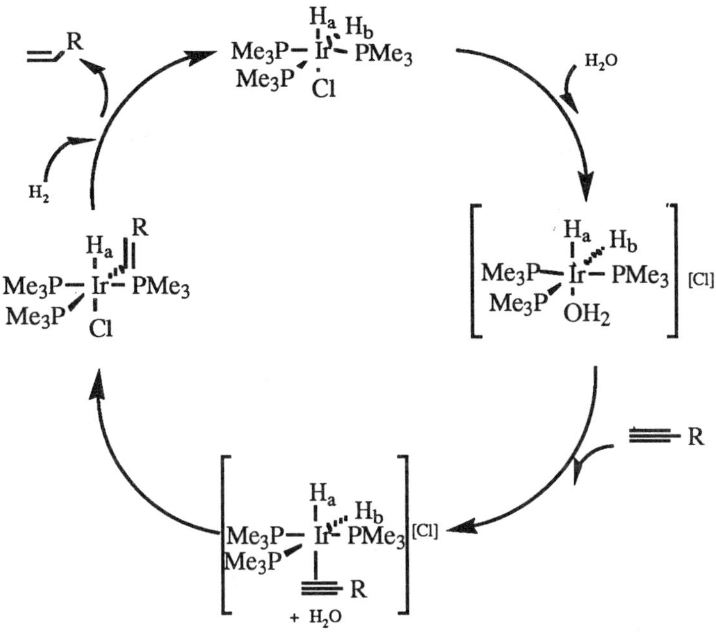

Figure 6. Proposed Catalytic Cycle for the Hydrogenation of Alkynes to Alkenes Using **5** as Catalyst.

6. Other Water-Soluble Iridium Phosphine Complexes

The tris-trimethylphosphine iridium compounds described above are all coordinatively saturated in their ground state. For catalytic activity of compounds such as **5** and **8**, an anionic ligand, either chloride or benzoate, must dissociate from the metal center. We were interested in synthesizing water-soluble iridium compounds with higher levels of unsaturation in order to obtain higher (or different) catalytic activity. We have approached this goal from two different directions. First, starting with complex **5**, we found that reaction with acid will convert the dihydride into a monohydride. For example, reaction between **5** and pyridinium chloride yields the cationic monohydrido pyridine iridium complex **11** (equation 16).

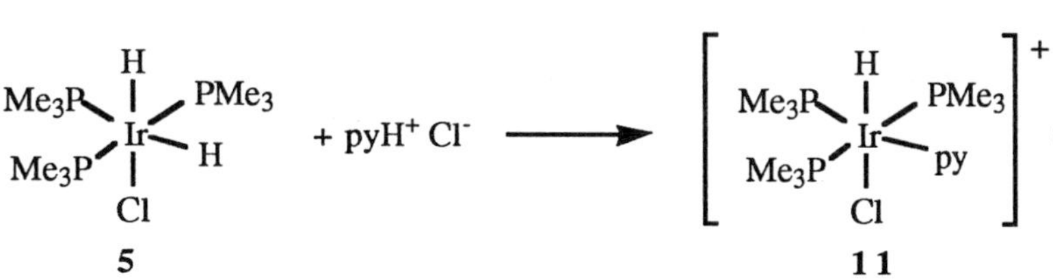 (16)

Complex **11** is water-soluble. While we are uncertain at this point of the nature of **11** in water, both the chloride and pyridine ligands are labile and this should lead to a very reactive iridium center.

The second approach to achieving higher coordinative unsaturation involves designing complexes with fewer phosphine ligands on the iridium. We were able to synthesize a bis-dimethylphosphinoethane (dmpe) complex of iridium, **12**, using the reaction shown in equation 16. This is a dmpe analog of some bis-diphenylphosphinoethane (dppe) compounds of iridium that have been synthesized by

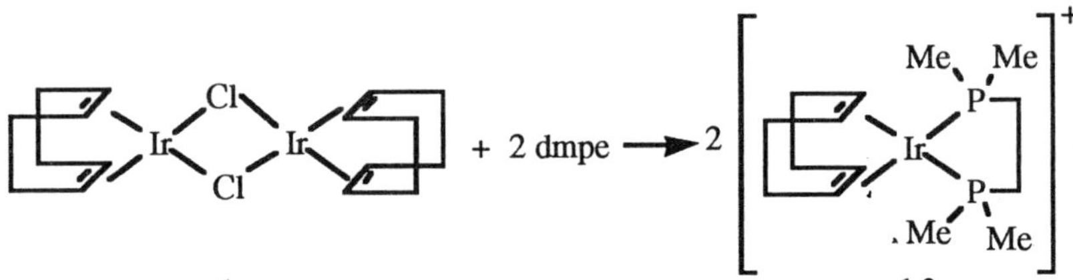

(17)

11 **12**

many others, but compound **12** is water soluble while the dppe compounds are not. Complex **12** is a square planar ionic compound as depicted in equation 17 when dissolved in water. In non-polar organic solvents or in the solid state, the chloride is bonded to the iridium yielding a 5-coordinate complex (figure 7).

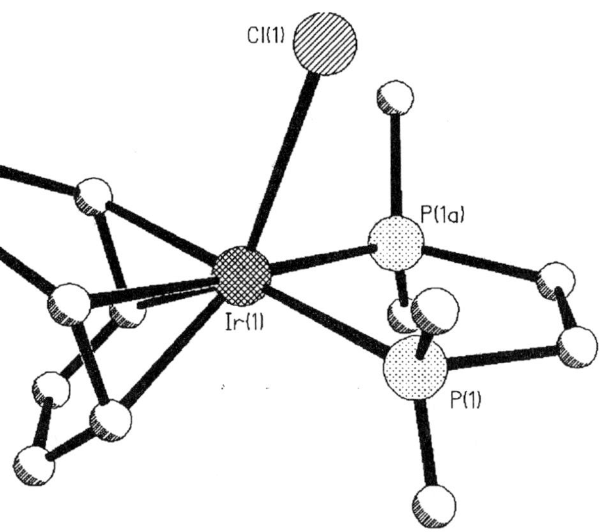

Figure 7. Crystal Structure of **12**.

That **12** is more reactive than **1** is evident from its reaction with hydrogen. While **1** requires elevated temperatures to react with H_2, **12** reacts with H_2 at room temperature to form the dihydrido complex **13**. Compound **13** is highly water-soluble and we are

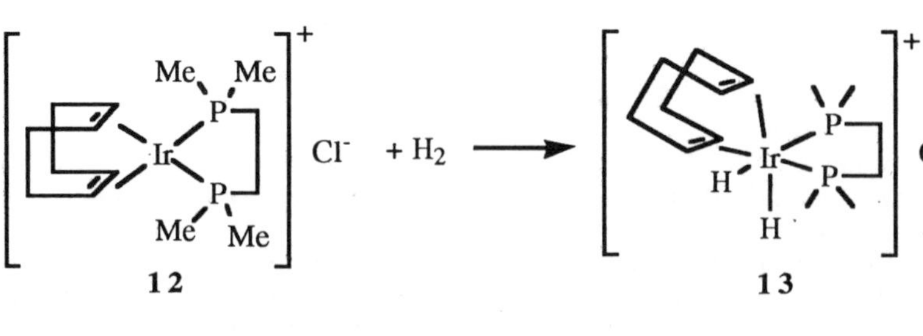

(18)

12 **13**

investigating its reactivity and possible catalytic activity.

7. Water-Soluble Organometallic Iridium Amine Complexes

In viewing these attempts to move the organometallic chemistry and catalysis of iridium into aqueous solutions, it seems that we are moving to bridge the gap between classical, Werner types of inorganic compounds and the organometallic compounds which have been studied more recently. To bridge that gap even further, it seemed as though organometallic complexes of iridium with amine ligands would be interesting targets. To make a direct analog of the dmpe complex, **12**, reaction between $[Ir(COD)Cl]_2$ and tetramethylethylenediamine (tmed) was studied. Instead of obtaining a complex analogous to **12**, we instead isolated the unusual salt $[(tmed)Ir(COD)][(COD)IrCl_2]$, **14** (equation 19 and figure 8).

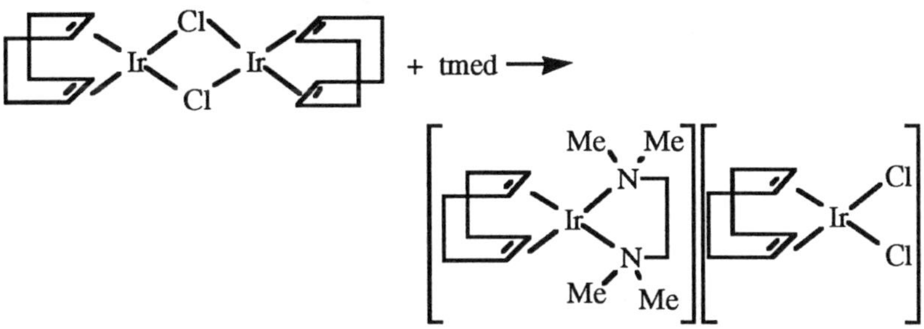

14 (19)

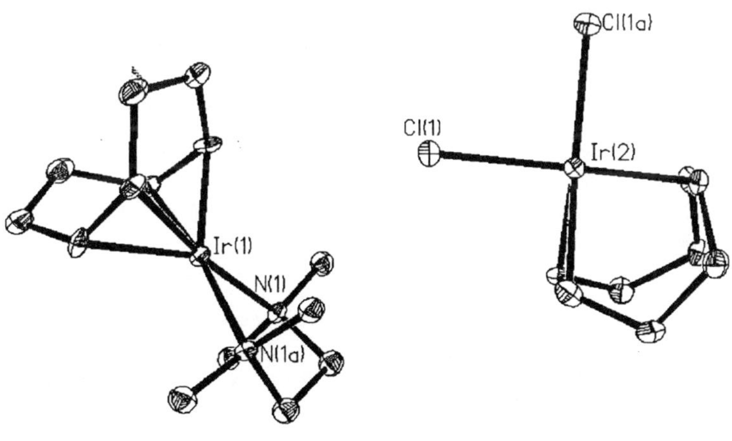

Figure 8. Crystal Structure of **14**.

The tmed ligand is somewhat labile and can be displaced by chloride. So, addition of chloride to solutions of **14** will drive the complex all the way to $[(COD)IrCl_2]^-$. Interestingly, tmed is the only diamine ligand that appears to be labile. Other chelating diamines, such as simple ethylenediamine, en, will form a complex that is analogous to **12** (equation 20). Like its phosphorus analog, complex **15** will add molecular hydrogen

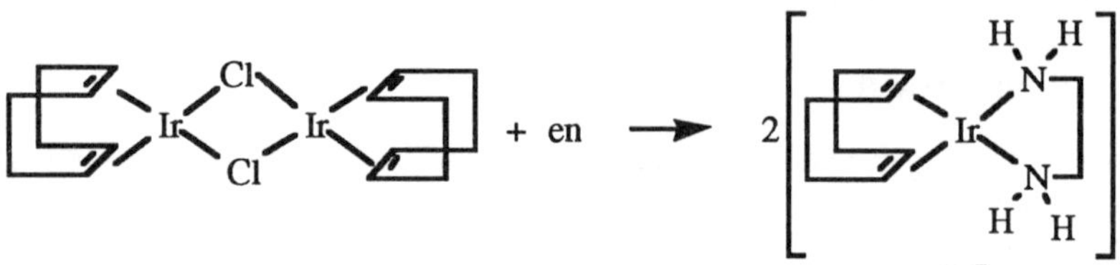

15

(20)

to yield the dihydrido complex **16** (equation 21).

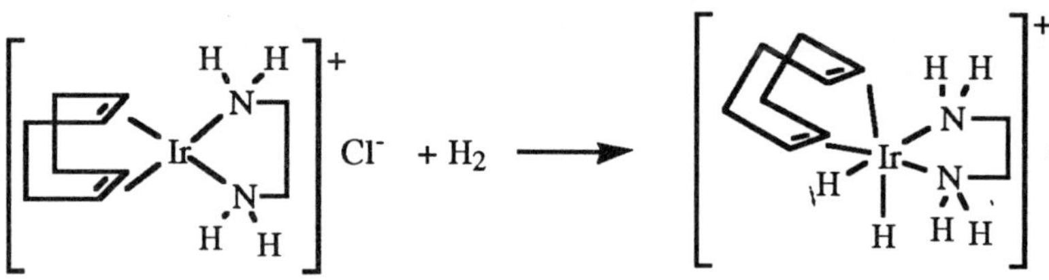

13

(21)

8. Conclusions

We have been able to synthesize and study a wide variety of organometallic iridium complexes that have significant water-solubility and a rich aqueous chemistry. The information that we have accumulated to date suggests that it will be possible to design and synthesize iridium complexes that can behave as catalysts for useful organic transformations in water. In fact, we have already demonstrated one such catalyst for hydrogenation chemistry. We plan to continue our investigations into the aqueous organometallic chemistry of iridium utilizing both phosphine and amine donor ligands.

9. Acknowledgments

We acknowledge financial support for this research from the National Science Foundation (NSF CHE 921-0427).

10. References

1. Herrman, W.A.; Kohlpaintner, C.W. *Angew. Chem. Int. Ed. Engl.* (1993) **32**, 1524.
2. See Barton, M.; Atwood, J. D. *J. Coord. Chem.* (1991) **24**, 43 and references therein.
3. a) Frazier, J. F.; Merola, J. S. *Polyhedron* (1992) **11**, 2917; b) Merola, J. S.; Kacmarcik, R. T. *Organometallics* (1989) **8**, 778.
4. Le, T.X.; Merola, J.S. *Organometallics* (1993) **12**, 3798.
5. Knorr, J. R.; Merola, J.S. *Organometallics* (1990) **9**, 3008.
6. a) Merola, J. S. *Organometallics* (1989) **8**, 2975; b) Selnau, H. E.; Merola, J. S. *J. Am. Chem. Soc.* (1991) **113**, 4008.
7. Ladipo, F. T.; Merola, J. S. *Inorg. Chem.* (1990) **29**, 4172.
8. Ladipo, F. T.; Kooti, M.; Merola, J. S. *Inorg. Chem.* (1993) **32**, 1681.

ORGANOMETALLIC RADICAL CHEMISTRY IN AQUEOUS SOLUTION

DAVID R. TYLER
University of Oregon
Department of Chemistry, Eugene, Oregon USA 97403

1. Introduction

Photochemically generated 17-valence electron metal radicals react with ligands to form 19-electron complexes, e.g., reaction 1 [1]. The 19-electron complexes are versatile reducing agents and we reported that they can be used to reduce a variety or organic, organometallic, and inorganic complexes, e.g., reaction 2. In addition to their versatility, two other features make the 19-electron complexes ideal choices as reducing agents for many applications. First, the complexes are easy to generate: the metal-metal bonded dimers are convenient precursors and the wavelength of irradiation ($\lambda > 500$ nm) generally precludes substrate absorption. Second, the 19-electron complexes are powerful reducing agents. In some cases, we estimate their oxidation potential to be 2 V or higher.

$$Cp_2Fe_2(CO)_4 \underset{}{\overset{h\nu}{\rightleftharpoons}} \underset{17\ e^-}{2\ CpFe(CO)_2} \underset{}{\overset{PR_3}{\rightleftharpoons}} \underset{19\ e^-}{CpFe(CO)_2PR_3} \quad (1)$$

$$(Cp = \eta^5\text{-}C_5H_5)$$

$$CpFe(CO)_2PR_3 + S \longrightarrow CpFe(CO)_2PR_3^+ + S^- \quad (2)$$

S = $CpMo(CO)_3Cl$, methylviologen (1,1'-dimethyl-4,4'-bipyridinium dication; MV^{2+}), Cp_2Co^+, $Ru_3(CO)_{12}$, *N*-n-butylpyridinium, $Mn_2(CO)_{10}$

One of our long-range goals is to explore the use of organometallic radicals in catalysis. In particular, we plan to exploit the reducing ability of 19-electron

I. T. Horváth and F. Joó (eds.), Aqueous Organometallic Chemistry and Catalysis, 47–60.

species for use as electron-transfer initiators and electron transfer relays in aqueous catalytic redox cycles (for example, in photochemical water-splitting schemes for solar conversion). However, a problem is that, because 19-electron complexes are generated from metal-metal bonded dimers, they cannot be used in aqueous solvents because all previously studied dimer complexes are soluble only in nonaqueous solvents. To overcome this problem, we began a research program to explore methods for generating organometallic radicals in aqueous solution.

Most organometallic complexes are water-solubilized by using appropriately functionalized phosphines as the water-solubilizing moiety; e.g., $HRh(CO)(PPh_3)_3$ is not soluble in water but $HRh(CO)(P(C_6H_4SO_3^-)_3)_3$ is soluble [2]. This approach, however, has its obvious limitations in the case of complexes without phosphines, such as the metal-metal bonded dimers. It was necessary therefore to develop other methods to render these complexes soluble. This paper reports on the various methods we have developed to water-solubilize the metal radicals and their precursors.

2. Generation of Radicals in Micelles

Our first attempts at generating and using metal radicals in aqueous solution were done in micellar solutions [3]. As is typical for micellar solutions, the key to the successful reduction of substrates with 19-electron complexes was the proper preparation of the solutions. In a typical preparation, $Cp_2Fe_2(CO)_4$ (5 mg) was dissolved in benzene (0.25 mL). This solution (20 μL) was then added to 3 mL of an aqueous solution containing the substrate to be reduced (typically about 10^{-3} M) and the surfactant CTAB (1.6 x 10^{-3} M; CTAB = hexadecyltrimethylammonium bromide). PBu_3 (5 μL) was then added to the solution. (PBu_3 could also be added to the benzene solution before mixing with the aqueous solution. In either case, an excess of PBu_3 must be avoided to prevent clouding the combined solutions.) The combined solutions, consisting of two phases, were then stirred or shaken for *ca.* 1 min., after which time the solution was homogeneous. The solution was then transferred to the appropriate spectroscopic cell or cuvette and irradiated ($\lambda > 500$ nm). When the substrate was MV^{2+}, esr and electronic absorption spectra of the reaction solution after irradiation clearly showed the formation of MV^+. The choice of phosphine ligand used to generate the 19-electron complex is not critical. Thus, for example, 1,2-bis(diphenylphosphino)ethane (dppe) gave results identical to those obtained above with PBu_3.

It is logical to propose that a 19-electron $CpFe(CO)_2PBu_3$ complex is the reducing agent in the reaction above. Unfortunately, IR detection of the

$CpFe(CO)_2PBu_3^+$ cation was impossible in the micellar solution. However, IR analysis of reactions run in MeCN showed the formation of $CpFe(CO)_2PBu_3^+$ [ν(C≡O) 2049 and 2003 cm^{-1}]. In addition, electronic absorption spectroscopic analysis of reactions run in MeCN also showed that *ca.* 2 moles of MV^+ were formed for every mole of $Cp_2Fe_2(CO)_4$ consumed, consistent with the stoichiometry in equations 1 and 2.

In addition to regular micelles, MV^{2+} was also reduced in reverse micellar solutions. The procedure was essentially identical to that above except that didodecyldimethylammonium bromide (DDAB) was used as the surfactant and the amounts of benzene and aqueous solution were reversed. MV^+ was again detected by electronic absorption and esr spectroscopy.

Other complexes that are soluble in water but insoluble in organic solvents can also be reduced in micellar solutions, e.g., $K_3Fe(CN)_6$. This complex was reduced by a procedure essentially identical to the one described above for MV^{2+}, the only difference being that sodium dodecyl sulfate was used as the surfactant. The product of the reduction was Prussian Blue.

In summary of this section, 17- and 19-electron complexes can be generated in micellar solutions and their chemistry exploited. However, a major disadvantage is that the amounts of material used are so small that it is hard to foresee any practical application of this technique.

3. Synthesis of Water-Soluble Dimers

In order to circumvent the problem of scale associated with the use of micelles, we decided to water-solubilize the radicals in a more direct fashion. The method we are pursuing is to functionalize the Cp rings of dimers containing Cp ligands with an ionizable or polar substituent. Two molecules synthesized thus far are $[(CpCH_2CH_2NH_3^+)_2Mo_2(CO)_6][NO_3^-]_2$ and $(CpCOOH)_2W_2(CO)_6$.

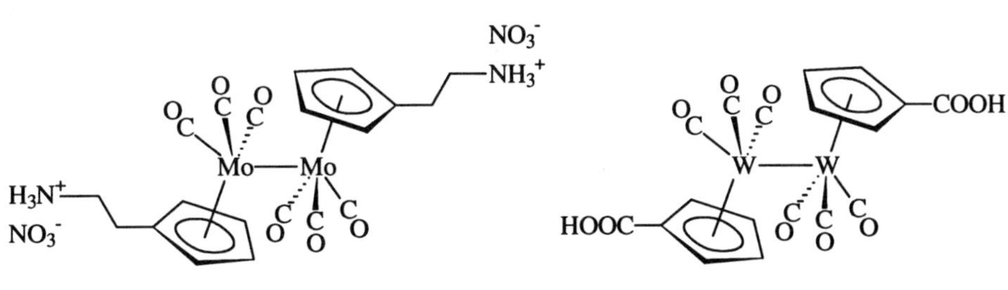

The $(CpCOOH)_2W_2(CO)_6$ dimer was synthesized by the route in Scheme I, which is a combination of the methods used by Rausch [6] for the substitution of a Cp ring and by Birdwhistell [7] for the synthesis of the $Cp_2M_2(CO)_6$

dimers. The synthesis of $[(CpCH_2CH_2NH_3^+)_2Mo_2(CO)_6][NO_3^-]_2$ is shown in Scheme II.

Scheme I. Synthesis of the $(CpCOOH)_2W_2(CO)_6$ Complex

$$Na^+[C_5H_5^-] \xrightarrow[\text{2. } W(CO)_6]{\text{1. } CO(OCH_3)_2} (C_5H_4C(O)OCH_3)W(CO)_3^-$$

$$(C_5H_4C(O)OCH_3)W(CO)_3^- \xrightarrow[\text{HOAc, } H_2O]{Fe(NO)_3)_3 \bullet 9H_2O,} (C_5H_4C(O)OCH_3)_2W_2(CO)_6$$

$$(C_5H_4C(O)OCH_3)_2W_2(CO)_6 \xrightarrow[\text{2. } CH_3COOH/H_2O]{\text{1. 2-propanol/}H_2O/KOH} (CpCOOH)_2W_2(CO)_6$$

Scheme II. Synthesis of the $[(CpCH_2CH_2NH_3^+)_2Mo_2(CO)_6][NO_3^-]_2$ Complex

$$C_5H_6 + C_2H_4O \xrightarrow[\text{TBOK}]{\text{TBOH}} C_5H_5CH_2CH_2OH$$

$$C_5H_5CH_2CH_2OH \xrightarrow[Et_3N,\ CH_2Cl_2]{TsCl} C_5H_5CH_2CH_2OTs$$

$$C_5H_5CH_2CH_2OTs \xrightarrow[\text{DMSO}]{NaN_3} C_5H_5CH_2CH_2N_3$$

$$C_5H_5CH_2CH_2N_3 \xrightarrow[Et_2O]{LiAlH_4} C_5H_5CH_2CH_2NH_2$$

$$C_5H_5CH_2CH_2NH_2 \xrightarrow[\text{2. } Mo(CO)_6]{\text{1. } Na^0} [(C_5H_4CH_2CH_2NH_2)Mo(CO)_3]^-$$

$$[(C_5H_4CH_2CH_2NH_2)Mo(CO)_3]^- \xrightarrow[H_2O]{Fe(NO_3)_3 \bullet 9H_2O} [(C_5H_4CH_2CH_2NH_3^+)_2Mo_2(CO)_6][NO_3^-]_2$$

The pK_a of the $(CpCOOH)_2W_2(CO)_6$ complex is 4.5, and the molecule is soluble in buffered aqueous solutions at pH 6 or higher. Measurements showed that the solubility of the dimer was at least 20 g/L (3×10^{-2} M) in buffered aqueous solution at pH 10 and 25 °C . Note that the dimer was deprotonated in the basic, buffered aqueous solvents employed in this study. The $[(CpCH_2CH_2NH_3^+)_2Mo_2(CO)_6][NO_3^-]_2$ complex is soluble in aqueous solutions at pH 8 or lower. Quantitative measurements showed that the dimer is soluble up to 2.0 g/L (2.8 mM) in pH 7 buffer at 25 °C. Note that the amine functionality will be protonated in all of the reactions described in this paper. The dimer with the nitrate counterion was insoluble in almost all common organic solvents except methanol. The NO_3^- counterion could be exchanged with PF_6^-, rendering the dimer soluble in THF.

The electronic spectra of the dimers are essentially identical to those of the unsubstituted parent complexes. Thus, the weak bands around 500 nm (≈2000 M^{-1} cm^{-1}) are assigned to $d\pi^* \rightarrow \sigma^*$ transitions and the intense bands at ≈370 nm (≈18000 M^{-1} cm^{-1}) are assigned to the $\sigma \rightarrow \sigma^*$ transitions. All of the photochemical reactions described below were carried out by irradiation into the low energy tail of the $d\pi^* \rightarrow \sigma^*$ bands ($\lambda > 525$ nm for general irradiations or $\lambda = 550$ nm for quantum yields). The dimers were also characterized by NMR and IR spectroscopy. The structure of the $(CpCOOH)_2W_2(CO)_6$ complex was also obtained by x-ray crystallography [5].

4. Photolysis of the Metal-Metal Bonds

In order to use the $[(CpCH_2CH_2NH_3^+)_2Mo_2(CO)_6][PF_6^-]_2$ and $(CpCOOH)_2$-$W_2(CO)_6$ complexes as a source of 17-electron radicals for the generation of 19-electron complexes, it was first necessary to establish that irradiation resulted in cleavage of the metal-metal bonds. This was accomplished by doing standard metal radical trapping experiments [8] in both aqueous and nonaqueous solutions. Thus, irradiation of $[(CpCH_2CH_2NH_3^+)_2Mo_2(CO)_6][PF_6^-]_2$ in 1 M CCl_4 in THF led to the quantitative formation of $[(CpCH_2CH_2NH_3^+)$-$Mo(CO)_3Cl][PF_6^-]$ (eq 3). Likewise, irradiation of $(CpCOOH)_2W_2(CO)_6$ gave $(CpCOOH)W(CO)_3Cl$ quantitatively. Metal chloride products such as these are taken as evidence of metal-metal bond homolysis in analogous systems.

$$[(CpCH_2CH_2NH_3^+)_2Mo_2(CO)_6][PF_6^-]_2 \xrightarrow[THF/CCl_4]{h\nu} 2\,[(CpCH_2CH_2NH_3^+)Mo(CO)_3Cl][PF_6^-] \quad (3)$$

Irradiation of *aqueous* solutions of $[(CpCH_2CH_2NH_3^+)_2Mo_2(CO)_6][NO_3^-]_2$ and CCl_3CH_2OH (a water-soluble chlorine atom donor) at pH 7 or pH 2 led to the quantitative formation of $[(CpCH_2CH_2NH_3^+)Mo(CO)_3Cl]^+$ (eq 4).

$$[(CpCH_2CH_2NH_3^+)_2Mo_2(CO)_6][NO_3^-]_2 \xrightarrow[\substack{H_2O\ (pH\ 2\ and\ 7)\\ CCl_3CH_2OH}]{h\nu} 2\,[(CpCH_2CH_2NH_3^+)Mo(CO)_3Cl][NO_3^-] \quad (4)$$

The quantum yield for disappearance of $[(CpCH_2CH_2NH_3^+)_2Mo_2(CO)_6]$-$[PF_6^-]_2$ in 1.0 M CCl_4 in THF (0.39 ± 0.04) is considerably smaller than for the analogous reaction in aqueous solution (0.78 ± 0.15 at pH 7). We explain these results as follows. In THF, the cationic dimer and its counterions probably exist as a closely associated ion pair. This may slow the rate of diffusion of the photogenerated geminate radical pair out of the cage, due to the increased mass and bulk of the diffusing species. Slower diffusion out of the cage will favor geminate recombination, resulting in lower quantum yields for nonaqueous systems. On the other hand, in aqueous systems, the dimer and its counterions probably exist as highly solvated ions, reducing the apparent mass and bulk, and thus favoring diffusion of the caged radical pair out of the solvent cage (relative to cage recombination). The effect of mass and sterics on the quantum yields is also apparent in the comparison of $[(CpCH_2CH_2NH_3^+)_2Mo_2$-$(CO)_6]$ to $(MeCp)_2Mo_2(CO)_6$: the quantum yield for disappearance of $[(CpCH_2CH_2NH_3^+)_2Mo_2(CO)_6]$ in 1.0 M CCl_4 in THF (0.39 ± 0.04) is smaller than that for disappearance of $(MeCp)_2Mo_2(CO)_6$ under identical conditions (0.55 ± 0.05). These same trends were observed in the quantum yields of the $(CpCOOH)_2W_2(CO)_6$ dimer.

5. Formation of 19-Electron Complexes

Our previous studies showed that 19-electron complexes are key intermediates in the photochemical disproportionation reactions of the metal-metal bonded carbonyl dimers [1]. Therefore, disproportionation of the dimers would strongly suggest the intermediate formation of 19-electron complexes. In fact, disproportionation did occur with both dimers in nonaqueous and aqueous solutions. For example, irradiation of $[(CpCH_2CH_2NH_3^+)_2Mo_2(CO)_6][PF_6^-]_2$ in the presence of PPh_3 in THF gave the disproportionation products shown in eq 5. Note that both $[(CpCH_2CH_2NH_3^+)Mo^{II}(CO)_3(PPh_3)]^{2+}$ and $[(CpCH_2CH_2$-$NH_3^+)Mo^{II}(CO)_2(PPh_3)_2]^{2+}$ form. Analogous mono- and di- substituted cationic products also form in the disproportionation reactions of the $Cp_2M_2(CO)_6$ complexes in nonaqueous solvents, so the formation of both products in eq 5 should not be construed as unusual. The disproportionation chain mechanism accounts for both products.

$$[(CpCH_2CH_2NH_3^+)_2Mo_2(CO)_6][PF_6^-]_2 \xrightarrow[PPh_3/THF]{h\nu}$$

$$[(CpCH_2CH_2NH_3^+)Mo^{II}(CO)_3PPh_3]^{2+}$$
$$[(CpCH_2CH_2NH_3^+)Mo^{II}(CO)_2PPh_3)_2]^{2+}$$
$$[(CpCH_2CH_2NH_3^+)Mo^0(CO)_3]^0 \qquad (5)$$

It is also important to note that the ammonium ion does not play a role in the overall disproportionation reaction when a good ligand such as a phosphine is present. However, in the absence of a better ligand, the ammonium functionality does play a role during photolysis of the dimer in nonaqueous solvent, and supramolecular chemistry is observed [9]. Analogous disproportionation chemistry is observed in nonaqueous solvents with the $(CpCOOH)_2W_2(CO)_6$ dimer.

The reaction of $[(CpCH_2CH_2NH_3^+)_2Mo_2(CO)_6][NO_3^-]_2$ in *aqueous* solution in the presence of the water-soluble phosphine PTA gave the disproportionation products shown in eq 6.

$$[(CpCH_2CH_2NH_3^+)_2Mo_2(CO)_6][NO_3^-]_2 \xrightarrow[PTA/H_2O\ pH\ 7]{h\nu}$$

$$[(CpCH_2CH_2NH_3^+)Mo^{II}(CO)_3PTA]^{2+}$$
$$[(CpCH_2CH_2NH_3^+)Mo^{II}(CO)_2(PTA)_2]^{2+}$$
$$[(CpCH_2CH_2NH_3^+)Mo^0(CO)_3]^0 \qquad (6)$$

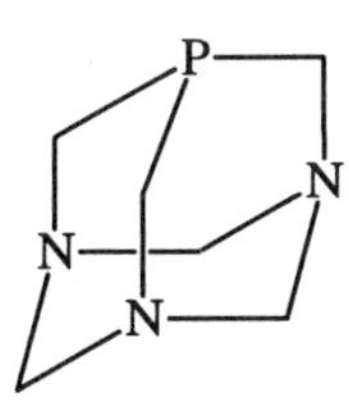

phosphatriazaadamantane (PTA)

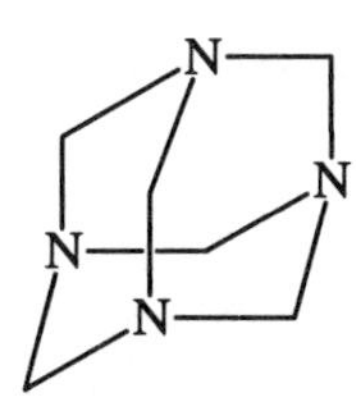

tetraazaadamantane $[N_4(CH_2)_6]$

Note again that both mono- and di- substituted cationic products $[(CpCH_2CH_2NH_3^+)Mo^{II}(CO)_3(PTA)^{2+}]$ and $[(CpCH_2CH_2NH_3^+)Mo^{II}(CO)_2(PTA)_2]^{2+}$ are formed, just as in nonaqueous solvents. Although the PTA ligand has nitrogen atoms that could conceivably coordinate to the molybdenum, it is more likely that coordination is through the phosphorus atom. Two points support this claim. First, previous work showed that the phosphorus atom in the PTA ligand preferentially binds to low valent group VI metals [10]. Second, in control experiments using tetraazaadamantane in place of PTA, no cationic products containing the $N_4(CH_2)_6$ ligand formed.

When no suitable ligands are present, H_2O will act as the disproportionating ligand, giving the products shown in eqs 7 and 8.

$$[(CpCH_2CH_2NH_3^+)_2Mo_2(CO)_6][NO_3^-]_2 \xrightarrow[H_2O\ pH\ 7]{h\nu} [(CpCH_2CH_2NH_3^+)Mo^{II}(CO)_3H_2O]^{2+} \quad [(CpCH_2CH_2NH_3^+)Mo^0(CO)_3]^0 \tag{7}$$

$$[(CpCH_2CH_2NH_3^+)_2Mo_2(CO)_6][NO_3^-]_2 \xrightarrow[H_2O\ pH\ 2]{h\nu} [(CpCH_2CH_2NH_3^+)Mo^{II}(CO)_3H_2O]^{2+} \quad [(CpCH_2CH_2NH_3^+)Mo^0(CO)_3H]^+ \tag{8}$$

These disproportionation reactions are analogous to the reactions in coordinating organic solvents in the absence of a ligand. Again, analogous disproportionation reactions were observed with the $(CpCOOH)_2W_2(CO)_6$ dimer in aqueous solution.

In summary of this section, because 19-electron species are generally accepted as intermediates in the disproportionation reactions, the disproportionation reactivity of the $[(CpCH_2CH_2NH_3^+)_2Mo_2(CO)_6][NO_3^-]_2$ and $(CpCOOH)_2W_2(CO)_6$ dimers strongly suggests that irradiation of these complexes in the presence of appropriate ligands results in the formation of 19-electron species:

$$(CpCH_2CH_2NH_3^+)_2Mo_2(CO)_6^+ \xrightarrow[\text{aqueous solution}]{h\nu} \underset{19\ e^-}{2\ (CpCH_2CH_2NH_3^+)Mo(CO)_3L} \tag{9}$$

$$L = PTA,\ H_2O$$

$$(CpCOO^-)_2W_2(CO)_6 + 2\ L \xrightarrow[\text{aqueous solution}]{h\nu} \underset{19\ e^-}{2\ (CpCOO^-)W(CO)_3L} \tag{10}$$

6. Reduction of Water-Soluble Substrates

The 19-electron complexes generated in aqueous solution can be used as photogenerated reducing agents. The substrates reduced were methyl viologen (MV^{2+}), ferricyanide, and cytochrome c. For each of these substrates, one-electron reduction occurred. Generalized reactions are shown in eqs 11 and 12.

$$[(CpCH_2CH_2NH_3^+)_2Mo_2(CO)_6]^{2+} + 2\ PTA + 2\ S \xrightarrow[H_2O\ pH\ 7]{h\nu}$$

$$2\ S^- + [(CpCH_2CH_2NH_3^+)Mo^{II}(CO)_3PTA]^{2+} \qquad (11)$$

$S = MV^{2+}$ (methyl viologen), cytochrome c^{III}, $Fe(CN)_6^{3-}$

$$(CpCOO^-)_2W_2(CO)_6 + 2\ PR_3 + 2\ S \xrightarrow[H_2O]{h\nu}$$

$$2\ S^- + (CpCOO^-)W(CO)_3(PR_3) \qquad (12)$$

For each dimer, the mass balance was studied for the reduction of MV^{2+}. In each case, in the presence of excess MV^{2+}, the reduction of MV^{2+} to $MV^{+\bullet}$ was quantitative, based on the amount of dimer present.

The electrochemical potentials of the $[(CpCH_2CH_2NH_3^+)Mo(CO)_3(PTA)]^0$ and $(CpCOO^-)W(CO)_3(PR_3)$ complexes can be estimated by their ability to reduce the substrates in reactions 11 and 12. The potential of the $MV^{2+}/MV^{+\bullet}$ couple (the most difficult to reduce substrate in eqs 11 and 12) is -0.46 V vs. SCE; thus, the potential of the 18-/19-electron complex couple is more negative than this value. In fact, the potential is probably much more negative than -0.46 V vs. SCE because one of the key steps in the disproportionation reaction chain mechanism is the reduction of the dimer reactant by a 19-electron complex (e.g., the reduction of $[(CpCH_2CH_2NH_3^+)_2Mo_2(CO)_6]^{2+}$ by $[(CpCH_2CH_2NH_3^+)Mo(CO)_3L)]^+$ (L = PPh_3, PTA, or H_2O) is a key step in the disproportionation). The electrochemical potential of the 18-/19-electron complex couple must, therefore, be more negative than the reduction potential of the dimer complexes. The potentials of the metal carbonyl dimers are less than -1 V vs. SCE, and thus the potential of the 18-/19-electron complex couple can be estimated as less than -1 V vs. SCE.

7. A New Water-Soluble Ligand

A limitation in the study of aqueous 19-electron complexes is the small number of appropriate water-soluble ligands, chelating phosphines in particular. Commonly used water-soluble phosphines include I ("triphenylphosphine-monosulfonate," TPPMS), II ("triphenylphosphinetrisulfonate," TPPTS), and various sulfonated-derivatives of dppe (1,2-bis(diphenylphosphino)ethane) and related chelating ligands (e.g., complex III).

$P(C_6H_5)_2(C_6H_4SO_3^-)$ — **I** — TPPMS

$P(C_6H_4SO_3^-)_3$ — **II** — TPPTS

$(^-O_3SH_4C_6)_2P{-}CH_2CH_2{-}P(C_6H_4SO_3^-)_2$ — **III**

Unfortunately, the use of these ligands can be problematic. For example, TPPMS is not soluble in aqueous solution at room temperature, and TPPTS and the sulfonated diphosphines are difficult to purify.

For our work with water-soluble 19-electron complexes, we needed a ligand that was readily soluble in aqueous solution at room temperature, that could be synthesized without complication in a pure state, and that was bidentate. For that reason, we synthesized the L_2 ligand, a new ligand that meets these requirements [11].

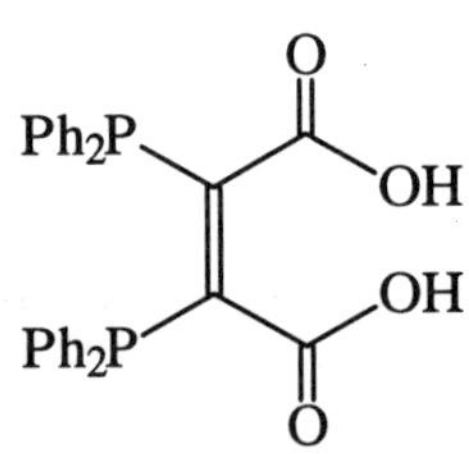

L_2

L_2 was synthesized by the route in Scheme III and was easily purified by extraction followed by crystallization from diethyl ether. The structure of L_2 was confirmed by X-ray crystallography. Ligand L_2 is readily soluble in aqueous solution at pH 5 or greater. For example, the solubility is >1 M at pH 7. (The ligand is, of course, deprotonated at this pH.)

Scheme III. Synthesis of L_2

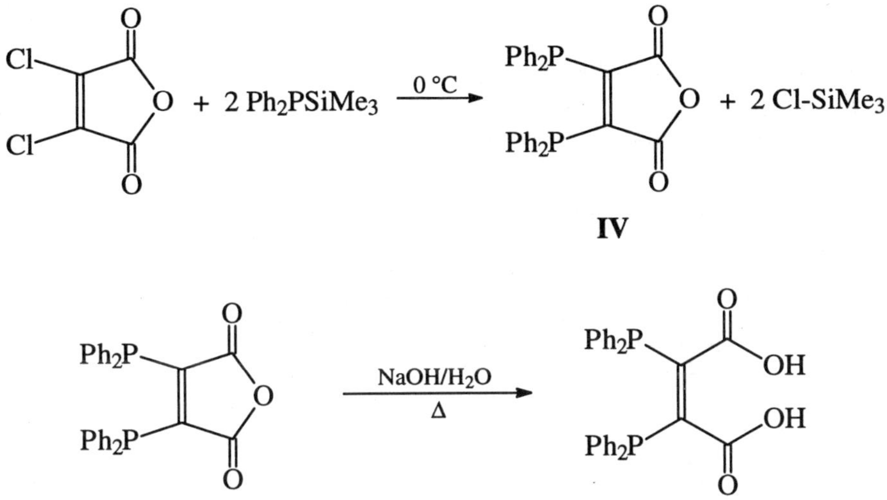

IV

L_2

To demonstrate the utility of L_2 in aqueous organometallic chemistry, we irradiated ($\lambda > 500$ nm) solutions of the water-soluble complexes $(CpCOOH)_2$-$W_2(CO)_6$ and $[(CpCH_2CH_2NH_3^+)_2Mo_2(CO)_6][NO_3^-]_2$ in the presence of L_2. The disproportionation reactions (eqs 13 and 14) proceeded analogously to the reaction of $Cp_2M_2(CO)_6$ (M = Mo or W) and dppe (1,2-bis(diphenylphosphino)-ethane) in benzene solution (eq 15).

$$[(CpCOO^-)_2W^I_2(CO)_6]^{2-} + L_2^{2-} \xrightarrow[H_2O\ pH\ 10]{h\nu}$$

$$[(CpCOO^-)W^{II}(CO)_2(L_2^{2-})]^{2-} + [(CpCOO^-)W^0(CO)_3]^{2-} + CO \qquad (13)$$

$$[(CpCH_2CH_2NH_3^+)_2Mo^I_2(CO)_6][NO_3^-]_2 + L_2^{2-} \xrightarrow[H_2O\ pH\ 7]{h\nu}$$

$$[(CpCH_2CH_2NH_3^+)Mo^{II}(CO)_2(L_2^{2-})]^0 +$$

$$[(CpCH_2CH_2NH_3^+)Mo^0(CO)_3]^0 + 2\ NO_3^- + CO \qquad (14)$$

L_2^{2-} = deprotonated L_2

$$Cp_2Mo_2(CO)_6 + dppe \xrightarrow{h\nu} CpMo^{II}(CO)_2dppe^+ + CpMo^0(CO)_3^- + CO \quad (15)$$

In the presence of reducible substrates, disproportionation was inhibited, and reduction of the substrate occurred as exemplified by the following reaction in which methyl viologen (MV^{2+}) is reduced:

$$Cp_2W_2(CO)_6 + 2\ dppe + 2\ MV^{2+} \longrightarrow 2\ CpW(CO)_2dppe^+ + 2\ MV^{+\bullet} + 2\ CO \quad (16)$$

An analogous reaction occurred with the L_2 ligand and $[(CpCOO^-)_2W_2(CO)_6]^{2-}$ in aqueous solution:

$$[(CpCOO^-)_2W_2(CO)_6]^{2-} + 2\ L_2^{2-} + 2\ MV^{2+} \xrightarrow[H_2O\ pH\ 10]{h\nu} 2\ [(CpCOO^-)W^{II}(CO)_2(L_2^{2-})]^{2-} + 2\ MV^{+\bullet} + 2\ CO \quad (17)$$

In summary, L_2 is an aqueous-soluble ligand that is easily synthesized and purified. In aqueous solution, L_2^{2-} reacts analogously to the dppe ligand in nonaqueous solvents; it should find a wide range of uses in the emerging area of aqueous organometallic chemistry.

8. Conclusions

Although metal radicals can be generated in aqueous solution by solubilizing them (or their precursors) with micelles, the low concentrations involved make this an impractical technique for most purposes. A better method for doing aqueous metal radical chemistry is to solubilize the molecules using appropriate ionic or polar functional groups on the ligands. The method discussed above involves substituting the Cp rings of metal-metal dimers with either the -COOH or $-CH_2CH_2NH_3^+$ groups. The $(CpCOOH)_2W_2(CO)_6$ complex is soluble in aqueous solution at pH 6 or higher, and the $[(CpCH_2CH_2NH_3^+)_2Mo_2(CO)_6][NO_3^-]_2$ dimer is soluble at pH 8 or lower. The photochemistry of these dimers in aqueous solvents is essentially identical with the photochemistry of the unsubstituted dimers in nonaqueous solvents. Thus, irradiation at wavelengths longer than 500 nm results in homolytic cleavage of the metal-metal bond to form 17-electron radicals. Evidence for these radicals comes from trapping reactions with aqueous-soluble radical traps. For example, CCl_3CH_2OH reacts with $(CpCH_2CH_2NH_3^+)Mo(CO)_3^{\cdot}$ to form $(CpCH_2CH_2NH_3^+)Mo(CO)_3Cl$. In the presence of a suitable water-soluble ligand, photochemical disproportionation occurs. Disproportionation reactions are taken as indicative of the formation of 19-electron complexes. In the presence of a reducible substrate, the disproportionation reactions are inhibited and substrate reduction occurs.

The synthesis of the chelating ligand L_2 was reported. This ligand is water soluble, and it reacts with 17-electron radicals in aqueous solution to form 19-electron complexes.

It is to be emphasized that, contrary to some expectations, the 17- and 19-electron radicals formed in these studies do not react with water. No evidence for metal oxides was observed, and the 19-electron complexes did not reduce water to form hydrogen. In all aspects, the chemistry of these radicals in aqueous solution appears to mimic their reactivity in nonaqueous solution. It is also noted that no evidence for involvement of the sidechains (either the -COOH or $-CH_2CH_2NH_3^+$ group) was found in any of the reactions.

Finally, the technique of using ionic sidechains on Cp rings to water-solubilize organometallic complexes is general. In another project in my lab, the compound below was synthesized, and it is water soluble.

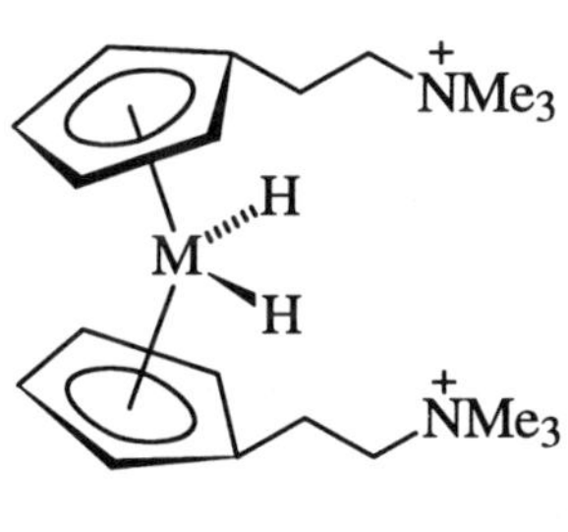

Acknowledgment is made to the U. S. National Science Foundation for the support of this research. Marcie Wood is acknowledged for her assistance in the preparation of this manuscript.

References

1. Tyler, D. R. (1991) 19-Electron organometallic adducts, *Acc. Chem. Res.* **24**, 325-331.
2. Barton, M. and Atwood, J. D. (1991) Aqueous soluble organometallic complexes, *J. Coord. Chem.* **24**, 43-67.
3. MacKenzie, V. and Tyler, D. R. (1987) Reduction of water-soluble substrates in micellar solutions using photochemically generated nineteen-electron organometallic complexes, *J. Chem. Soc., Chem. Commun.* 1783-1784.

4. Avey, A. and Tyler, D. R. (1992) Synthesis and photochemistry of the aqueous-soluble $[(CpCH_2CH_2NH_3^+)_2Mo_2(CO)_6]$ complex. Generation of 19-electron complexes in aqueous solution, *Organometallics* **11**, 3856-3863.
5. Avey, A., Tenhaeff, S. C., Weakley, T. J. R., and Tyler, D. R. (1991) Organometallic photochemistry in aqueous solution. Synthesis, crystal and molecular structure, and photochemistry of the $(CpCOOH)_2W_2(CO)_6$ complex. Generation of 19-electron organometallic complexes in aqueous solution and their use as reducing agents. Photochemical production of hydrogen, *Organometallics* **10**, 3607-3613.
6. Macomber, D. W. and Rausch, M. D. (1993) Functionally substituted cyclopentadienyltricarbonylmethyl derivatives of chromium, molybdenum, and tungsten, *J. Organomet. Chem.* **258**, 331-341.
7. Birdwhistell, R., Hackett, P., and Manning, A. R. (1978) A simple and effective preparation of $(\eta\text{-}RC_5H_4)_2M_2(CO)_6$ complexes (M = Cr, Mo and W), *J. Organomet. Chem.* **157**, 239-241.
8. Meyer, T. J. and Caspar, J. V. (1985) Photochemistry of metal-metal bonds, *Chem. Rev.* **85**, 187-218.
9. Avey, A., Weakley, T. J. R., and Tyler, D. R. (1993) Supramolecular photochemistry of the $[(CpCH_2CH_2NH_3^+)_2Mo_2(CO)_6][PF_6^-]_2$ complex. Chemical ramifications of a tentacle ligand covalently bonded nearby to a reactive metal center, *J. Am. Chem. Soc.* **115**, 7706-7715.
10. Darensbourg, M. Y. and Daigle, D. (1975) Synthesis and spectral properties of metal carbonyl derivatives of the tetrabasic, nonchelating phosphatriazaadamantane ligand, *Inorg. Chem.* **14**, 1217-1218.
11. Avey, A., Schut, D. M., Weakley, T. J. R., and Tyler, D. R. (1993) A new water-soluble phosphine for use in aqueous organometallic systems. Products for the reactions of 2,3-bis(diphenylphosphino)maleic anhydride with water and oxygen, *Inorg. Chem.* **32**, 233-236.

THE USE OF 1,3,5-TRIAZA-7-PHOSPHAADAMANTANE (PTA), A WATER-SOLUBLE , AIR STABLE LIGAND, IN ORGANOMETALLIC CHEMISTRY AND CATALYSIS

DONALD J. DARENSBOURG, TARA J. DECUIR, AND JOSEPH H. REIBENSPIES
Department of Chemistry
Texas A&M University
College Station, Texas 77843, USA

Interest in organometallic chemistry in aqueous media is increasing dramatically because of its far-reaching applications in large-scale industrial processes [1], fine organic synthesis [2-4], and the biological sciences [5]. Water's highly protic, very polar nature and its ability to form strong hydrogen bonds and act as a donor ligand make it a unique and interesting solvent. It is economical, non-toxic, and environmentally acceptable and can also exert a positive influence on regioselectivity and activity [6]. Solubility of transition metal complexes in aqueous media is usually accomplished by the attachment of charged or polar substituents (e.g. $-SO_3H$, -COOH, -OH, $-NH_2$) to phosphines [1a]. Electrostatic forces are also important in solubilizing highly charged complex ions in water.

In 1974, Daigle and co-workers synthesized the water soluble phosphine 1,3,5-triaza-7-phosphaadamantane (PTA, **1**) and its oxide (PTAO), and methyl-ammonium (PTA-Me^+) derivatives [7]. This ligand was subsequently shown to form stable complexes with Group 6 metal carbonyls [8] and mercury salts [9]. Recently PTA has been used in the development of water-soluble catalysts [10]. In this paper comparisons of the electronic and steric requirements will be made between PTA and the more commonly employed water-soluble

I. T. Horváth and F. Joó (eds.),
Aqueous Organometallic Chemistry and Catalysis, 61–80.

sulfonated phosphine ligand. The use of the PTA ligand in several biphasic catalytic processes will be described. These include regio-selective catalytic hydrogenation of unsaturated aldehydes.

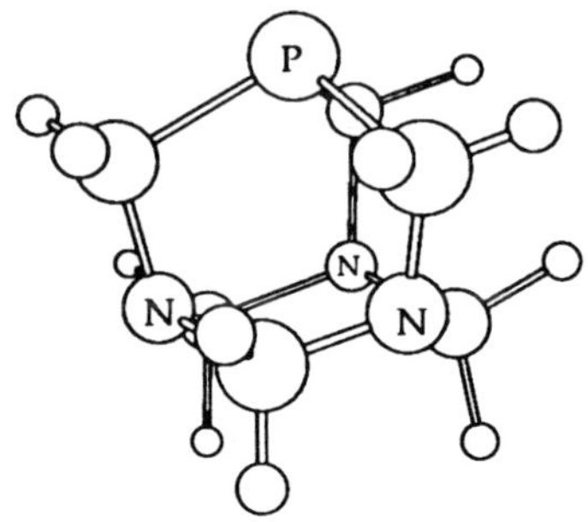

1

Our initial entry into the investigation of the organometallic chemistry associated with water-soluble phosphines was aimed at defining the electronic and steric differences between PPh_3 and its *tris*-sulfonated analog, TPPTS. Although spectroscopic studies have revealed these ligands exert similar influences on the $\nu_{(CO)}$, ^{31}P and ^{13}C NMR spectral properties of comparable derivatives [11], we felt it was necessary to obtain structural comparisons, as well as to examine relative ligand binding *via* dissociative ligand substitution processes. In this regard the X-ray structure of [Na-kryptofix-221]$_3$[$W(CO)_5P\{C_6H_4$-*m*-$SO_3\}_3$] was determined [12]. It was necessary to encapsulate the Na^+ with kryptofix-221 since all attempts to grow crystals of the complex in its absence were unsuccessful. Figure 1 contains a ball and stick representation of the [$W(CO)_5P\{C_6H_4$-*m*-$SO_3\}_3$]$^{-3}$ anion. The TPPTS ligand is situated 2.554(11)Å from the tungsten metal center, a distance quite similar to that observed in $W(CO)_5PPh_2H$ [13]. Of greater importance was the relative steric requirements of the *tris*-sulfonated phosphine ligand compared to that of PPh_3. On the basis of a molecular model from the structural data, TPPTS appears to be about 20% larger than triphenylphosphine, or in terms of Tolman's cone angle, a value of about 170° would be appropriate. Further studies have corroborated this estimate of the steric influence of TPPTS [14].

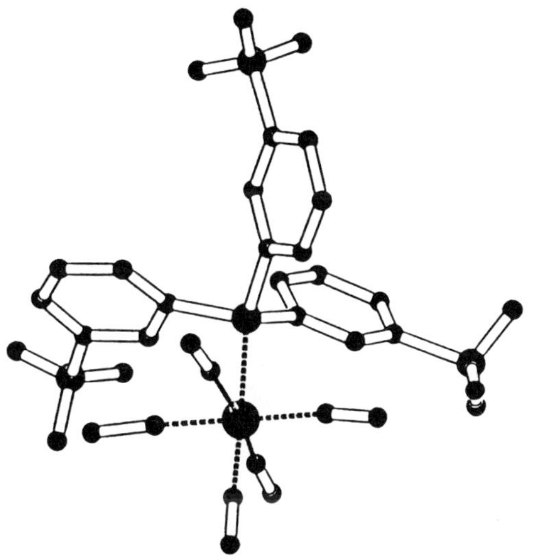

Figure 1. Ball and stick structure of $[W(CO)_5P\{C_6H_4\text{-}m\text{-}SO_3\}_3]^{-3}$ anion.

The crystal structure of [Na-kryptofix-221]$_3$$[W(CO)_5P\{C_6H_4\text{-}m\text{-}SO_3\}_3]$ provides a good model for the steric environment of an isolated metal-bound TPPTS ligand. That is, in this instance the shortest $Na^+\cdots O_3S^-$ distance is 5.47Å and there are no waters of solvation present. However, in general the metal coordinated sulfonated phosphine ion is involved in a wide range of contact ion-pairs or solvent separated ion-pairs with the cations. In addition hydrogen-bonding interactions with water are significant. Hence, it would be advantageous to define the structure of metal TPPTS complexes crystallized from water, where the cations are not encapsulated. We have accomplished this for the $Fe(CO)_4$TPPTS derivative [15]. The Fe-P distance was determined to be 2.25(2)Å, which is quite similar to that noted for $Fe(CO)_4PPh_3$ where a bond length of 2.244(1)Å was defined [16]. Figure 2 displays the structure of the anion, where the $-SO_3^-$ groups are shown to minimize their steric influence. That is, the ligand's cone angle is greatly reduced from that seen in the encapsulated Na^+ derivative described in Figure 1. This is the result of intermolecular interligand interactions as depicted in the packing diagrams shown in Figure 3a and 3b.

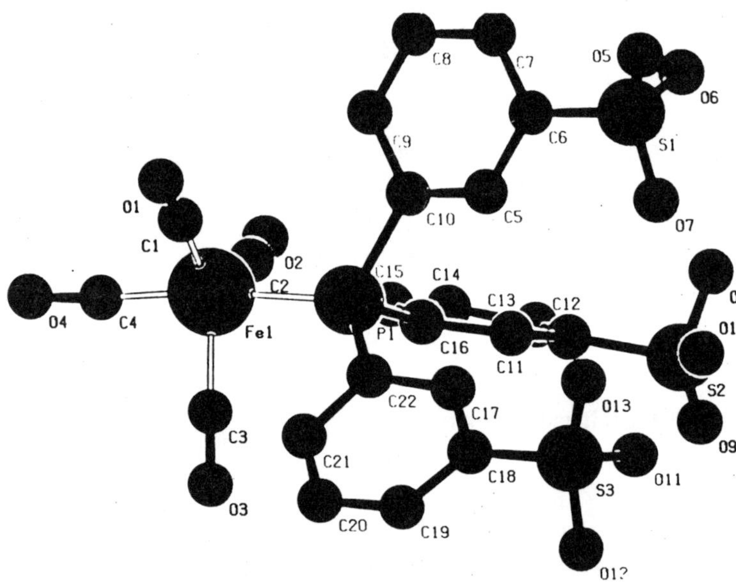

Figure 2. Ball and stick structure of $[Fe(CO)_4P\{C_6H_4\text{-}m\text{-}SO_3\}_3]^{3-}$ ion.

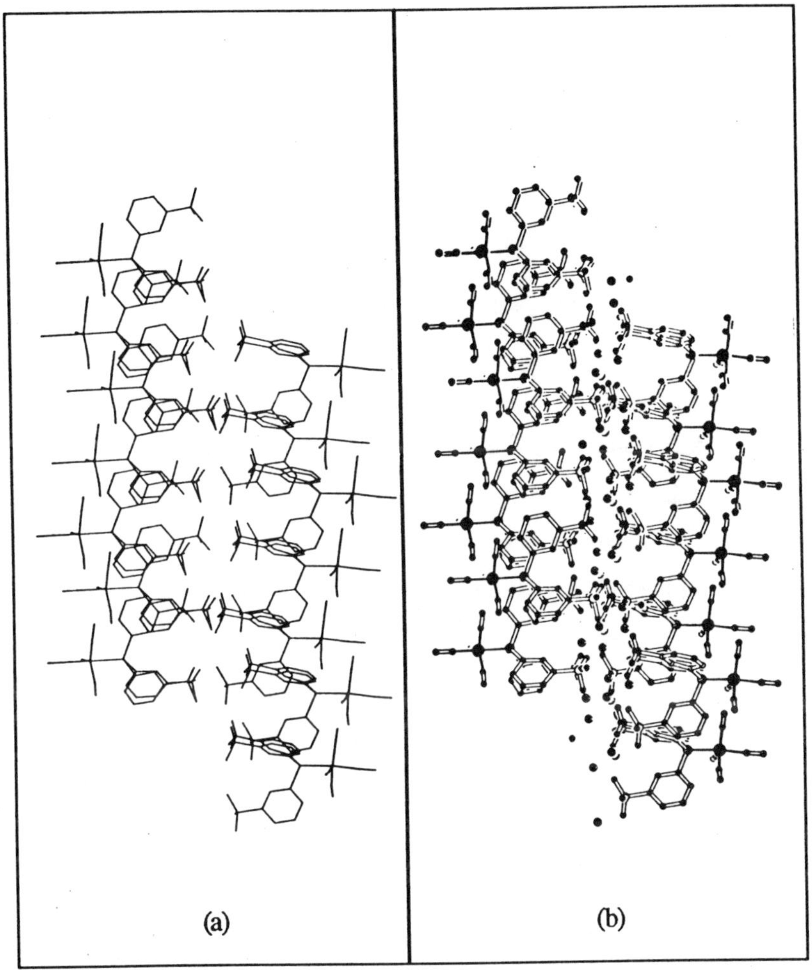

Figure 3. Packing diagrams for $Na_3[Fe(CO)_4P\{C_6H_4\text{-}m\text{-}SO_3\}_3]$ *(a)* stick structure without counter-ions and *(b)* ball and stick structure with counter-ions.

In order to have a more direct comparison of the binding energies of the trisulfonated triphenylphosphine *vs* triphenylphosphine to a metal center we have investigated ligand substitution reactions in appropriately substituted group 6 metal carbonyl derivatives. Reaction (1) was employed in theses studies [17]. Carbon monoxide, a poor nucleophile, was used as incoming ligand in an effort to insure a dissociative (**D**) process. However, the drawback of this procedure is that CO has a very low solubility in water. This was overcome by carrying out ***in situ*** experiments at high presssures of CO (>500 psi) using CIRCLE infrared techniques [18].

$$Mo(CO)_5L + CO \xrightarrow{k_1} M(CO)_6 + L \qquad (1)$$

($L=PPh_3$ or TPPTS and M=Mo or W)

In summary the kinetic parameters for TPPTS dissociation from $Mo(CO)_5TPPTS$ derivatives were shown to be quite similar to those determined for the analogous processes involving the PPh_3 ligand in the same solvent system. For example, the k_1 values for PPh_3 and TPPTS dissociation from the $\{Mo(CO)_5\}$ center were determined to be 8.60×10^{-4} sec^{-1} and 7.22×10^{-4} sec^{-1} at 135°C, respectively. Furthermore, there were only small effects on the rate of reactions noted upon going from aqueous to organic solvents for these dissociative processes. On the contrary, the greater steric requirements of the TPPTS ligand *vs* PPh_3 as derived from X-ray structural data were evident upon examining phosphine dissociation from [Na-kryptofix-221]$_6$[*cis*-$Mo(CO)_4\{P(C_6H_4SO_3)_3\}_2$], where the sulfonated phosphine ligand dissociates much faster than its unsulfonated counterpart. Alternatively, it was observed that in the absence of an encapsulating reagent for the Na^+ ions a slight stabilizing effect was noted for the two cis TPPTS ligand in *cis*-$Mo(CO)_4[TPPTS]_2$. Finally, we have demonstrated a greater than 10:1 selectivity of the pentacoordinate fragment, [$Mo(CO)_4$TPPTS] (Figure 4), for carbon monoxide *vs* TPPTS in 2-methoxyethanol. This latter observation is attributed to electrostatic repulsion between the pentacoodinate fragment and an incoming TPPTS ligand, as well as enhanced solvation of the free TPPTS ligand *vs* the metal bound.ligand.

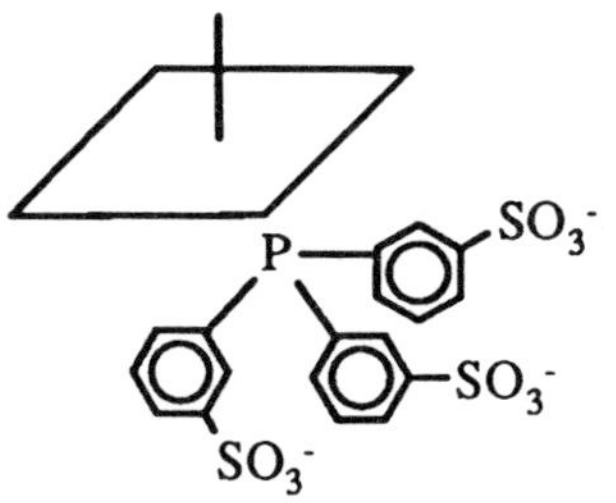

Figure 4. Pictoral representation of pentacoordinate fragment.

Alternative ligands which are electronically and sterically quite different from the sulfonated triphenylphosphines deserve judicious scrutiny. One such ligand is the PTA ligand, which is a small (cone angle 102°) [19], basic (similar to PMe_3) [8], and yet very water-soluble and air-stable phosphine. This tertiary phosphine ligand is readily prepared from the reaction of formaldehyde, tris(hydroxymethyl)phosphine, and hexamethylenetetramine and is easily purified by recrystallization from ethanol [7]. Herein, we will focus on the structural chemistry of the ruthenium derivatives of PTA, although these observations are found to be common for a variety of metal complexes.

The synthesis of the $RuCl_2(PTA)_4$ complex and its diprotonated analog is depicted in **Scheme 1** [20,21]. X-ray quality crystals were obtained from saturated water solutions in the absence or presence of HCl. Figure 5 contains structural representations of both derivatives. In either instance the ruthenium atom possesses a distorted octahedral geometry with cis chloride ligands. The two axial Ru-P bond lengths in **2** at 2.370(2)Å are longer than the two equitorial Ru-P bond lengths (2.260(2) Å) which are trans to the chloride ligands. The two Ru-Cl bond lengths have an average value of 2.496(2)Å with a Cl-Ru-Cl bond angle of 84.2(1)°. Noteworthy are the N-C distances in the four PTA ligands in **2** which are quite similar, displaying a range of values of 1.450(9)-1.485(9)Å with an average value of 1.467(8)Å.

These intraligand bond distances are essentially the same as those found in the free ligand, where the N-C bond distances have a spread of 1.461(4)-1.465(4)Å [22].

Scheme 1

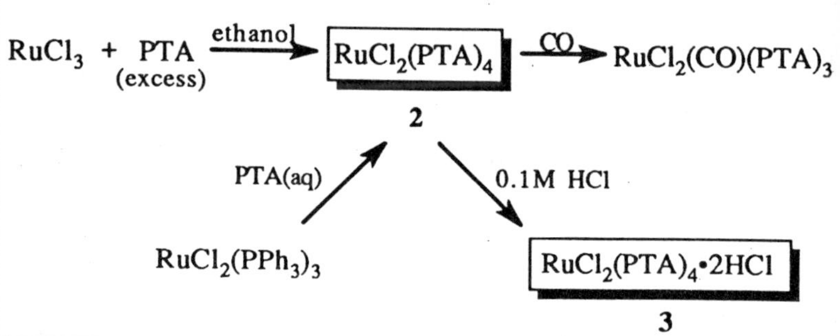

In complex **3** as in complex **2** the two mutually trans Ru-P bond lengths are longer by 0.097Å, 2.379(1) *vs* 2.282(1)Å, and the Ru-Cl bond distances are 2.457(2)Å, with a Cl-Ru-Cl bond angle of 86.0(1)°. Hence, the coordination geometry about the Ru(II) center is not significantly affected by protonation of the PTA ligands. *This is a common feature of all the metal-PTA complexes we have thusfar investigated.* This is of importance since the solubility of metal-PTA derivatives in water is often pH dependent. The major structural difference between complexes **2** and **3** is that the two mutually trans phosphine lighands are protonated at one of the nitrogen atoms. These quarternary nitrogens possess N-C average bond distances of 1.578(7)Å, slightly longer than their tertiary counterparts in the same ligand, 1.461(7)Å, or in the neutral PTA ligands in complexes **2** and **3**.

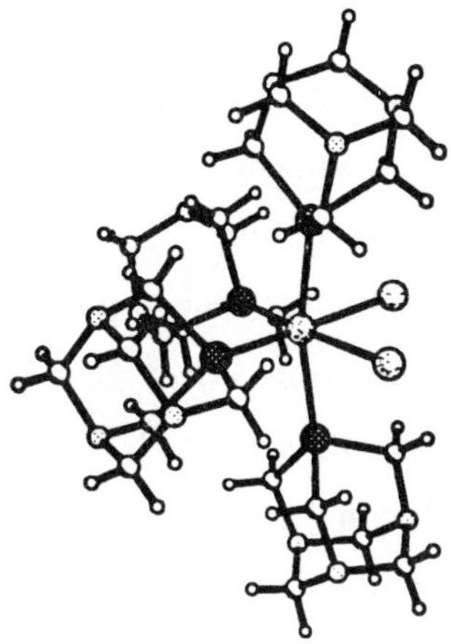

(a)

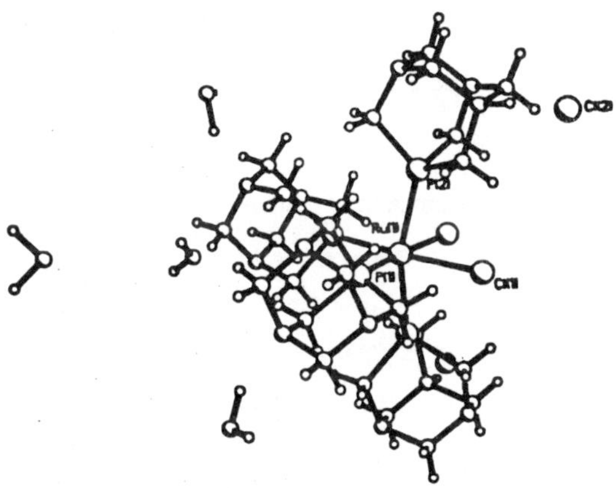

(b)

Figure 5. Ball-and-stick representation for the crystals (a) **2** and (b) **3**.

The minimization of ring strain in the adamantane structure of the PTA is vital to the ligand's overall stability. *Ab initio* calculations predicts that the successive protonation of the nitrogen atoms of the PTA molecule will introduce asymmetry in the intramolecular distances, whereas the distance between protonated nitrogens and the adjacent alpha carbon atoms will lengthen and the distances between the alpha carbon and the unprotonated beta nitrogen will shorten. Unfortunately resolution of these distances for any given structure determination is normally beyond the precision of the experiment. To detect the asymmetry in the intramolecular distances we have statistically examined the distribution of carbon-nitrogen bond lengths, known as the (N)C-N bond, in the six-membered ring of the adamantane structure which does not contain the phosphorus atom as one of its members. The examination of 228 distances from 37 individual structure determinations containing protonated and/or unprotonated PTA molecules, indicates that the overall (N)C-N bond length distribution is the convolution of three normal distributions (see Fig. 6) with the modes of these distributions equal to 1.44, 1.47 and 1.52. The statistical examination for the (N)C-N bond length distribution verifies the *ab initio* calculations and indicates that successive protonation of the PTA ligand will result in asymmetry in its intramolecular distances.

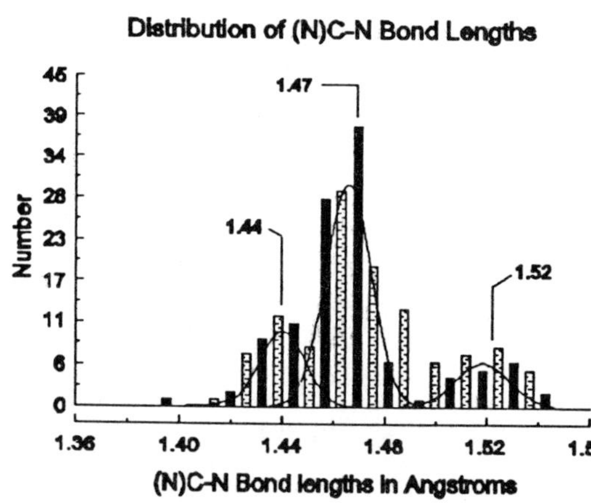

Figure 6. Histrogram of the Distribution of (N)C-N bond lengths for protonated and/or unprotonated PTA molecules.

PTA can be used as a suitable ligand for Ru(II) and Rh(I) complexes which catalyze the hydrogenations of aldehydes and olefins respectively (**Scheme 2**) [20, 21, 23]. These processes are analogous in many ways to the catalytic systems employing TPPMS (*meta*-sulfonatophenyl-diphenylphosphine) [24] or TPPTS [25] as ligands.

Although the catalytic processes are similar for the respective metal species of the given ligands, there are fundamental differences, most notably that pertaining to the requirements of metal/ligand ratios.

Scheme 2

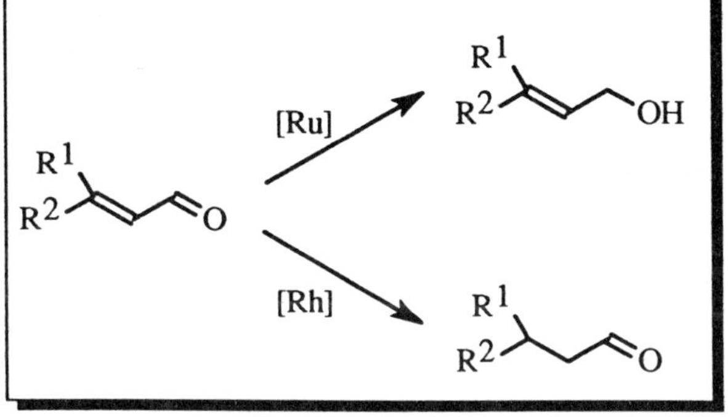

As with the Ru(II)/TPPMS system, Ru(II)/PTA species selectively catalyze the reduction of aldehydes to alcohols, where depending on the substrate being reduced selectivities can be as high as 98%. Specifically, when *cis*-$RuCl_2(PTA)_4$ is used as a biphasic catalyst in the presence of sodium formate or dihydrogen as the hydrogen source for the substrates 2-butenal (crotonaldehyde) and 3-phenyl-2-propenal (cinnamaldehyde), only the corresponding crotyl and cinnamyl alcohols are produced. In fact, for *cis*-$RuCl_2(PTA)_4$, the selectivity toward CO functionality reduction is so pronounced that the Ru(II)/PTA complex does not reduce such primary olefins as 1-decene. However, as the chain length of the aldehydic substrate increases, the percent conversion to the respective alcohol decreases, which is generally thought to be due to a decrease in miscibility of the organic substrate in the aqueous phases.

In contrast to the ruthenium(II)/PTA biphasic systems, analogous rhodium(I)/PTA species, synthesized according to equation (2), selectively catalyze the reduction of C-C double bonds;

Rh(I)/PTA systems catalyze the hydrogenation of unsaturated aldehydes to saturated aldehydes.

$$RhCl_3 + \underset{(excess)}{PTA} \xrightarrow{ethanol} \boxed{RhCl(PTA)_3} \xrightarrow{0.1M\ HCl} \boxed{RhCl(PTA)_2 \cdot 2HCl} \qquad (2)$$

Superficially, Rh(I)/PTA and Rh(I)/TPPMS complexes catalyze similar reactions. As an example of the selectivity of the Rh(I) system, when Rh(PTA)3Cl, in the presence of sodium formate or H_2, is used as a hydrogenation catalyst for *trans*-cimmamaldehyde, hydrocinnamaldehyde is the major product at 93.9% conversion, as illustrated in equation (3).

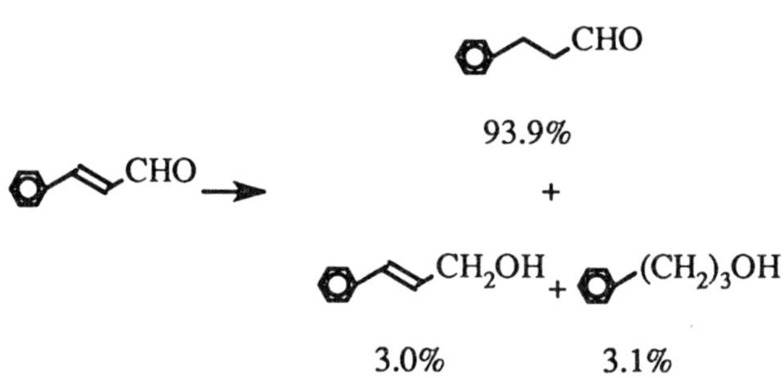

(3)

Although Rh(PTA)3Cl is similar to Wilkinson's catalyst, it does not readily decarbonylate aldehydes. The same rhodium complex also catalyzes isomerizations of the alkene bond. With allylbenzene as the substrate, hydrogenation is accompanied by isomerization of the double bond to afford *cis*- and primarily *trans*-propenylbenzene, as indicated in equation (4). This isomerization reaction is enhanced when sodium formate is used as the hydrogen source, as opposed to H_2 gas.

(4)

By altering the reaction temperature, the amount of the respective products can be controlled, where the isomerization products are prevalent at ambient temperatures.

Though the PTA complexes of Ru(II) and Rh(I) catalyze similar reactions as their sulfonated phosphine derivatives, there are differences in the overall systems. The most striking deviation for both Ru(II)/PTA and Rh(I)/PTA systems is that the presence of excess PTA is found to extensively, if not to completely, inhibit the reaction. This observation is in contrast to the corresponding TPPMS systems, where an excess of TPPMS not only enhances the catalytic activity of the metal species, but is required for efficient rates and high turnovers. Studies show that in the systems involving TPPMS, and even TPPTS in the presence of water, the formation of phosphonium salts occur, as indicated in **Scheme 3** [26, 27].

Scheme 3

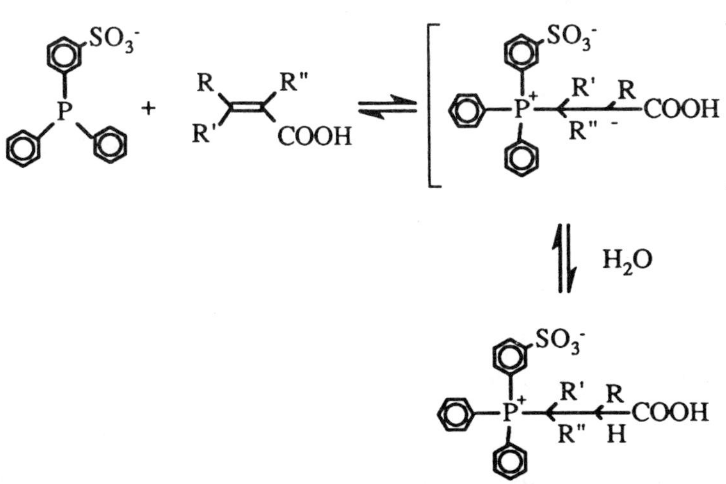

The formation of these phosphonium salts serves as the means of consuming the excess phosphine present in the system. To date such an interaction between substrate and PTA has not been observed. Without the removal of excess PTA, the phosphine binds to and hence saturates the metal site causing catalytic expiration.

The catalytically active species of Ru(II) requires at least three phosphine ligands. Where the processes utilizing $RuCl_2(TPPMS)_2$ and $RuCl_2(TPPTS)_2$ utilize excess phosphine present to attain this ligand requirement, $RuCl_2(PTA)_4$ achieves the same through ligand dissociation. Hence, the divergent behavior exhibited by the PTA and sulfonated phosphine catalysts is, at least in part, a manifestation of steric differences. [PTA has a cone angle of 102°, where the sulfonated phosphines have a cone angle > 165°.]

Another difference in the PTA and TPPMS (and TPPTS) catalytic systems lies in the reaction rates. For $RuCl_2(PTA)_4$ the rate of hydrogenation is approximately 30 times slower than for the analogous sulfonated (TPPMS and TPPTS) species. This phenomena is attributed to the surface-active properties of both TPPMS and TPPTS, in which a more rapid migration of the substrate into the aqueous phase is possible.

One very important criteria of any catalytic process, especially those employing costly transition metal species, is catalyst recycling and recovery. Biphasic catalysis is one of the answers to the question of ease of recovery which is an ever-present difficulty in both heterogeneous and homogeneous catalysis. In utilizing biphasic catalysis, ideally separation of substrate from catalyst is easier in that the catalyst remains in the aqueous phase and the substrate in the organic phase, a simple phase separation technique. In the system where the catalytic species is $RuCl_2(PTA)_4$, the acitivity of the catalyst did not decrease after three consecutive cycles. Likewise for $Rh(PTA)_3Cl$ recycling shows a minimal loss of activity until the fourth cycle, where the loss of activity was shown not to be due to rhodium leaching into the organic phase. Hence, both Ru(II) and Rh(I) systems discussed do indeed meet the criteria of catalyst recovery and cycling.

For the hydrogenation of benzaldehyde to benzyl alcohol, where the Ru(II)/PTA complex acts as the catalytic species, the following mechanism, based on experimental results, has been proposed, illustrated in **Scheme 4**.

Scheme 4

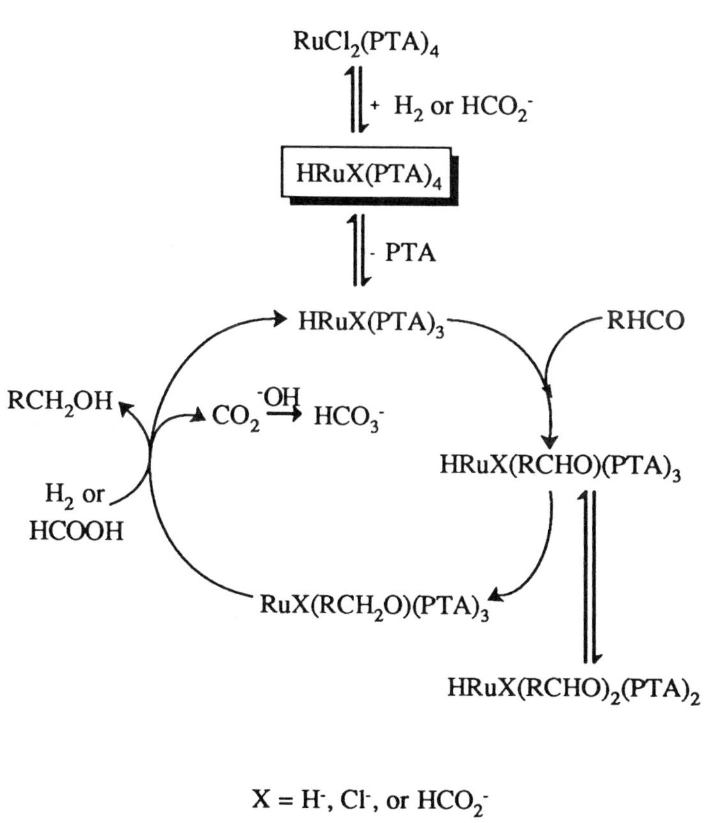

This cycle is in agreement with the experimental data obtained, thus far. In order to determine the hydrogen source, as either sodium formate or water, investigations were carried out in solution mixtures of (a) D_2O/NaO_2CH, (b) H_2O/NaO_2CD, and (c) D_2O/NaO_2CD. Mass spectral analysis demonstrated that the level of deuterium incorporation in the benzyl alcohol was considerably greater for experiment *b* as compared to *a*. Therefore the hydrogen source must be sodium formate.

The described process for $RuCl_2(PTA)_4$ was found to be first order in catalyst and aldehyde concentration (at low aldehyde concentrations) and independent of formate at high formate concentrations. It is assumed that the rate limiting step is the reaction of the metal-bound aldehyde with the metal hydride to afford the metal alkoxide intermediate. It is noteworthy that the cycle has many common features to that of the analogous ruthenium(II)/TPPMS system using sodium formate as the hydrogen source with the exception that the ruthenuim(II)/PTA system is inhibited by the presence of excess phosphine, as previously noted.

In the catalytic process utilizing $Rh(PTA)_3Cl$, experimental evidence has not been able to distinguish between several possible mechanisms. It is believed that complications are due, at least in part, to the reaction being catalyzed to some extent by bulk or finely divided metal colloids. Therefore it is likely that both homogeneous and heterogeneous reaction pathways are operative in this catalytic cycle.

For the rhodium(I)/PTA system, phosphine oxidation (PTA to PTAO) has been observed, whereas for the related rhodium(I)/TPPTS system, the presence of TPPTS oxide has not been noted. The oxygen incorporated into the phosphine originates from the water, as indicated by $H_2{}^{18}O$ studies, where a shift is seen for the P=O stretch in the infrared from 1151 cm^{-1} to 1120 cm^{-1}. The following reaction (equation (5)) is believed to be occurring, though the rhodium hydride species has not been directly detected.

$$ClRh^{I}(PTA)_3 \overset{H_2O}{\rightleftharpoons} ClRh^{II}(H)(OH)(PTA)_3 \searrow [HRh^{I}(PTA)_2]^{+} + PTAO + HCl \quad (5)$$

Despite differences in the requirements of amount of phosphine for the two analogous systems, PTA versus sulfonated phosphines, for the respective metals, it should be acknowledged that fundamentally the systems are very similar. Therefore, the potential use of 1,3,5-triaza-7-phosphaadamantane (PTA), a small, basic, air-stable, water-soluble phosphine, as a ligand for biphasic catalysis is attractive as a suitable alternative to the common sulfonated phosphines.

Acknowledgments:

The authors are most grateful to the National Science Foundation, whose support under grants (CHE91-19737, INT90-08227 and INT93-13951) has made possible their contributions to the research described herein. They are likewise extremely appreciative to all of their coworkers mentioned in the references, in particular, A. Bényei, C.J. Bischoff, D.J. Daigle, F. Joó, M. Kannisto, A. Kathó, and J.N.W. Stafford. Lastly, they would like to express their appreciation to Sheila Eng, for aiding in the use of the data base.

References

1. (a) Herrmann, W.A., Kulpe, J.A., Kellner, J., Riepl, H., Bahrmann, H. and Konkol, W. (1990) *Angewante Chemie, International Edition in English*, **29**, 391-95. (b) Escaffre, P.,

Thorez, A., and Kalck, P. (1987) *Nouveau Journal of Chimie,* **11**, 601. (c) Kalck, P., Escaffre, P., Serein-Spirau, F., and Thorez, A. (1982) *Nouveau Journal of Cheimie,* **12**, 687. (d) Bahrmann, H. and Bach, H. (1987) *Phosphorus and Sulfur*, **30**, 611. (e) Kuntz, E.G. (1987) *Chemtech*, 570. (f) Russell, M.J.H. (1988) *Platimum Metals Review*, **32(4)**, 179, (g) Larpent, C., Dabard, T. and Patin, H. (1987) *Inorganic Chemistry*, **26**, 2922-24.

2. (a) Joó, F. and Tóth, Z.J. (1980) *Journal of Molecular Catalysis*, **8**, 369-83. (b) Dror, Y. and Manassen, J. (1977) *Journal of Molecular Catalysis*, **2**, 219-22. (c) Nuzzo, R.G., Feitler, D. and Whitesides, G.M. (1979) *Journal of the American Chemical Society*, **101**, 3683-5. (d) Willner, I. and Maidan, R.J. (1988) *Journal of the Chemical Society, Chemical Communications*, 876-7. (e) Okano, T., Moriyama, Y., Konishi, H. adn Kiji, J. (1986) *Chemistry Letters* , 1463. (f) Kiji, J., Okana, T., Nishiumi, W., and Konishi, H. (1988) *Chemistry Letters*, 957.

3. (a) Amrani, Y. and Sinou, D.J. (1984) *Journal of Molecular Catalysis*, **24**, 231. (b) Sinou, D. and Amrani, Y. (1986) *Journal of Molecular Catalysis*, **36**, 319-27. (c) Amrani, Y., Lecomte, L., Sinou, D., Bakos, J., Tóth, I., and Heil, B. (1989) *Organometallics*, **8**, 542-7. (d) Tóth, I., Hanson, B., and Davis, M.E. (1990) *Catalysis Letters*, **5**, 183-8.

4. Chachoyan, A.A. and Sagradyan, L.I. (1987) *Biologicheskii Zhurnal Armenii,* **40**, 399-400.

5. (a) Vigo, C., Goni, F.M., Quinn, P.J. and Chapman, D. (1978) *Biochimica Biophysica Acta*, **508**, 1. (b) Madden, T.D., Peel, W.E., Quinn, P.J., and Chapman, D. (1980) *Journal of Biochemical and Biophysical Methods*, **2**, 19. (c) Farin, F., Van Gaal, H.L.M., Bonting, S.L., and Daemen,

F.J.M. (1982) *Biochim. Biophys. Acta,* **711**, 336-?. (d) Quinn, P.J., Joó, F., and Vigh, L. (1989) *Prog.ress in Biophysical Molecular Biol.ogy*, **53**, 71. (e) Willner, I., Maidan, R., and Shapiro, M. (1990) *Journal of the Chemical Society, Perkin Transactions II*, 559.

6. (a) Joó, F. unpublished manuscript. (b) Fauche, E., Senocq, F., Santini, C., and Basset, J.M. (1990) *Journal of the Chemical Society, Chemical Communications*, 1776.

7. Daigle, D.J., Pepperman, A.B.,Jr., and Vail, S.L. (1974) *Journal of Heterocyclic Chemistry*, **11**, 407.

8. (a) Darensbourg, M.Y. and Daigle, D.J., (1975) *Inorganic Chemistry,* **14**, 1217-8. (b) DeLerno, J.R., Trefonas, L.M., Darensbourg, M.Y., and Majeste, R.J. (1976) *Inorganic Chemistry*, **15**, 816-9. (c) Darensbourg, D.J. and Graves, A.H. (1979) *Inorganic Chemistry,* **18**, 1257-61.

9. Alyea, E.C., Fisher, K.J., and Johnson, S. (1989) *Canadian Journal of Chemistry*, **67**, 1319.

10. Kannisto, M. (1994) Ph.D. Dissertation, Texas A&M University.

11. (a)Herrmann, W.A., Kellner, J., and Riepl, H. (1990) *Journal of Organometallic Chemistry*, **389**, 103. (b) Horváth, I., Kastrup, R., Oswald, A., and Mozeleski, E. (1989) *Catalysis Letters*, **2**, 85.

12. Darensbourg, D.J., Bischoff, C.J., and Reibenspies, J.H. (1991) *Inorganic Chemistry*, **30**, 1144-47.

13. Reibenspies, J.H., Darensbourg, D.J., and Atnip, E. (1994) *Zeitshrift fur Kristallographie*, **209**, 759-60.

14. Bartik, T., Bartik, B., Hanson, B.E., Guo, I., and Tóth, I. (1993) *Organometallics,* **12**, 164-70.

15. (a) Darensbourg, D.J., Bischoff, C.J., and Reibenspies, J.H., unpublished observations. (b) Similar structural results have

been obtained for $Co_2(CO)_6(TPPTS)_2$. Bartik, T., Bartik, B., Hanson, B.E., Whitmire, K.H., and Guo, I. (1993) *Inorganic Chemistry*, **32**, 5833-37.

16. Riley, P.E. and Davis, R.E. (1980), *InorganicChemistry*, **19**, 159-65.

17. Darensbourg, D.J. and Bischoff, C.J. (1993) *Inorganic Chemistry*, **32**, 47-53.

18. Bischoff, C.J. (1991) Ph.D. Dissertation, Texas A&M University.

19. Tolman, C.A. (1977) *Chemistry Reviews*, **77**, 313-48.

20. Darensbourg, D.J., Joó, F., Kannisto, M., Kathó, A., and Reibenspies, J.H. (1992) *Organometallics*, **11**, 1990-93.

21. Darensbourg, D.J., Joó, F., Kannisto, M., Kathó, A., Reibenspies, J.H., and Daigle, D.J. (1994) *Inorganic Chemistry*, **33**, 200-208.

22. Fluck, E., Forster, J.E., Weidlein, J., and Hadicke, E.Z. (1977) *Zeitshriff furNaturforsch.*, **32B**, 499.

23. Darensbourg, D.J., Stafford, N.W., Joó, F., and Reibenspies, J.H. (in press) *Journal of Organometallic Chemistry*.

24. (a) Joó, F. and Bényei, A. (1989) *Journal of Organometallic Chemistry*, **363**, C19. (b) Bényei, A. and Joó, F. (1990) *Journal of Molecular Catalysis*, **58**, 151-63.

25. Grosselin, J.M., Mercier, C., Allmang, G., and Grass, F. (1991) *Organometallics*, **10**, 2126-33.

26. Bényei, A., Stafford, J.N.W., Kathó, A., Darensbourg, D.J., and Joó, F. (1993) *Journal of Molecular Catalysis*, **84**, 157-63.

27. Darensbourg, D.J., Joó, F., Kathó, A., Stafford, J.N.W., Bényei, A., and Reibenspies, J.H. (1994) *Inorganic Chemistry*, **33**, 175-177.

RECENT DEVELOPMENTS IN METAL CATALYZED BIPHASIC REACTIONS

VLADIMIR GRUSHIN and HOWARD ALPER*
University of Ottawa
Department of Chemistry
10 Marie Curie
Ottawa, Ontario, Canada K1N 6N5

Organic reactions catalyzed by transition metal complexes, under aqueous, biphasic, or phase transfer catalysis conditions, have attracted considerable attention in recent years.[1,2] Of particular note are metal catalyzed carbonylation reactions of halides to form carboxylic acids.[3] Examples include the biscarbonylation of vinylic dibromides, catalyzed by $Pd(diphos)_2$ in aqueous base-organic two-phase media containing a phase transfer agent (eq. 1)[4], and the

$$\mathrm{ArCH{=}CBr_2 + CO} \xrightarrow[\substack{\mathrm{PhCH_2N(C_2H_5)_3^+Cl^-} \\ \mathrm{t\text{-}AmOH,\ 5N\,NaOH} \\ \mathrm{r.t.,\ 1\ atm.}}]{\mathrm{Pd(diphos)_2}} \mathrm{ArCH = C(COOH)_2} \qquad (1)$$

conversion of iodoarenes to carboxylic acids in water or aqueous N,N-dimethylformamide catalyzed by palladium (II) complexes in the presence of a base (eq. 2)[5], or the nickel cyanide phase transfer catalyzed carbonylation reaction.[6] Iodoxyarenes,

I. T. Horváth and F. Joó (eds.), Aqueous Organometallic Chemistry and Catalysis, 81–95.

$$\mathrm{ArI} + \mathrm{CO} \xrightarrow[\substack{K_2CO_3 \text{ or NaOAc} \\ 2.\ H_3O^+}]{1.\ H_2O/DMF,\ Pd(II)} \mathrm{ArCOOH} \tag{2}$$

readily generated from iodoarenes using commercial laundry bleach (sodium hypochlorite), are excellent reactants for the tetrachloropalladate catalyzed hydrocarboxylation in water, under gentle conditions, affording carboxylic acids in 48-89% yield (eq)[7]

$$\mathrm{ArIO_2} + \mathrm{CO} \xrightarrow[2.\ H_3O^+]{\substack{1.\ Na_2PdCl_4,\ Na_2CO_3 \\ H_2O}} \mathrm{ArCOOH} \tag{3}$$

More attractive than the carbonylation of either iodoxy or iodoarenes (or bromoarenes) is the production of acids from appreciably cheaper chloroarenes. This paper is concerned with recent developments in the author's laboratory on the synthetic utility of the biphasic palladium (II) catalyzed carbonylation of chloroarenes, and experiments, using chloro- and iodoarenes as reactants, designed to gain insight into the reaction pathway.

The conventional protocol for the carbonylation of halides of various classes involves the use of carbon monoxide. It is possible, to substitute chloroform and aqueous base, for carbon monoxide for the conversion of several kinds of halides to acids. Examples utilizing this methodology are presented below.

SCOPE OF THE BIPHASIC PALLADIUM (II) CATALYZED CARBONYLATION REACTION OF CHLOROARENES

Treatment of neat chloroarenes with one atmosphere of carbon monoxide and 20% aqueous KOH, using $(Cy_3P)_2PdCl_2$ as the catalyst, affords carboxylic acids in fine yields.[8,9] The ratio of substrate to catalyst was 400:1, with good turnover numbers attained in some cases (Table 1). Electron-donating groups activate the chloroarenes to carbonylation,

Table 1. Carbonylation of Chloroarenes Catalyzed by $[(Cy_3P)_2PdCl_2]$ under Biphasic Conditions[a]

$$ArCl + CO \xrightarrow[{[(Cy_3P)_2PdCl_2],\ reflux}]{20\%\ KOH} ArCOOK$$

entry	Ar	reactn time, h	catalytic turnover no.[b]
1	C_6H_5	72	89
2	1-$C_{10}H_7$	72	103
3	*p*-$CH_3C_6H_4$	72	116
4	*p*-$CH_3C_6H_4$	48	88
5	*m*-$CH_3C_6H_4$	48	90
6[c]	*m*-$CH_3C_6H_4$	48	85
7	*o*-$CH_3C_6H_4$	48	65
8	*p*-$CH_3OC_6H_4$	48	105
9[d]	*m*-$C_6H_5COC_6H_4$	48	9
10	*p*-$CF_3C_6H_4$	48	5
11	*p*-$CF_3C_6H_4$	72	7
12	*p*-$O_2NC_6H_4$	24	0

[a]Reaction conditions: chloroarene (40 mmol), $[(Cy_3P)_2PdCl_2]$ (0.1 mmol), 20% KOH (10 mL), CO (1 atm). [b] Catalytic turnover numbers, not yields are listed, since the chloroarenes were used as both the reagents and the organic phase. [c]In the presence of 18-crown-6 (0.3 mmol). [d] 10 mmol of *m*-chlorobenzophenone.

while electron-attracting groups reduce the reactivity of the organic substrate. A p-nitro group completely inhibits reaction. Note that while bis(triphenylphosphine)palladium dichloride is inactive for the carbonylation of chloroarenes, it is an effective catalyst for the biphasic carbonylation of bromo- and iodoarenes.

DELINEATION OF THE PATHWAY FOR THE CARBONYLATION OF HALOARENES

Palladium (II) is reduced to a palladium (0) species in the biphasic carbonylation reaction of haloarenes. How does this occur? It is the hydroxide ion which induces disproportionation of palladium (II) tertiary phosphine complexes to palladium (0) and phosphine oxides. It was found that bis(tricyclohexylphosphine)palladium dichloride reacts with chlorobenzene and potassium hydroxide in a water-benzene mixture to give a σ-phenylpalladium dimer (1), containing hydroxo bridges, and tricyclohexylphosphine oxide (eq. 4). Similarly, 1, L=PPh_3 was formed,

$$2L_2PdCl_2 + 2PhX + 6KOH \xrightarrow{C_6H_6,\ H_2O}$$

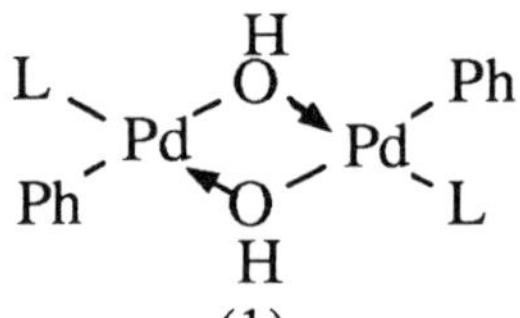

$$+ 2LO + 2KI + 4KCl + 2H_2O \qquad (4)$$

(1)

$L = PCy_3$, $X = Cl$
$L = PPh_3$, $X = I$

together with triphenylphosphine oxide, when bis(triphenylphosphine) palladium dichloride was treated with iodobenzene and potassium hydroxide. Note that when the latter process was repeated without the haloarene, palladium metal was formed together with the phosphine oxide.

The stereochemistry of the disproportionation reaction was determined using bis[(S)-benzymethylphenylphosphine]palladium dichloride for the reaction with hydroxide ion and iodobenzene (eq. 5). (S)-Benzylmethylphenylphosphine

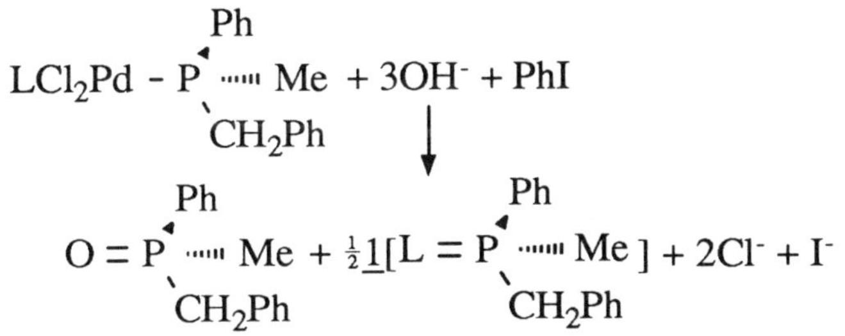

(5)

oxide was isolated in 99.9-99.6% optical purity (50-80% chemical yields). Consequently, this reaction takes place with retention of configuration at phosphorus. The mechanism, outlined in Scheme 1, is consistent with the stereochemistry and other results described above.

Scheme 1

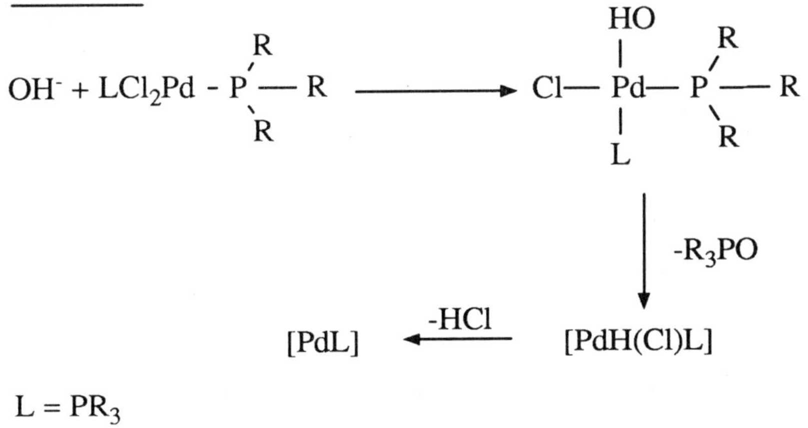

In contrast to the usual practice of effecting L_2PdCl_2 reactions with added excess phosphine, the present study was done in the absence of any added ligand. From our findings, one can state that *palladium (II) phosphine complexes that are used as catalysts for carbonylation and other reactions, occuring in the presence of alkali, are reduced to a Pd(O) species by their own coordinated phosphine ligands which are oxidized to phosphine oxides, with hydroxide ion functioning as a specific promoter.*

The disproportionation reaction resulting in the reduction of palladium (II) to palladium (0) is likely of broad applicability to other Pd(II) catalyzed reactions including Heck arylation, the Suzuki reaction, and other coupling processes.[10] These reactions are conventionally run by either using excess inorganic (e.g. NaOH, Na_2CO_3) or organic (e.g. R_3N, OR^-) bases to neutralize acid released during the course of the reaction. For example, tertiary amines contain enough water to react by reversible deprotonation to give the hydroxide anion which can then induce the reduction of Pd(II) to Pd(0).

The formation of the dimeric palladium complexes in the reaction of L_2PdCl_2 with KOH and either iodo- or chlorobenzene can be accounted for by the pathway outlined in Scheme 2 (illustrated for PhI), i.e. oxidative addition of the haloarene to the Pd(O) intermediate affording a palladium (II) complex which can dimerize to 2 and form 1 by reaction with hydroxide ion.

Scheme 2

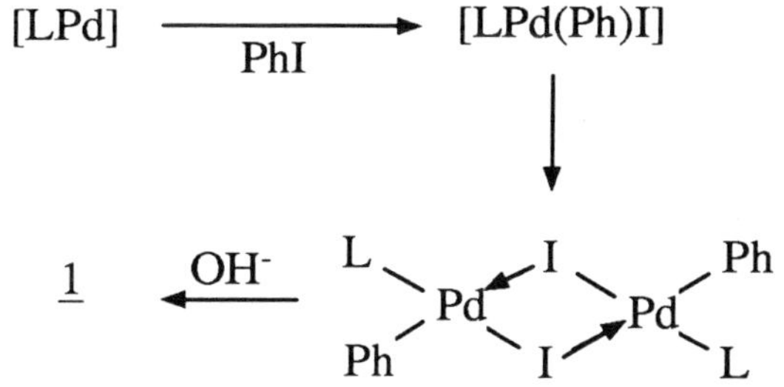

(2)

The next matter which needs to be addressed is the pathway(s) for formation of the carboxylic acid. There are two routes by which the acid is formed from the haloarene. The centerpiece for both routes is the dimeric palladium complex, 1, with benzoic anhydride being another key intermediate. When, 1, L=PPh_3 is reacted with iodobenzene and carbon monoxide in toluene, the benzoylpalladium complex 3, resulting from carbonyl insertion into the Ph-Pd bond, was obtained in 98% yield (eq. 6). Benzoic anhydride (4) was the accompanying product of the reaction, reproducibly formed in 95 ± 5% yield. When 1 L = PPh_3, was reacted with CO alone, benzoic anhydride (4) was obtained, together with a

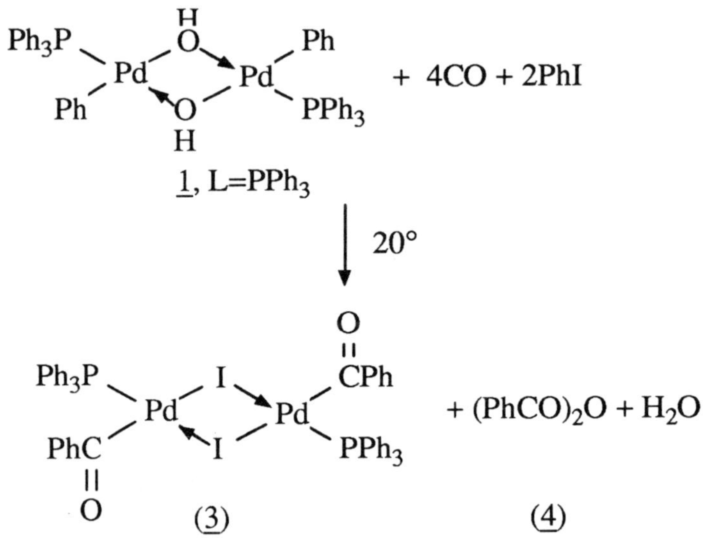

(6)

palladium cluster having the presumed formula $Pd_3(CO)_3(Ph_3P)_3$.

Attempts to identify intermediates of the reaction of 1, L=PPh_3 and CO failed, even at low temperature. However, it was gratifying to be able to isolate such intermediates using 1, L=PCy_3, as reactants. Carbonylation of the latter afforded trans-benzoatophenylbis(tricyclohexylphosphine)palladium (II) (5) (eq. 7)

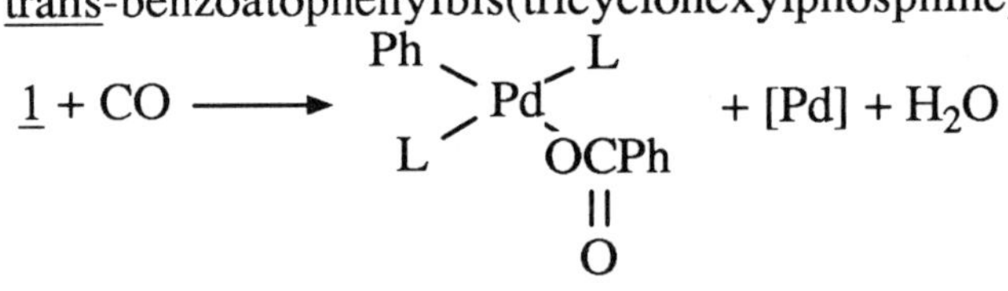

(7)

5, L=PCy_3

Accompanying 5 was a palladium cluster, of unknown composition, which contains carbonyl and phosphine ligands. One can imagine that insertion of CO into either the Pd-C or Pd-O bond of 1, followed by reductive elimination, would give benzoic acid which on exposure to as yet unreacted 1, can be converted to 5. Facile reaction of 1, L=PCy_3 or

PPh_3, does occur with benzoic acid to form dimeric benzoate complexes 6 (eq. 8). The latter, on treatment with two equivalents of L, gave the mononuclear complexes 5 (eq. 9).

1 + 2PhCOOH →

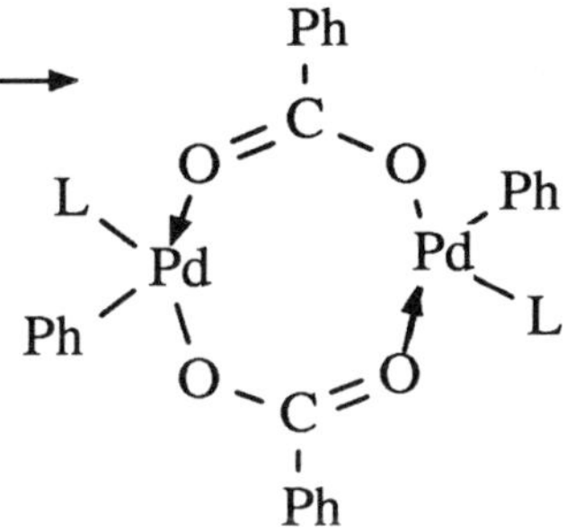

(6) (8)

$$\underline{6} + 2L \rightarrow \underline{5} \qquad (9)$$

$L = PPh_3, PCy_3$

Carbonylation of 5, obtained from 1 or 6, occurs slowly (7-10 days) to form the trans-(benzoato)benzoylpalladium complexes 7 (eq. 10). Note that 7, L=PCy_3, can be formed in excellent yield by oxidative addition of benzoic anhydride to bis(tricyclohexylphosphine)palladium (0).

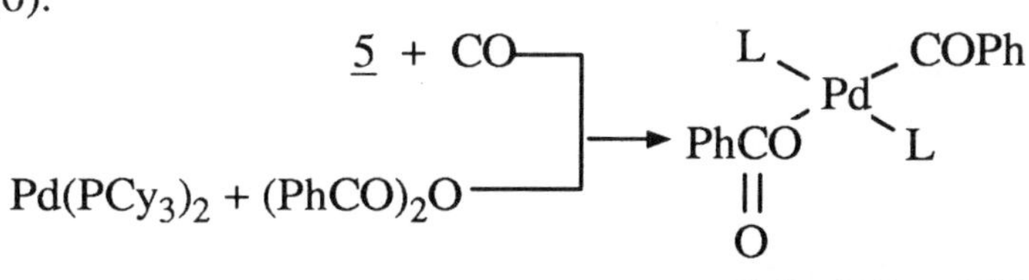

(10)

7, L=PCy_3, PPh_3

Benzoic anhydride was produced when either of 5, 6 (eq.11) or 7 (L=PPh_3) was treated with CO and iodobenzene

$$\underline{6}, L{=}PPh_3 + 4CO + 2PhI \rightarrow \underline{5}, L{=}PPh_3 + 2(PhCO)_2O \qquad (11)$$

(4)

The results described herein establish for the first time the occurence of two pathways for the transformation of haloarenes to carboxylic acids, the major one likely involving ArPd and ArCOPd containing intermediates (Scheme 3), and the other proceeding via aromatic anhydrides.

SCHEME 3

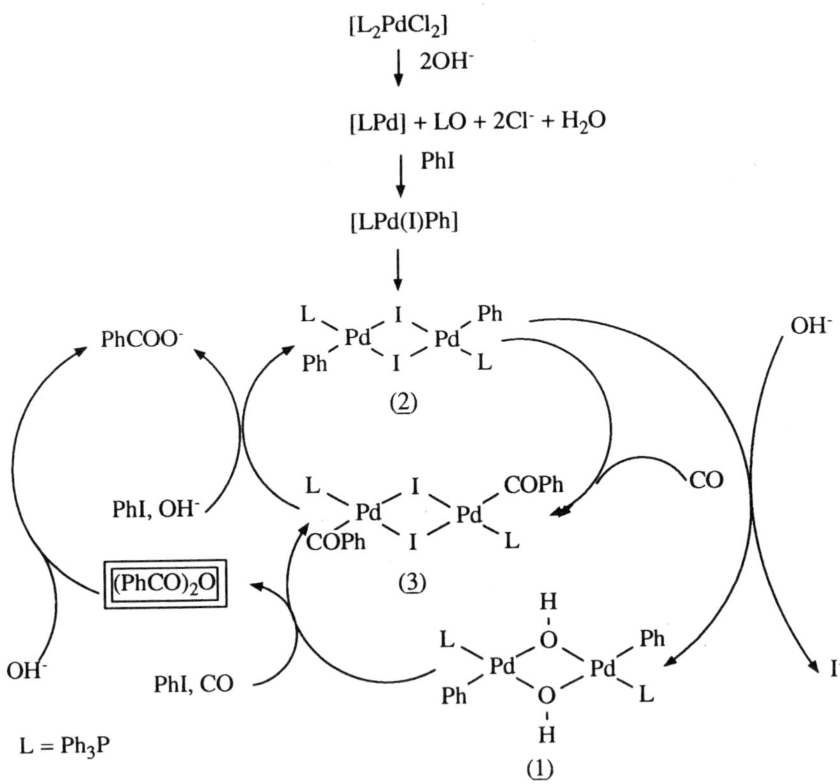

CARBONYLATION OF HALIDES BY CHLOROFORM AND ALKALI

While carbon monoxide is an exceptionally useful reactant for the preparation of an extensive range of organic products, it is highly toxic and requires storage in stainless steel tanks and cylinders. An alternative approach to conducting reactions with carbon monoxide is to generate it *in situ* from inexpensive, easy-to-handle reagents. We have found that an excellent method for achieving carbonylation is by use of a biphasic chloroform and alkali system.[11]

Treatment of iodoarenes with chloroform, 60% KOH, and bis(triphenylphosphine)palladium dichloride as the catalyst (20:1 ratio of substrate: catalyst), at room temperature for 24 hours, affords carboxylic acids in 43-92% yields (Table 2). Vinylic bromides also undergo reaction (eq. 12), as do

$$PhCH{=}CHBr + CHCl_3 + KOH \xrightarrow[(PPh_3)_2PdCl_2]{PPh_3 \text{ then } H_3O^+} PhCH{=}CHCOOH \qquad (12)$$

benzylic bromides and, to a small extent, benzyl chloride (when $CeCl_3.6H_2O$ is a co-catalyst) (Table 3).

Table 2. Biphasic Carbonylation of Iodoarenes with Chloroform, Catalyzed by $[(Ph_3P)_2PdCl_2]$ [a]

$$ArI + CHCl_3 \xrightarrow[(2)\ H_3O^+]{(1)\ [Pd],\ OH^-} ArCOOH$$

entry	Ar	isolated yield, %[b]	entry no.	Ar	isolated yield, %[b]
1	C_6H_5	54-72	6	$3\text{-}CH_3OC_6H_4$	91-92
2	$4\text{-}CH_3C_6H_4$	54-76	7	$4\text{-}ClC_6H_4$	54-65
3	$3\text{-}CH_3C_6H_4$	77-81	8	$3\text{-}ClC_6H_4$	76-89
4	$2\text{-}CH_3C_6H_4$	43-45	9	$1\text{-}C_{10}H_7$	78-81
5	$4\text{-}CH_3OC_6H_4$	60-73			

[a]Reaction conditions: 1.5 mmol of iodoarene, 5 g of 60% KOH, 1 mL of $CHCl_3$, 0.075 mmol of $[(Ph_3P)_2PdCl_2]$, N_2, 22°C, 24 h. [b]After purification by vacuum sublimation.

Table 3. Biphasic Carbonylation of Benzylic Halides with Chloroform, Catalyzed by $[(Ph_3P)_2PdCl_2]$ [a]

$$ArCH_2X + CHCl_3 \xrightarrow[(2)\ H_3O^+]{(1)\ [Pd],\ OH^-} ArCH_2COOH$$

entry	substrate	additive (amt, equiv)	isolated yield, %[b]
1	$C_6H_5CH_2Br$		60-70
2	$C_6H_5CH_2Cl$		<0.5
3	$C_6H_5CH_2Cl$	KBr (1)	1
4	$C_6H_5CH_2Cl$	$CeCl_3{\bullet}6H_2O$ (0.01)	10
5	$C_6H_5CH_2Cl$	$CeCl_3{\bullet}6H_2O$ (0.1)	11.5
6	$4\text{-}CH_3C_6H_4CH_2Br$		59-68
7	$3\text{-}CH_3C_6H_4CH_2Br$		54-60
8	$2\text{-}CH_3C_6H_4CH_2Br$		43-47
9	$2\text{-}C_{10}H_7CH_2Br$		42-49
10	$4\text{-}F\text{-}C_6H_4CH_2Br$		26-45

[a]Reaction conditions: 1.5 mmol of benzylic halide, 4 g of 50% KOH, 1.2 mL of $CHCl_3$, 0.015 mmol of $[(Ph_3P)_2PdCl_2]$, N_2, 22°C, 24 h. [b] After purification by vacuum sublimation.

Use of $^{13}CHCl_3$ for the reaction of iodobenzene gave $Ph^{13}COOH$ in >98% isotopic purity, demonstrating that CO arises from chloroform. One can account for the conversion of halides to acids by the pathway illustrated in Scheme 4. The key intermediate is the palladium dichlorocarbene complex 8 which reacts with alkali to form the palladium complex 9.

Scheme 4

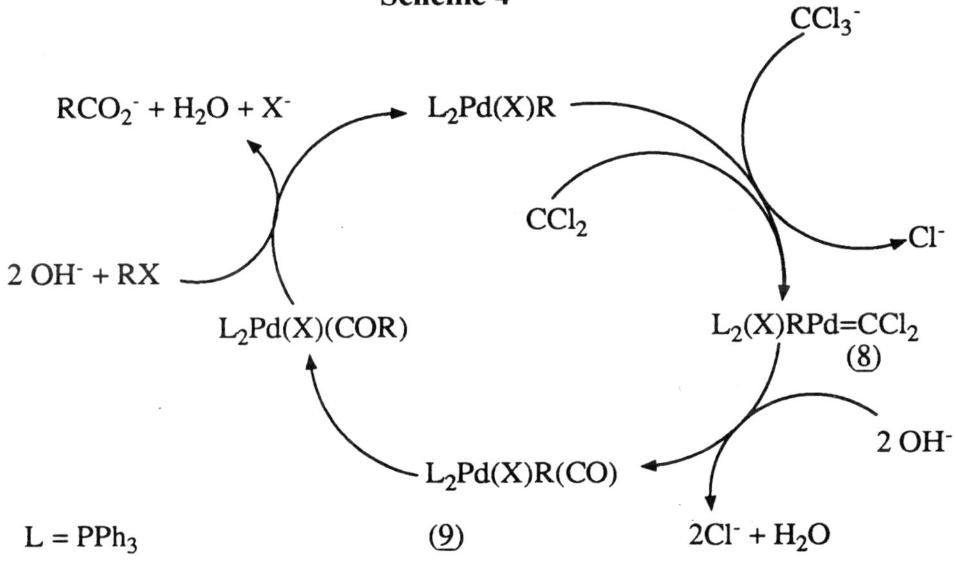

In conclusion, methodology has been developed for the biphasic carbonylation of chloroarenes to carboxylic acids, and there are two routes by which the acid is formed. The chloroform-alkali system is of considerable potential for the exceptionally mild *in situ* generation of carbon monoxide and its application to catalytic organic reactions.

Acknowledgments

We are grateful to the Natural Sciences and Engineering Research Council of Canada for support of this research.

References

1. Goldberg, Y. (1992) Phase Transfer Catalysis. *Selected Problems and Applications*, Gordon and Beach Science Publishers, Reading, U.K.
2. Dehmlow, E.V. and Dehmlow, S.S. (1993) *Phase Transfer Catalysis*, Third Edition, VCH Publishers Inc., New York.
3. Grushin, V.V. and Alper, H. (1994) Transformations of Chloroarenes, Catalyzed by Transition-Metal Complexes, *Chem. Rev.,* **94**, 1047-1062.
4. Galamb, V., Gopal, M. and Alper, H. (1983) Synthesis of Diynes, α,β-Unsaturated Monoacids, and Diacids by the Selective Palladium (O) Catalyzed and Phase Transfer Catalyzed Reactions of Vinylic Bromides, *Organometallics*, **2**, 801.
5. Bumagin, N.A., Nikitin, K.V. and Beletskaya, I.P. (1988) Palladium Catalyzed Carbonylation of Aryl Iodides in Aqueous Media, *J. Organomet. Chem.*, **358**, 563-565.
6. Amer, I. and Alper, H. (1988) Nickel Cyanide and Phase Transfer Catalyzed Carbonylation of Aryl Iodides in the Absence of Light, *J. Org. Chem.*, **53**, 5147-5149.

7. Grushin, V.V. and Alper, H. (1993) Simple and Efficient Palladium-Catalyzed Carbonylation of Iodoxyarenes in Water Under Mild Conditions, *J. Org. Chem.*, **58**, 4794-4795.

8. Grushin, V.V. and Alper, H. (1993) Alkali Induced Disproportionation of Palladium (II) Tertiary Phosphine Complexes, [L_2PdCl_2], to LO and Palladium (O). Key Intermediates in the Biphasic Carbonylation of ArX Catalyzed by [L_2PdCl_2], *Organometallics*, **12**, 1890-1901.

9. Grushin, V.V. and Alper, H. (1992). An Exceptionally Simple Biphasic Method for the Metal Catalyzed Carbonylation of Chloroarenes, *J. Chem. Soc., Chem. Commun.*, 611-612.

10. Colquhoun, H.M., Thompson, D.J., and Twigg, M.V. (1991). *Carbonylation*, Plenum Press, New York.

11. Grushin, V.V. and Alper, H. (1993) Novel Palladium Catalyzed Carbonylation of Organic Halides by Chloroform and Alkali, *Organometallics*, **12**, 3846-3850.

COUNTER PHASE TRANSFER CATALYSIS BY WATER-SOLUBLE PHOSPHINE COMPLEXES

T. OKANO
Department of Materials Science, Faculty of Engineering
Tottori University, Tottori 680, Japan

1. Introduction

The phase transfer catalysis is one of the big subjects in the latter half of the 20th century, and a large number of studies have been reported. The phase transfer catalysts, which are crown ethers or onium salts, transport anions from aqueous phase into organic phase, and accelerate reactions of organic substrates with the anions. Hypothetically, it is possible to exist a functional molecule capable of transferring lipophilic molecules from organic phase into aqueous phase. However, such a molecule was not reported until recently. In 1986, Mathias et al. and we reported the first examples of such functional molecules. Mathias et al. [1] reported that 4-dimethylaminopyridine reacts with acid halides in organic phase to form pyridinium salts, which transfer into aqueous phase and react with amino acids in the aqueous phase to give amides. This catalysis was named "inverse phase transfer catalysis". We found that water-soluble palladium complexes, reacting with allyl halides in organic phase, transport them into aqueous

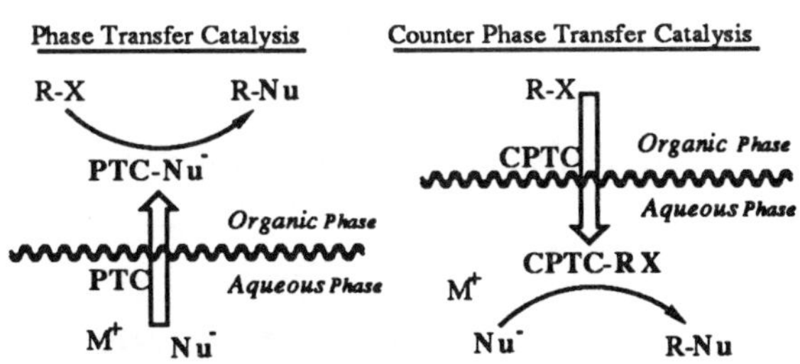

Figure 1. Normal phase transfer and counter phase transfer catalyses

I. T. Horváth and F. Joó (eds.), Aqueous Organometallic Chemistry and Catalysis, 97–110.

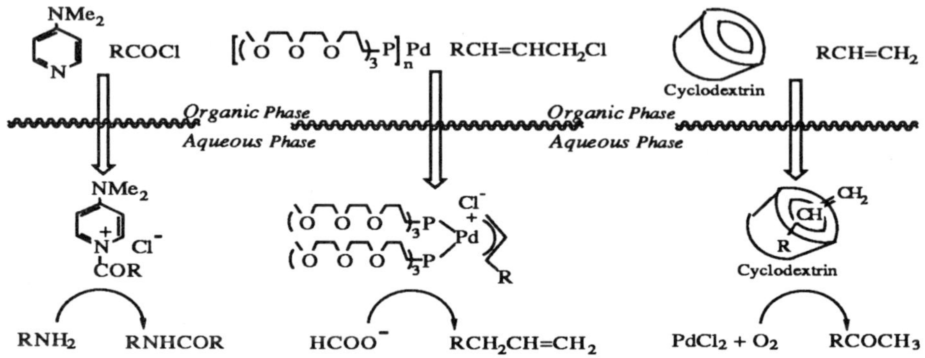

Figure 2. Counter or Inverse Phase Transfer Catalysts

phase, and catalyze the reduction of the allyl compounds with sodium formate in the aqueous phase [2]. We called this catalysis "counter phase transfer catalysis". Recently, Harada [3] reported cyclodextrins to act as a counter phase transfer catalyst. However, we believe the applicability of the water-soluble phosphine complexes to be superior to the others. Because, the pyridine derivatives are limited to the transportation of acid halides, and the cyclodextrins are limited in size of the cavity. On the other hand, transition metal complexes are able to interact with a wide variety of organic compounds. Additionally, the hydrophilic metal complexes are able to perform the double function as counter phase transfer catalyst and metal catalyst. Therefore, we expect a wide applicability and high functions of the hydrophilic transition metal complexes. However, the studies on the catalyses of hydrophilic complexes have not progressed so far. Here, I would like to introduce some catalytic reactions of organic halides using hydrophilic palladium phosphine complexes under two-phase conditions.

2. Counter Phase Transfer Catalytic Reactions of Allyl Halides

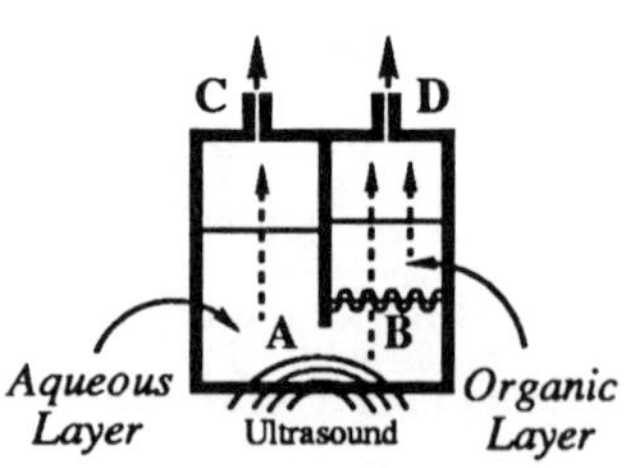

Figure 3. Partitioned reactor

It is difficult to verify the counter or inverse phase transfer catalysis strictly, because such a catalyst more or less acts as a surfactant as well as a normal phase transfer catalyst [4]. Therefore, we examined whether the reduction of allyl acetate with sodium formate under two-phase conditions evolves the propene gas from the

$$\underset{Org.}{CH_2{=}CHCH_2OAc} + \underset{Aq.}{NaOOCH} \xrightarrow[H_2O\text{-Heptane}]{PdCl_2L_2} \underset{Gas}{CH_2{=}CHCH_3} \uparrow$$

Table 1. Catalytic Reduction of Allyl Acetate with Sodium Formate

Catalyst	Yield/% of C_3H_6	From Aq. : From Org. and Int.
$PdCl_2[P(n\text{-}Bu)_3]_2$	38	12 : 88
$PdCl_2[P\{CH_2(CH_2CH_2O)_3CH_3\}_3]_2$	55	98 : 2

Conditions; $PdCl_2L_2$ (0.1 mmol) : C_3H_5OAc : NaOOCH = 1:200:400, Heptane (4 ml), H_2O (7 ml), 80°C, 0.5 h.

aqueous phase or the organic phase [2]. We used a partitioned reactor as shown in Fig. 2. Part A was charged with an aqueous solution of catalyst and sodium formate, and a heptane solution of allyl acetate was placed in Part B. The propene gas evolved from Part A of the aqueous layer was collected from Outlet C, and the gas from the heptane layer, the interface, and Part B of the aqueous layer was collected from Outlet D. As the volume ratio of the aqueous solution in Part A to that in Part B is known, the gas amount from each phase can be calculated. The reduction with hydrophobic catalyst, $PdCl_2(PBu_3)_2$, evolved the gas mostly from the organic layer and/or the interface, whereas the reaction with hydrophilic catalyst, $PdCl_2[P\{CH_2(CH_2CH_2O)_3CH_3\}_3]_2$, evolved the propene gas mostly from the aqueous layer. This fact proves that the hydrophilic complex transports the allyl acetate from the organic phase into the aqueous phase and catalyzes the reduction to propene with sodium formate in the aqueous phase.

Palladium-catalyzed allylation of aldehydes with allyl compounds and stannous chloride was found recently [5], and is an important reaction for organic synthesis.

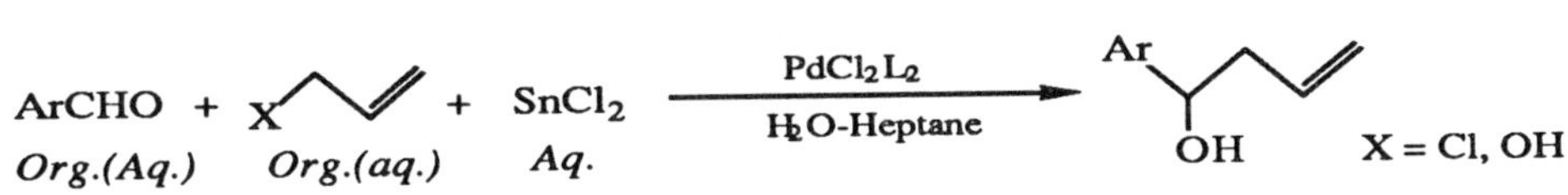

Table 2. Allylation of aldehydes with allyl compounds with $SnCl_2$

Catalyst	Aldehyde	Allyl compound	Yield/%
$PdCl_2(PPh_2(C_6H_4SO_3Na))_2$	PhCHO	$CH_2{=}CHCH_2Cl$	98
	PhCHO	$CH_2{=}CHCH_2OH$	98
	p-$NaOOCC_6H_4CHO$	$CH_2{=}CHCH_2Cl$	94
$PdCl_2(P(C_6H_4CH_3)_3)_2$	PhCHO	$CH_2{=}CHCH_2Cl$	60
	PhCHO	$CH_2{=}CHCH_2OH$	13

$PdCl_2L_2$ (0.1mmol) : PhCHO : $CH_2{=}CHCH_2Cl$: $SnCl_2$ = 1:100:200:200, H_2O (10ml), Heptane (10ml), 35°C, 20 h.

However, aldehydes are hydrophobic, while stannous chloride is lipophobic. Accordingly, amphiphilic solvents such as DMF and DMSO were necessary for the homogeneous reaction. To this reaction, we tried to apply a sulfonated phosphine complex, $PdCl_2\{PPh_2(m\text{-}C_6H_4SO_3Na)\}_2$ (**1**), which is soluble in water but does not have the function as normal phase transfer catalyst. The reaction even in a heptane-water two-phase system gave the allylated product in an excellent yield. On the other hand, hydrophobic catalyst, $PdCl_2\{P(p\text{-}C_6H_4CH_3)_3\}_2$ (**2**), was less active in this system. When allyl alcohol which is soluble in water was used instead of allyl chloride, the catalytic activity of the hydrophobic catalyst (**2**) was seriously decreased, while the water-soluble catalyst still showed a high reactivity. This water-soluble catalyst was also applicable to the allylation of a water-soluble aldehyde, which gave the corresponding alcohol in 94% yield.

Table 3. Transportation of allyl chloride from organic phase into aqueous phase

$PdCl_2L_2$	Conversion/% of C_3H_5Cl
L = $PPh_2(C_6H_4SO_3Na)$	51
$P(C_6H_4CH_3)_3$	22

$PdCl_2L_2$ (0.1 mmol) : C_3H_5Cl : $SnCl_2$ = 1:200:200, H_2O (10 ml), Heptane (10 ml), at 35°C for 1 h

The palladium-demanded step in this allylation is the reaction of allyl compounds with stannous chloride to form allyl stannic compounds. Therefore, the consumption rate of allyl chloride in the heptane phase was checked in the biphasic reaction without the aldehyde. When the water-soluble catalyst (**1**) was used, the conversion of ally chloride was 51% at 35°C for 1 h. However, the conversion was only 22% in the reaction with the hydrophobic catalyst (**2**). This result suggests that the high efficiency of the water-soluble catalyst on the allylation is attributed to the fast transfer of allyl chloride from the organic phase into the aqueous phase via a hydrophilic π-allyl palladium compound, which yields the allyl stannic compounds.

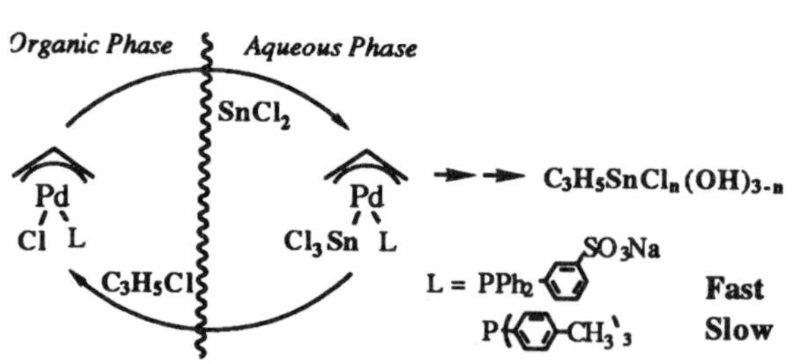

Figure 4. Possible mechanism for the transportation

The biphasic allylation of benzaldehyde in the presence of **1** showed an unusual solvent effect. When a heptane solution was used as an organic phase, the product yield was 54% for 7 h. However, the reactions in toluene and in anisole gave the product in 40% and 33% yields, respectively.

Table 4. Solvent Effect on the Allylation of Benzaldehyde

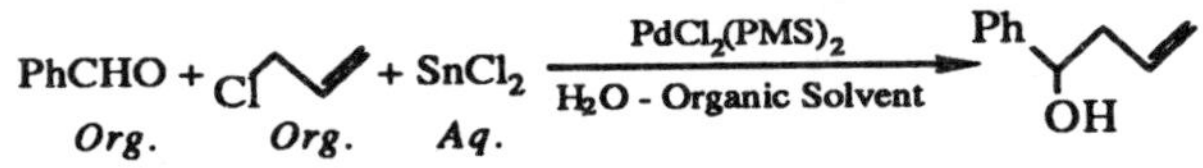

Organic Solvent	Yield/%
Heptane	54
Toluene	40
Anisole	33

$PdCl_2(PMS)_2$ (0.1mmol):PhCHO:C_3H_5Cl:$SnCl_2$=1:100:200:200, H_2O (10ml), Organic solvent (10ml), at 35°C for 7 h.

Generally, the efficiency of normal phase transfer catalysts is higher in polar solvents than that in nonpolar solvents. This is the inverse of the solvent effect on this allylation. An analogous solvent effect is also observed in the counter phase transfer catalytic reduction of allyl halides with sodium formate. Therefore, the inverse solvent effect is considered to be characteristic of the counter phase transfer catalysis.

A similar solvent effect was also observed in a biphasic carboxylation of cinnamyl chloride with NaOH and CO at atmospheric pressure. The rate of CO absorption of the

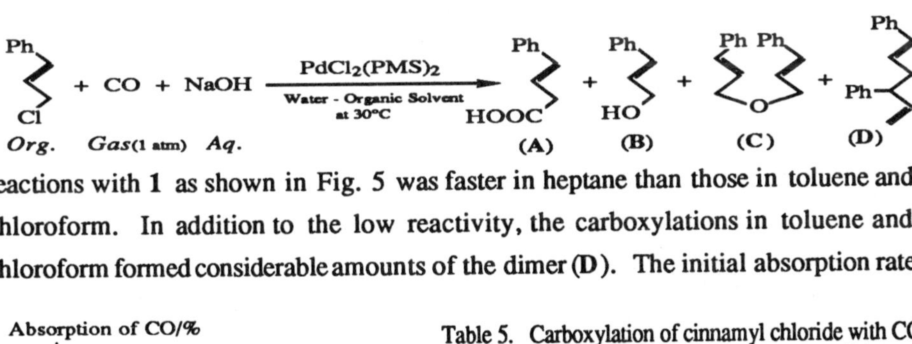

(A) (B) (C) (D)

reactions with **1** as shown in Fig. 5 was faster in heptane than those in toluene and in chloroform. In addition to the low reactivity, the carboxylations in toluene and in chloroform formed considerable amounts of the dimer (**D**). The initial absorption rate in

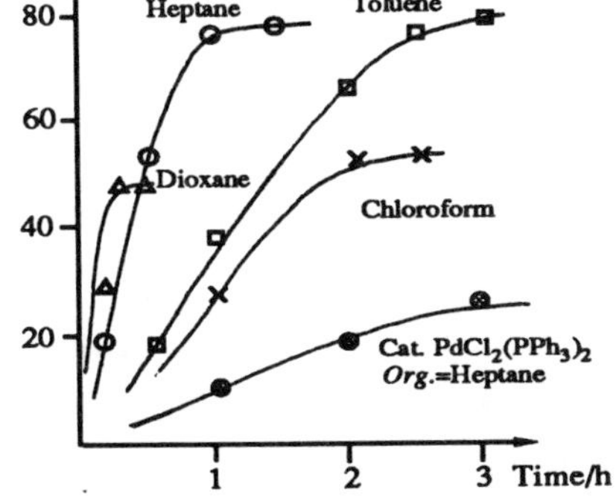

Figure 5. Rate of CO absorption

Table 5. Carboxylation of cinnamyl chloride with CO and NaOH under two-phase conditions

Organic Phase (Time)	Yields/%		
	A	B + C	D
Heptane (1.5 h)	47 (58)*	15	21
Toluene (4 h)	49 (53)*	7	37
Chloroform (2.5 h)	38 (40)*	17	40
Dioxane (0.5 h)	26 (39)*	30	11

* Selectivity (A/(A+B+C+D) x 100)
$PdCl_2\{PPh_2(m\text{-}C_6H_4SO_3Na)\}_2$(0.05mmol):PhCH=CHCH$_2$Cl
NaOH=1:100:1000, Heptane(10ml),H_2O(50ml).

Table 6. Counter Phase Transfer Catalytic Carboxylation of Allyl Halides

Allyl Halides	Reaction Time (h)	Product	Yield (%)[a]
	0.5		82
	1.2 [b]		63
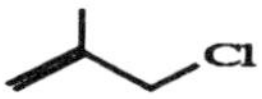	0.5 [c]		73
	0.5		73
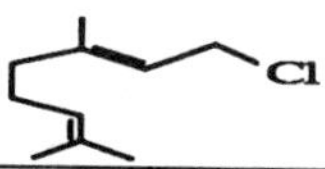	1.2	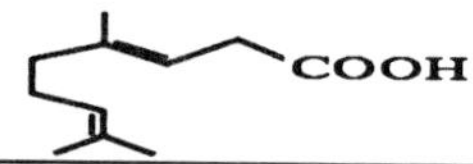	64

$PdCl_2\{PPh_2(m\text{-}C_6H_4SO_3Na)\}_2$ (0.05 mmol) : $PhCH{=}CHCH_2Cl$: NaOH = 1:100:200, Heptane (10 ml), H_2O (50 ml), 30°C. a) Isolated yield.
b) H_2O (10 ml), NaOH (25 mmol) c) NaOH (50 mmol)

dioxane was the fastest, but large amounts of cinnamyl alcohol and the ether (C) were formed. This fact indicates that simple nucleophilic substitutions occurred in a solvent miscible with water like dioxane. Therefore, it is concluded that heptane is the best solvent for the carboxylation with the water-soluble catalyst.

Since the synthesis of β,γ–unsaturated carboxylic acids is rather difficult by normal organic reactions, we tried the carboxylation of several allyl halides under the counter phase transfer catalytic conditions. Though the carboxylation of cinnamyl chloride gave many products, these allyl halides gave the corresponding carboxylic acids in considerable yields in short reaction times under mild conditions.

One of the advantages of this counter phase transfer catalytic reaction is easy separation of the product. Simple acidification of the separated aqueous solution, which contains the product and the sulfonated phosphine complex, gave almost pure carboxylic acid, because the sulfonated phosphine complex is soluble even in the acidic aqueous solution.

3. Biphasic Carboxylation of Benzyl Halides

The hydrophilic palladium complex (1) was also a good catalyst for the carboxylation of benzyl halides under water-heptane two-phase conditions. In the case of benzyl chloride, the carboxylation with the water-soluble catalyst could be completed at 50°C within 5 h to give phenyl acetic acid in a 89% isolated yield. However, the hydrophobic catalyst, $PdCl_2(PPh_3)_2$, was inactive in this two-phase solvent system. Benzyl bromide was more reactive than the chloride. The carboxylation with 1 proceeded at 30°C to give the

Table 7. Carboxylation of benzyl halides under two-phase conditions

$$\underset{Org.}{PhCH_2X} + \underset{Gas(1atm)}{CO} + \underset{Aq.}{NaOH} \xrightarrow[H_2O\text{ - Heptane}]{PdCl_2L_2\ (1\ mol\%)} PhCH_2COOH + PhCH_2OH$$

$PhCH_2X$ X=	$PdCl_2L_2$ L=	Temp. (°C)	Time (h)	Yields (%) $PhCH_2COOH$[a]	$PhCH_2OH$[b]
Cl	$PPh_2(m\text{-}C_6H_4SO_3Na)$	50	5	89	1
	PPh_3	50	12	2	6
Br [c]	$PPh_2(m\text{-}C_6H_4SO_3Na)$	30	5	89	0
	$P(p\text{-}C_6H_4CH_3)_3$	30	7	34	14
	$P(p\text{-}C_6H_4CH_3)_3 + Bu_4NBr$[d]	30	7	46	20
	$P(p\text{-}C_6H_4CH_3)_3$ in EtOH	30	2	37	58

$PdCl_2L_2$ (0.1 mmol) : $PhCH_2X$: NaOH = 1:100:250, Heptane (8.9 ml), H_2O (10 ml).
a) Isolated yield. b) GC yield. c) Heptane (10 ml) d) Bu_4Br (2 mmol)

carboxylic acid in a high yield, while the reaction with **2** was slow, and formed a considerable amount of benzyl alcohol. The addition of normal phase transfer catalyst like Bu_4NBr improved the catalytic activity of **2**. However, the acceleration effect was not enough, and the addition of Bu_4NBr was ineffective in retarding the formation of benzyl alcohol. Though the carboxylation with **2** in ethanol was very fast, a large amount of benzyl alcohol was formed by the nucleophilic substitution. These results indicate that the water-soluble transition metal complex is very efficient for the metal-catalyzed reaction of organic substrates susceptible to nucleophilic substitution.

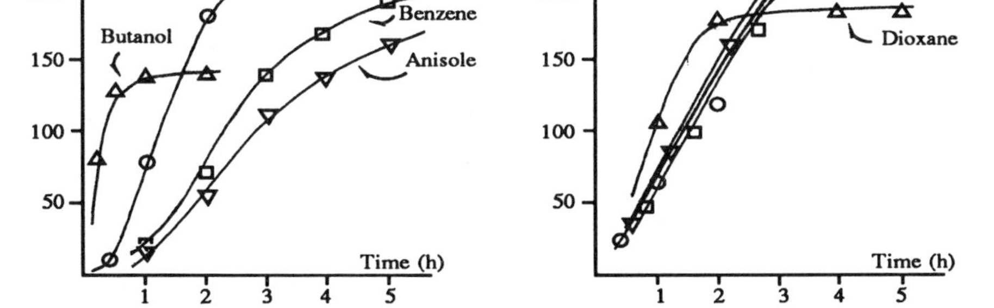

$PdCl_2[PPh_2(m\text{-}C_6H_4SO_3Na)]_2$ (0.1 mmol), $PhCH_2X$ (10 mmol), NaOH (25 mmol), CO (1 atm), H_2O (10 ml), Organic solvent (X=Cl, 8.9 ml and X=Br, 10 ml).

Figure 6. Carboxylation of benzyl halides under various two-phase conditions

Since the amounts of CO absorption were almost comparable to the yields of phenyl acetic acid in the biphasic carboxylation of benzyl halides with **1**, the solvent effect was estimated on the basis of the CO absorption. Though unexpected short induction periods were observed in the reactions of benzyl chloride at 50°C, the rate in heptane was faster than those in benzene and in anisole as shown in Fig. 6. However, the rate of the carbonylation of benzyl bromide in heptane was almost equivalent to those in toluene and in anisole. This solvent effect was different from those on the other reactions with **1** as mentioned above. However, we regard this carboxylation as a counter phase transfer catalytic reaction, because the solvent effect is also different from that on the normal phase transfer catalytic reactions.

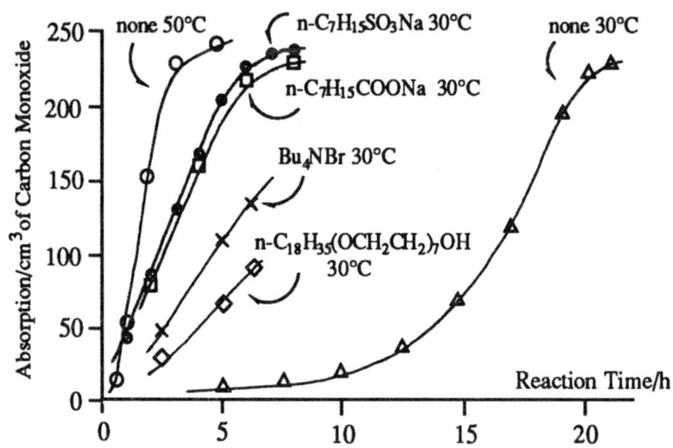

Additive (2 mmol), $PdCl_2[PPh_2(m\text{-}C_6H_4SO_3Na)]_2$ (0.1 mmol), $PhCH_2Cl$ (10 mmol), aq-NaOH (2.5N, 10 ml), heptane 10 ml), CO (1 atm)

Figure 7. Effect of additives on carboxylation of $PhCH_2Cl$ catalyzed by $PdCl_2[PPh_2(m\text{-}C_6H_4SO_3Na)]_2$ under two-phase conditions

The induction period of the carboxylation of benzyl chloride in the heptane-water system was very long at 30°C as shown in Fig. 7, while the rate in the latter half of the reaction was rather fast. This result implies that the accumulation of sodium phenyl acetate accelerate the carboxylation. In fact, the addition of sodium phenyl acetate greatly improved the reactivity. Accordingly, the long induction period is supposed to be the time for accumulating the sodium carboxylate. Therefore, the effect of various additives was tested. Sodium heptanesulfonate, sodium octanoate, tetrabutylammonium bromide, and n-$C_{18}H_{35}(OCH_2CH_2)_7OH$ were found to be effective in accelerating the carboxylation. Tetrabutylammonium bromide and n-$C_{18}H_{35}(OCH_2CH_2)_7OH$ have two functions as a phase transfer catalyst as well as a surfactant. However, the carboxylate and the sulfonate only have a function as a surfactant. Therefore, the acceleration effect of these additives may be due to an action as a surfactant.

Table 8. Effect of n-$C_7H_{15}SO_3Na$ on the carboxylation of $PhCH_2Cl$

Catalyst	Organic Phase	Additive (mmol)	Absorption (%) of CO
$PdCl_2(PPh_3)_2$	Toluene	0	3
		2	3
$Pd(PPh_3)_4$	Toluene	0	4
		2	4
$PdCl_2\{PPh_2(m\text{-}C_6H_4SO_3Na)\}_2$	Toluene	0	5
		2	59
$PdCl_2\{PPh_2(m\text{-}C_6H_4SO_3Na)\}_2$	Heptane	0	20 (17)*
		2	100 (93)*

$PhCH_2Cl$ (10 mmol), aq-NaOH (2.5N, 10 ml), Organic solvent (10 ml), Catalyst (0.1 mmol), CO (1 atm), 30°C, 8 h. * Isolated yield of $PhCH_2COOH$.

The acceleration effect of heptanesulfonate was observed in the carboxylations in toluene as well as in heptane. However, the addition had no effect on the reactions with hydrophobic two-valent catalyst, $PdCl_2(PPh_3)_2$, nor with zero-valent catalyst, $Pd(PPh_3)_4$. This fact suggests that the acceleration effect is not due to a simple action of the surfactant. One of the explanations for the effect is as follows (see Fig. 8). The hydrophilic group of the sulfonated phosphine complex in organic phase is strongly associated with heptanesulfonate between the ionic groups. The lipophilic phenyl group is associated with the alkyl chain of the sulfonate in aqueous phase. Easy exchange of these associations results in easy phase transfer of the catalyst, and accelerates the carboxylation. On the basis of this mechanism, the inefficiency of sulfonate in the reaction with the lipophilic catalysts can be explained.

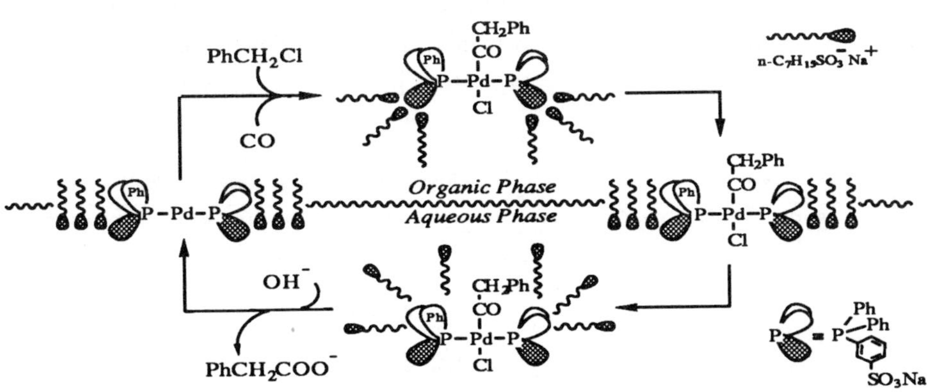

Figure 8. Probable mechanism for the carboxylation

4. Biphasic Reactions of Aryl Halides

Phenyl iodide could also be carboxylated with NaOH and CO at atmospheric pressure in the presence of the water-soluble phosphine catalyst under biphasic conditions, and gave benzoic acid in high isolated yields. Though the solvent effect was neutral in this reaction as in the carboxylation of benzyl bromide, the efficiency of the water-soluble phosphine catalyst was far higher than that of the hydrophobic complex (**2**). The reaction with **2** was promoted by the addition of a phase transfer catalyst, but the

Table 9. Carboxylation of phenyl iodide under two-phase conditions

$$\underset{Org.}{PhI} + \underset{Gas(1atm)}{CO} + \underset{Aq.}{NaOH} \xrightarrow[H_2O\text{ - Organic Solvent}]{PdCl_2L_2\ (1\ mol\%)} PhCOOH$$

$PdCl_2L_2$	Organic Phase	Isolated yield/%
L= 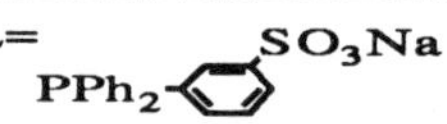	Heptane	88
	Toluene	84
	Anisole	81
	Dioxane	90
L= 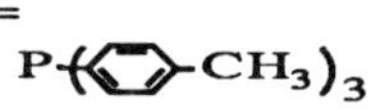	Heptane	5
	Anisole	2
	Anisole (Bu_4NBr)[a]	24

$PdCl_2L_2$ (0.1 mmol) : L : PhI : NaOH = 1:5:100:250, H_2O (10 ml), Organic Solvent (10 ml), at 50°C for 4 h. a) Bu_4NBr (1 mmol)

reaction was still slower than the carboxylation with **1**. Alper et al. have also reported that the addition of phase transfer catalysts is not always effective in accelerating the biphasically metal-calalyzed reactions [6].

The carboxylation of some aryl halides was summarized in Table 10. Aryl iodides were carboxylated at 50°C to give the corresponding benzoic acids in considerable yields. Though aryl bromides were less reactive than the iodides, aryl bromides having electron-withdrawing groups were carboxylated at 80°C to give the carboxylic acids in good

Table 10. Carboxylation of aryl halides catalyzed by $PdCl_2\{PPh_2(m\text{-}C_6H_4SO_3Na)\}_2$

Aryl iodide	Time/h	Yield/% of product	Aryl halide	Time/h	Yield/% of product
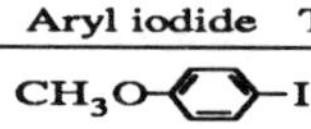	5	54	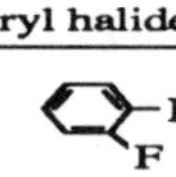	27	78
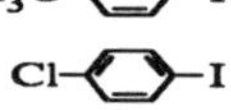	5	95		24 (80°C)	47
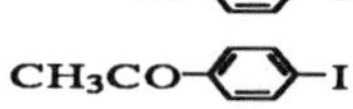	5	91	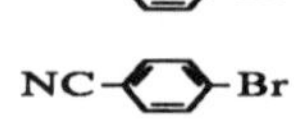	5 (80°C)	92
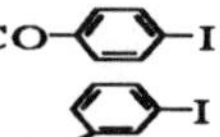	7	77			
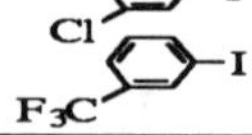	5	76	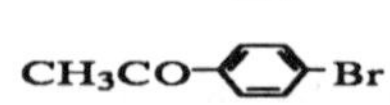	7 (80°C)	83

$PdCl_2L_2$ (0.1 mmol):L:PhI:NaOH = 1:5:100:250, H_2O (10 ml), Organic Solvent (10 ml), at 50°C.

Table 11. Cyanation of PhI in the presence of $PdCl_2\{PPh_2(m\text{-}C_6H_4SO_3Na)\}_2$

$$\underset{Org.}{C_6H_5I} + \underset{Aq.}{MCN} \xrightarrow[\text{in Water - Heptane Reflux for 1 h}]{PdCl_2(PMS)_2\ (1\ mol\%)} C_6H_5CN$$

MCN (MCN/PhI=1.3 moles/mol)	$NaBH_4$ ($NaBH_4$/Pd moles/mol)	$ZnCl_2$ ($ZnCl_2$/MCN moles/mol)	Yield/% of PhCN [a)]
NaCN	2	0.38	99
	none	0.38	<1
	2	0	0
	2	0.23	0
	2	0.27	71
	2	0.77	51
KCN	2	0.38	98
$Na_2Zn(CN)_4$ [b)]	2	Zn/CN = 0.25	52
$K_2Zn(CN)_4$ [b)]	2	Zn/CN = 0.25	15

$PdCl_2(PMS)_2$ (0.1 mmol) : PhI : MCN = 1:100:130, H_2O (10 ml), Heptane (10 ml).
a) GC yield. b) $M_2Zn(CN)_4$ (5 mmol).

yields. As aryl chlorides were inactive for the carboxylation, the selective carboxylation of chloro iodo arenes could be achieved.

Cyanation of phenyl iodide with NaCN was also catalyzed by the water-soluble phosphine complex under two-phase conditions. This reaction was very fast in the presences of $NaBH_4$ and zinc chloride, and gave benzonitrile almost quantitatively. In the absence of $NaBH_4$, however, the reaction did not proceed. This fact indicates that the catalytically active species is a zero-valent palladium complex. Interestingly, without zinc chloride, no benzonitrile was obtained. When the ratio of zinc chloride to NaCN was 0.23, the reaction did not occurred. The cyanation was effected by the use of zinc chloride more than 0.25 moler ratio. This result suggests that tetracyano zincate is an active cyanide source. In fact, these zincate complexes were efficient as a cyanide source.

The solvent effect on the cyanation with the sulfonated phosphine complex was

Table 12. Solvent effect on the cyanation by $PdCl_2(PMS)_2$

Organic Phase	Yield/%* of PhCN Reflux for 1 h	for 30 min
Heptane	99	89
Dibutyl ether	98	89
Toluene	99	70
Anisole	98	53
Dioxane	0	

$PdCl_2(PMS)_2$ (0.1 mmol):$NaBH_4$:$ZnCl_2$:PhI:NaCN = 1:2:50:100:130, Water (10 ml), Organic solvent (10 ml). * GC yield.

Table 13. Cyanation of PhI in the presence of various palladium catalysts

Catalyst	Additives (mmol)	Yield/% of PhCN[a)]
$PdCl_2(PPh_2\text{-}C_6H_4(SO_3Na))_2$	$NaBH_4$ (0.2)	99
$PdCl_2(PPh_2\text{-}C_6H_4(COONa))_2$	$NaBH_4$ (0.2)	23
$PdCl_2(PPh_2\text{-}C_6H_4(CH_2NMe_3Br))_2$	$NaBH_4$ (0.2)	5
$PdCl_2\{P(CH_2CH_2OCH_2CH_2OCH_2CH_2OCH_3)_3\}_2$	$NaBH_4$ (0.2)	0
$PdCl_2\{P(C_6H_4\text{-}CH_3)_3\}_2$ b)	$NaBH_4$ (0.2) Bu_4NI (1)	0
	$NaBH_4$ (0.2) n-$C_7H_{15}SO_3Na$ (1)	1
$Pd(PPh_3)_4$ b)	none	2

Palladium complex (0.1 mmol) : PhI : NaCN : $ZnCl_2$ = 1:100:130:50, heptane (10 ml), H_2O (10 ml), reflux for 1 h. a) GC yield. b) Anisole (10 ml) was used instead of heptane.

characteristic of the counter phase transfer catalysis. The catalytic activity was higher in nonpolar solvents than that in polar solvents.

The hydrophobic catalyst (2) was totally inactive in this biphasic cyanation. The addition of a phase transfer catalyst or a surfactant was ineffective. The zero-valent palladium complex, $Pd(PPh_3)_4$, had no catalytic activity. On the other hand, carboxylated phosphine complex, $PdCl_2\{PPh_2(m\text{-}C_6H_4COONa)\}_2$, was active for the cyanation, and phosphine complex having an ammonium group, $PdBr_2\{PPh_2[m\text{-}C_6H_4CH_2N(CH_3)_3Br]\}_2$, also showed a catalytic activity. However, the polyether phosphine complex, $PdCl_2[P\{CH_2(CH_2CH_2O)_3CH_3\}_3]_2$, was not active, while the complex is soluble in water. This fact suggests that the phosphines having ionic groups were necessary for the occurrence of the catalytic action, though the mechanism

Table 14. Counter phase pransfer catalytic cyanation of aryl halides

Aryl halide	Time (h)	Product/Yield(%)		Aryl halide	Time (h)	Product/Yield(%)[a)]	
I-C_6H_4-OCH_3	8	NC-C_6H_4-OCH_3	92	I-C_6H_4-CF_3	5	NC-C_6H_4-CF_3	99 (85)[b)]
I-C_6H_4-Cl	5	NC-C_6H_4-Cl	90	I-C_6H_4-Cl	3	NC-C_6H_4-Cl	99
I-C_6H_4-Br	5	NC-C_6H_4-Br	96	I-C_6H_4-OCH_3	24	NC-C_6H_4-OCH_3	79
I-C_6H_4-$COCH_3$	2	NC-C_6H_4-$COCH_3$	75	Br-C_6H_5	5	NC-C_6H_5	0
I-C_6H_4-COOEt	2	NC-C_6H_4-COOEt	93				

$PdCl_2(PMS)_2$ (0.1 mmol) : $NaBH_4$: $ZnCl_2$: PhI : NaCN = 1:2:50:100:130, H_2O (10 ml), Heptane (10 ml).
a) GC yield. b) Isolated yield.

for the cyanation is not clear.

The biphasic cyanation with **2** was applied to several aryl halides (see Table 14). Aryl iodides were readily cyanated to the corresponding aryl cyanides in excellent yields. Since aryl bromides and chlorides were not cyanated under these conditions, the selective cyanation of chloro or bromo iodo benzenes could be achieved.

5. Conclusion

As mentioned above, several biphasic catalytic reactions of organic halides using the sulfonated phosphine palladium complex was introduced. We considered all of these as the counter phase transfer catalytic reactions. One of the characteristics of the counter-phase transfer catalyst is the high efficiency and the high selectivity. The efficiency and the selectivity are higher than that of the mixed system of hydrophobic metal catalyst and normal phase transfer catalyst. The high efficiency and the high selectivity arose from the easy transportation of organic halides by complexation into the aqueous phase in which nucleophiles exist. Although the use of amphiphilic solvents or normal phase transfer catalysts is efficient in enhancing the catalytic activity of hydrophobic catalysts, that is ineffective in retarding the side-reactions of organic substrate susceptible to nucleophilic substitution such as the reactions of allyl halides and benzyl halides. Because the increase in the concentration of nucleophiles in the organic phase containing organic halides results in increase in the rate of nucleophilic substitutions.

Another characteristic is the solvent effect. The catalytic activity of the hydrophilic complexes is higher in nonpolar solvent than that in polar solvent, or is almost unchanged in various solvents immiscible with water. This is the inverse of the solvent effect on the normal phase transfer catalysis. This difference may be ascribed to the transportation step, which is the transporting organic substrates from organic phase into aqueous phase in the counter phase transfer catalysis, whereas that is the transporting inorganic reagents from aqueous phase into organic phase in normal phase transfer catalysis.

In conclusion, the counter phase transfer catalytic reactions using sulfonated phosphine complex have several advantages as follows. The water-soluble metal complex is able to perform the double function as a counter phase transfer catalyst and as a metal catalyst concertedly. The efficiency of the counter phase transfer catalyst is generally higher than that of a mixture of lipophilic metal catalyst and normal phase transfer catalyst. Additionally, the water-soluble metal catalyst has the advantage of high selectivity. The last is the easy separation. If the productis in organic phase, the catalyst in aqueous phase can easily be separated. Even if the product and the catalyst exist in aqueous phase, the product could be isolated, because the sulfonated phosphine complex has a high affinity to water in a wide range of pH.

Transition metal complexes are able to interact with a wide variety of organic compounds, we expect the great development of the counter phase transfer catalysis in the near future.

6. References

1. Mathias, J.S. and Vaidya, R.A. (1986) Inverse phase transfer catalysis. First report of a new class of interfacial reactions, *J. Am. Chem. Soc.* **108**, 1093-1094.
2. Okano, T., Moriyama, Y., Konishi, H., and Kiji, J. (1986) Counter phase transfer catalysis by water-soluble phosphine complexes. Catalytic reduction of allyl chlorides and acetates with sodium formate in two-phase systems, *Chem. Lett.* 1463-1466.
3. Harada, A. (1990) Organic reactions in aqueous solutions. Inverse phase transfer reactions, *J. Syn. Org. Chem. Jpn.* **48**, 517-521
4. Wasmser, C.C. and Yates, J.A. (1989) Kinetics and mechanisms for the two-phase reaction between aqueous aniline and benzoyl chloride in chloroform, with and without pyridine catalysis, *J. Org. Chem.* **54**, 150-154.
5. Takahara, J.P., Masuyama, Y., and Kurusu, Y. (1992) Palladium-catalyzed carbonyl allylation by allylic alcohols with $SnCl_2$, *J. Am. Chem. Soc.* **114**, 2577-2586.
6. Grushin, V.V. and Alper, H. (1993) Alkali-induced disproportionation of palladium(II) tertiary phosphine complexes, $[L_2PdCl_2]$, to LO and palladium(0). Key intermediates in the biphasic carbonylation of ArX catalyzed by $[L_2PdCl_2]$, *Organometallics*, **12**, 1890-1901.

METAL-CATALYSED HYDROPHOSPHINATION AS A ROUTE TO WATER-SOLUBLE PHOSPHINES

PAUL G. PRINGLE DAVID BREWIN,.
MARTIN B. SMITH, and KERRY WORBOYS
School of Chemistry, University of Bristol
Cantocks Close, Bristol, UK.

1. Introduction

It is difficult to exaggerate the impact that phosphine ligands have made on coordination chemistry, organometallic chemistry and homogeneous catalysis over the past 30 years. It is therefore not surprising that hydrophilic phosphines have been at the forefront in the development of aqueous organometallic chemistry and biphasic catalysis.

Most of the water-solubilising phosphines studied to date have been derivatives of arylphosphines such as PPh_3, dppe, chiraphos [1]. Great impetus for this work was provided by the success of the Ruhrchemie propene hydroformylation process which employs a Rh-tppts complex as the biphasic catalyst.

The greater basicity of trialkylphosphines such as PMe_3; PEt_3 and $Me_2PCH_2CH_2PMe_2$, (dmpe) makes them better ligands than arylphosphines for metals in high oxidation states and for when high complex nucleophilicity is required, e.g. in the activation of N_2 or CH_4. Hydrophilic trialkylphosphines may hold the key to transferring this chemistry to aqueous media. We have been interested in the synthesis and chemistry of hydrophilic alkylphosphines which can be made by routes which exploit catalytic hydrophosphination of alkenes and aldehydes [2-7]:

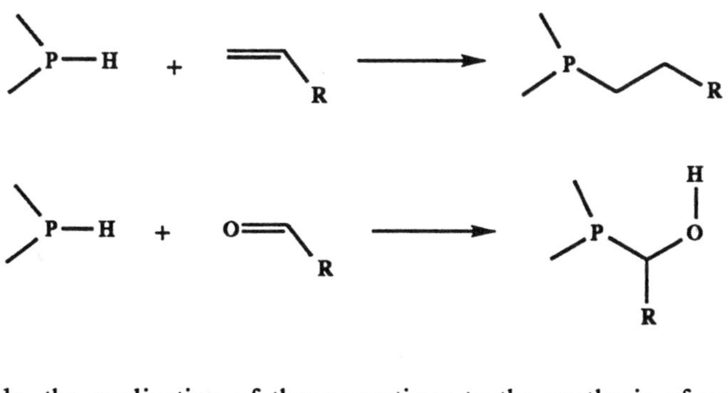

In this article, the application of these reactions to the synthesis of water-soluble monophosphines and diphosphines will be presented.

I. T. Horváth and F. Joó (eds.), Aqueous Organometallic Chemistry and Catalysis, 111–122.

2. ω-Hydroxyalkylphosphines

2.1 TRIS(HYDROXYMETHYL)PHOSPHINE

The acid-catalysed addition of PH_3 to formaldehyde is a large scale industrial process to produce the flame-retardant salts of the phosphonium ion $[P(CH_2OH)_4]^+$. In 1958, it was reported in a German patent [8] that tris(hydroxymethyl)phosphine (thmp) was the product when the same reaction is carried out using a $PtCl_2$ catalyst instead of a protic acid. We confirmed these observations and investigated the homogeneous catalysis in detail [6]. The active species is most likely a platinum(0) complex which is efficient (0.1 mol% of Pt catalyst is used and the only product is thmp) and not inhibited by the large amounts of free thmp present in the reaction mixture.

Thmp and PEt_3 are similar is size and mass but have very different physical and chemical properties. Thmp is a white, moderately air-stable solid (m.p. 58 °C) which is very soluble in water and methanol but insoluble in most common organic solvents whereas PEt_3 is a volatile, very air-sensitive liquid (m.p. -86 °C) which is miscible with all common organic solvents. These differences are attributed to the presence of hydrogen-bonding in thmp.

The coordination chemistry of thmp with nickel(II), palladium(II) and platinum(II) is, in many respects, as expected for a small trialkylphosphine [5, 9]. The ^{31}P NMR spectroscopic parameters for thmp complexes of platinum(II) closely resemble those for analogous triethylphosphine complexes (see Table 1) indicating that the bonding of thmp and PEt_3 to the metal is similar. It is the water-solubility of the thmp complexes that sharply distinguishes them from the PEt_3 analogues.

TABLE 1. Comparison of the ^{31}P coordination chemical shifts ($\Delta\delta$) and coupling constants (J) for selected complexes of thmp (in D_2O) and triethylphosphine analogues (in $CDCl_3$)

	$\Delta\delta$ / ppm	^1J(PtP) / Hz
cis-$[PtCl_2(thmp)_2]$	33.5	3431
cis-$[PtCl_2(PEt_3)_2]$	29.1	3516
trans-$[PtI_2(thmp)_2]$	17.4	2162
trans-$[PtI_2(PEt_3)_2]$	20.4	2273
cis-$[Pt(OH_2)_2(thmp)_2]^{2+}$	28.6	3647
cis-$[Pt(OH_2)_2(PEt_3)_2]^{2+}$	29.4	3737
$[Pd(thmp)_4]$	36.6	
$[Pd(PEt_3)_4]$	18.9	

The zerovalent complexes $[M(thmp)_4]$ (M = Ni, Pd, Pt) were studied in detail because they are rare examples of water-soluble metal(0) complexes, they are surprisingly air-stable, and we purport that they are the catalyst precursors in the hydrophosphination of CH_2O. The crystal structure of $[Pd(thmp)_4]$ (see Figure 1) revealed a network of hydrogen-bonds.

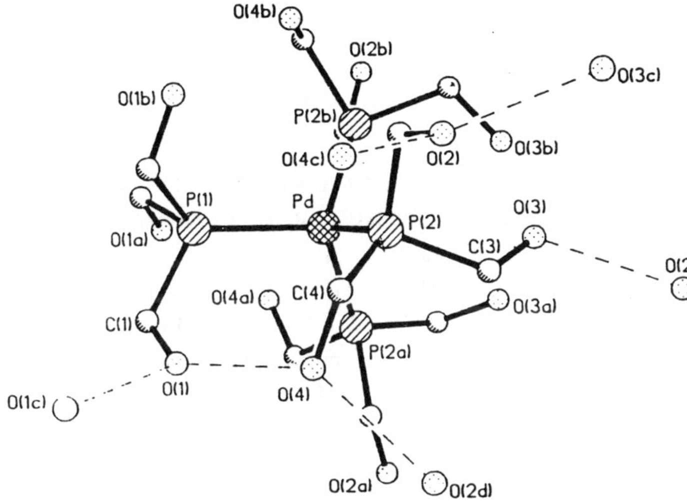

Figure 1. Molecular structure of $[Pd(thmp)_4]$

The intramolecular hydrogen-bonding, shown schematically in Figure 2, stitch together the four ligands to make them effectively a tetradentate ligand. The multiple chelate effect resulting from this hydrogen-bonded P_4Pd cage can explain the high kinetic stability of the metal(0) species, the absence of dissociation or rapid exchange detectable by ^{31}P NMR, and the unusual $\Delta\delta$ value for $[Pd(thmp)_4]$ (see Table 1).

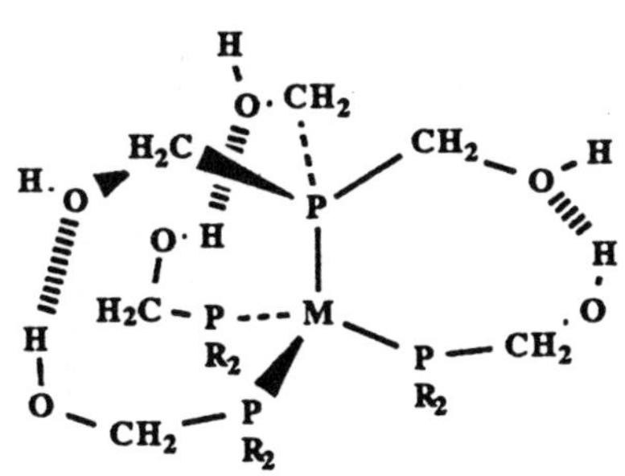

Figure 2. Schematic depiction of the intramolecular hydrogen bonding in $[M(thmp)_4]$

Aqueous solutions of the [Pt(thmp)$_4$] are strongly alkaline because in water the platinum(0) species is in equilibrium with [PtH(thmp)$_4$]OH:

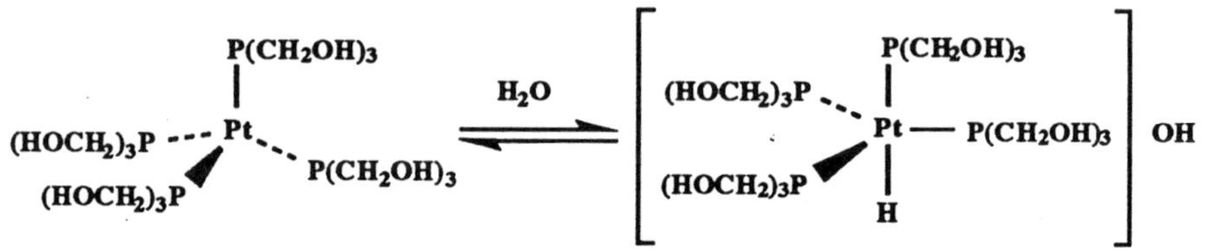

The uniqueness of this chemistry is not that water adds to the platinum(0) since the analogous PEt_3 complex does likewise [10], but that the product is five-coordinate even at ambient temperatures (as is clearly shown by ^{31}P and ^{1}H NMR spectroscopy [5]) whereas the ostensibly similar $[PtH(PEt_3)_4]^+$ dissociates to $[PtH(PEt_3)_3]^+$ above -137 °C [10]. This difference can be explained in terms of the hydrogen-bonding present in the Pt(thmp)$_4$ unit stabilising trigonal bipyramidal coordination in a manner similar to tripodal ligands such as $P(CH_2CH_2PPh_2)_3$ [11].

2.2 TRIS(HYDROXYPROPYL)PHOSPHINE

We wanted to investigate the chemistry of $P(CH_2CH_2CH_2OH)_3$ (thpp) to see whether the unusual properties of thmp were general to tris(ω-hydroxyalkyl)phosphines. Thpp, which was made by the radical-initiated addition of PH_3 to allyl alcohol [12], is a moderately air-stable liquid. It is more thermally stable than thmp since it can be distilled at 196-198 °C (1 mmHg) while thmp, above 130 °C, decomposes exothermically, releasing gaseous PH_3. Thpp is soluble in water but, unlike thmp, it is also soluble in most common organic solvents.

The water-soluble complexes [PtX$_2$(thpp)$_2$] are fluxional in solution due to the presence of the equilibria:

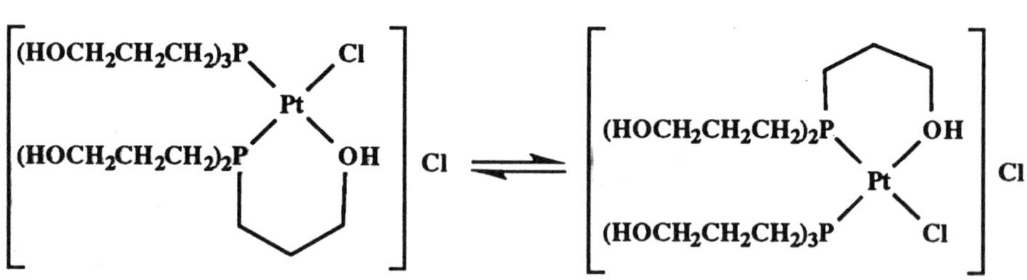

We have previously reported [13, 14] similar fluxionality with platinum(II) complexes of the ligands $Ph_2P(CH_2)_nOH$ (n = 2 or 3).

In its platinum(0) chemistry, thpp resembles a conventional PR_3 more than thmp. Hence in dmso, at ambient temperatures, $[Pt(thpp)_3]$ exchanges rapidly on the NMR timescale with free thpp and in water it forms the square planar $[PtH(thpp)_3]OH$.

In summary, the ^{31}P NMR parameters for the platinum coordination complexes of thmp and thpp suggest that they bond to metals much like other trialkylphosphines with the hydroxyl groups conferring water solubility. The presence of hydrogen-bonding has a much more profound effect on the coordination chemistry of thmp than thpp.

2.3 HYDROXYMETHYL DIPHOSPHINES

Electron-rich bis(dialkylphosphino)alkanes such as dmpe are important chelating ligands for hard and soft metal ions. We reasoned that diphosphines containing bis(hydroxymethyl)phosphino groups would be as versatile as dmpe and have the novel features of thmp chemistry. Recently we have found that formaldehyde adds readily to diprimary phosphines, without need of a catalyst, to give the bidentate, air-stable, water-soluble ligands dhmpe, dhmpp and dhmpb as the only products:

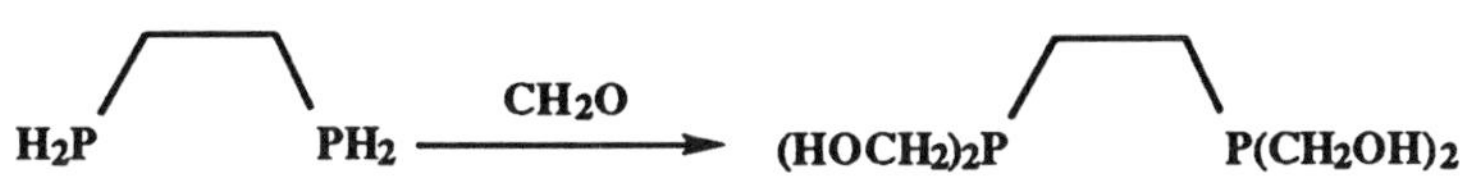

thmpe

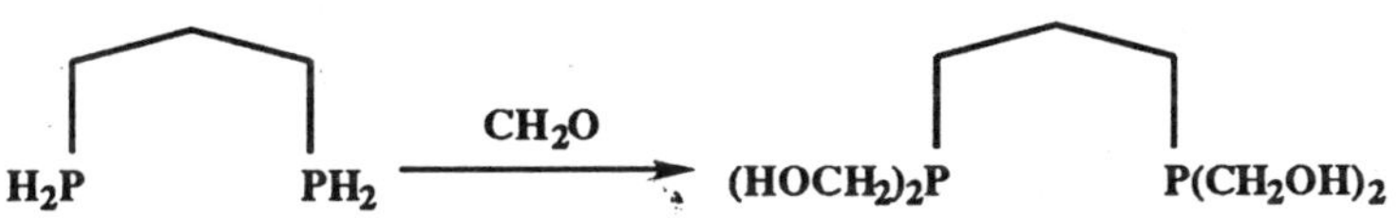

thmpp

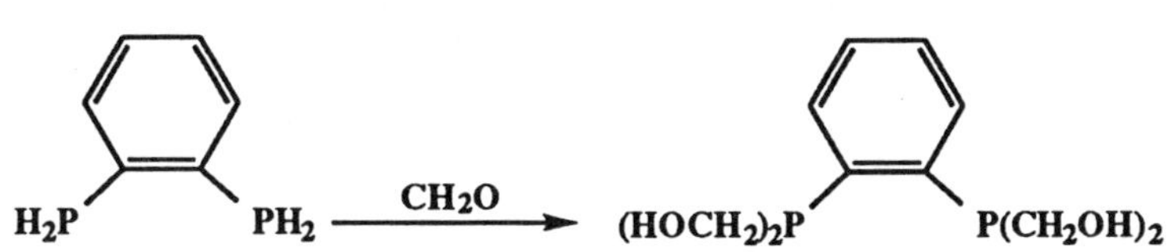

thmpb

From our preliminary results with thmpb and platinum(II), unexpected chemistry has already emerged. We have been unable to form $[PtCl_2(thmpb)]$ because the bis chelate $[Pt(thmpb)_2]Cl_2$ is the only complex detected, even when a ratio of 5:1 of $[PtCl_2(cod)]$:thmpb is used in the synthesis. The complex $[Pt(thmpb)_2]Cl_2$ was recrystallised from boiling water and the crystal structure of $[Pt(thmpb)_2]Cl_2$ (Figure 3) shows that in the solid state extensive intermolecular hydrogen-bonding is present.

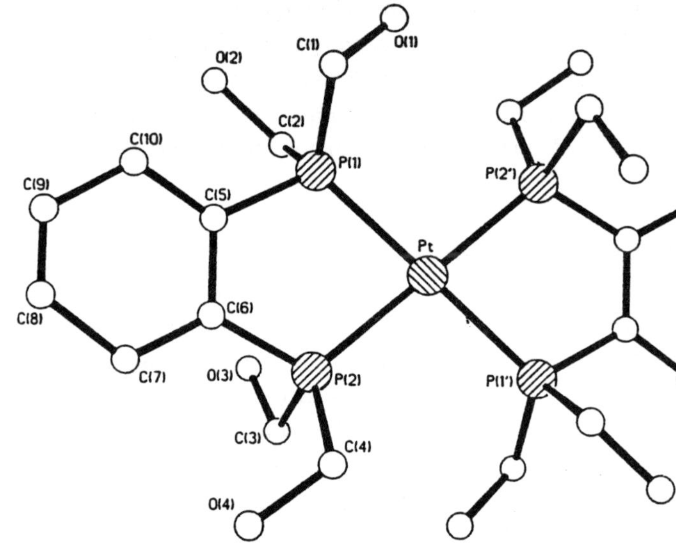

Figure 3 Molecular structure of $[Pt(thmpb)_2]^{2+}$

We believe that in solution the formation of the bis chelate is driven by the formation of an intramolecularly hydrogen-bonded macrocycle, shown schematically below:

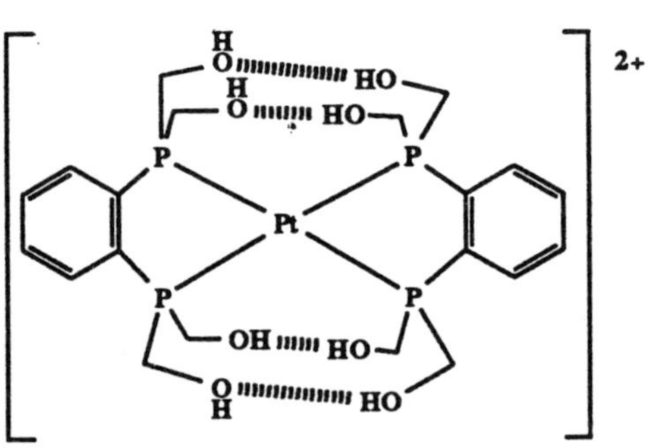

3. 2-Carboxyethyl Phosphines

3.1 TRIS(2-CARBETHOXYETHYL)PHOSPHINE

Neither the triester phosphine $P(CH_2CH_2CO_2Et)_3$ nor the highly water-soluble $P(CH_2CH_2CO_2Na)_3$ have been much studied as ligands [15, 16]. They can be easily made from the commercially available $P(CH_2CH_2CN)_3$:

$$P(CH_2CH_2CN)_3 \xrightarrow{HCl\ /\ EtOH} P(CH_2CH_2CO_2Et)_3 \xrightarrow{NaOH} P(CH_2CH_2CO_2Na)_3$$

We previously showed that platinum(0) complexes will catalyse the addition of PH_3 to CH_2=CHCN to give $P(CH_2CH_2CN)_3$ as the only product [4] although this platinum-catalysed synthesis is of purely academic interest since it offers no advantages over the radical-initiated or base-catalysed alternatives [17]. The chemo- and regioselectivity advantages of platinum(0) catalysis are spectacularly demonstrated in the reaction of PH_3 with CH_2=$CHCO_2Et$. The radical-initiated addition yielded mainly telomers and the desired product, $P(CH_2CH_2CO_2Et)_3$, was formed in only 2% yield. In contrast the platinum(0)-catalysed additions gave $P(CH_2CH_2CO_2Et)_3$ in 85% yield.

A mechanism for the hydrophosphination of Michael-activated alkenes is proposed in the Scheme.

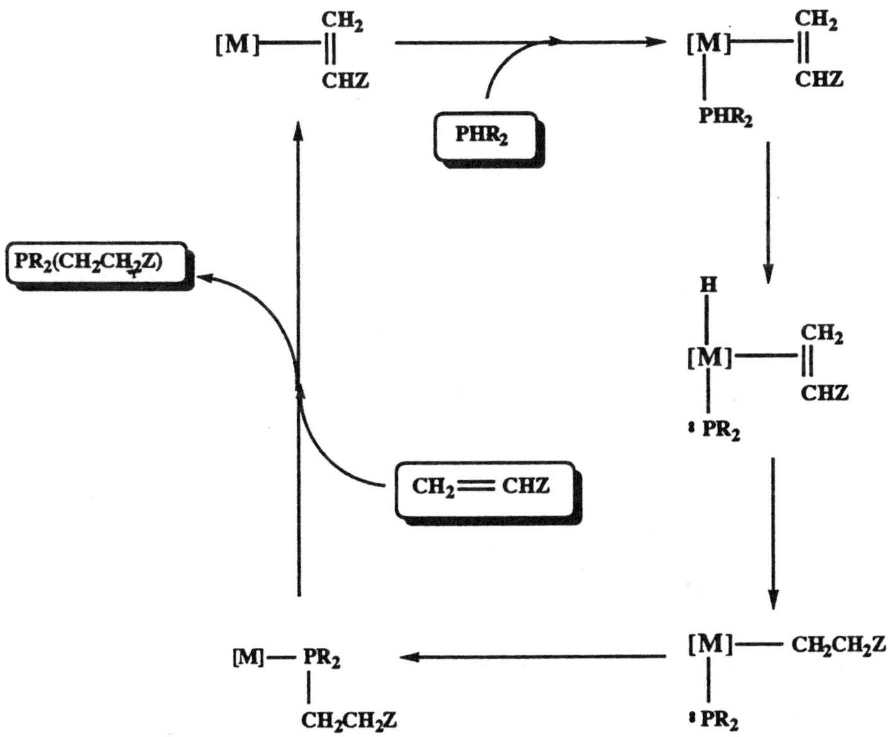

Scheme. Possible mechanism for the hydrophosphination of activated alkenes

A clue to the explanation of why the catalysis appears not to be inhibited by the formed phosphine is that platinum(0) complexes of $P(CH_2CH_2Z)_3$ (Z = CN or CO_2Et) are invariably unsaturated; for example, the ^{31}P NMR signal for the 16-electron species

[Pt{P($CH_2CH_2CN)_3$}$_3$] in the presence of 10 equivalents of P($CH_2CH_2CN)_3$, as shown in Figure 4, is broadened but not shifted, indicating that some exchange takes place but there is no evidence for the formation of the 18-electron species [Pt{P($CH_2CH_2CN)_3$}$_4$]. Of course these observations beg the question, why do the rather remote β-substituents on the alkyl group stabilise coordinative unsaturation on the platinum?

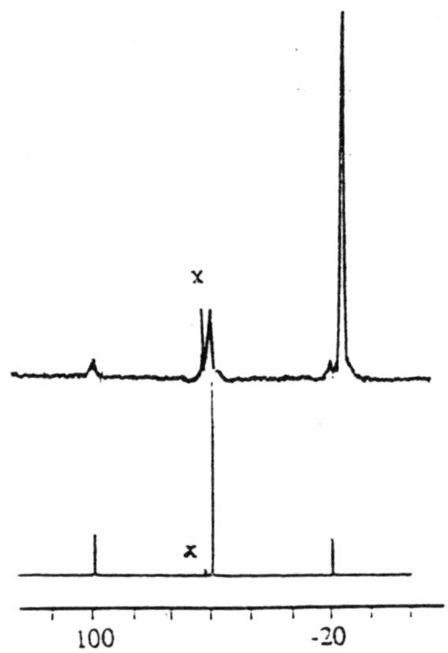

Figure 4 ^{31}P NMR spectra of [Pt{P($CH_2CH_2CN)_3$}$_3$]: in the absence of P($CH_2CH_2CN)_3$ (lower spectrum); in the presence of 10 equivalents of P($CH_2CH_2CN)_3$ (upper spectrum).

3.2 2-CARBOXYETHYL DIPHOSPHINES

Ethylenediaminetetraacetate (edta) is one of the most studied ligands in coordination chemistry with diverse applications ranging from corrosion prevention to medicine. In contrast, the only studies of the diphosphine analogue of edta have been carried out by Podlahová and coworkers [18-20].

Our strategy for the synthesis of diphosphine tetraacetates was to add diprimary phosphines to CH_2=$CHCO_2Et$ followed by saponification with NaOH:

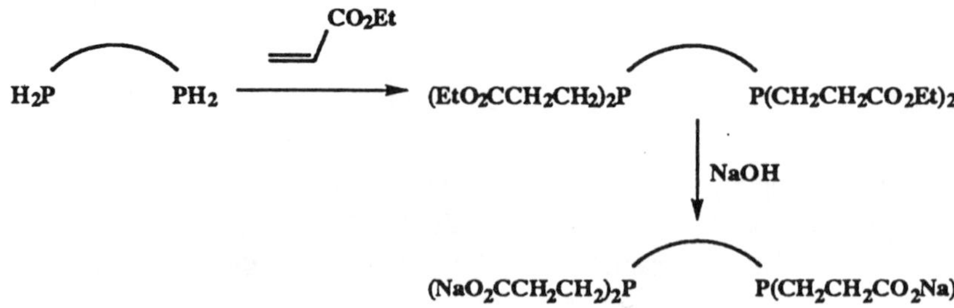

The first step works even better than we had hoped. For example, in the presence of the platinum(0) complex [Pt(norbornene)$_3$], 1,2-diphosphinobenzene gave the desired linear tetraester as the only P-containing product observed by ^{31}P NMR. In the absence of platinum(0), no reaction takes place and in the presence of radical initiators, at least three major tertiary phosphine products were formed. The chelating diphosphine product would be expected to be a powerful inhibitor and therefore it is remarkable that this catalysis occurs at all let alone with excellent regioselectivity. A ^{31}P NMR study of the stoichiometric reaction sequence [Pt(norbornene)(diphosphine tetraester)] plus 4 equivalents of CH_2=$CHCO_2Et$ followed by $H_2PCH_2CH_2CH_2PH_2$ sheds some light on what is going on in these reactions.

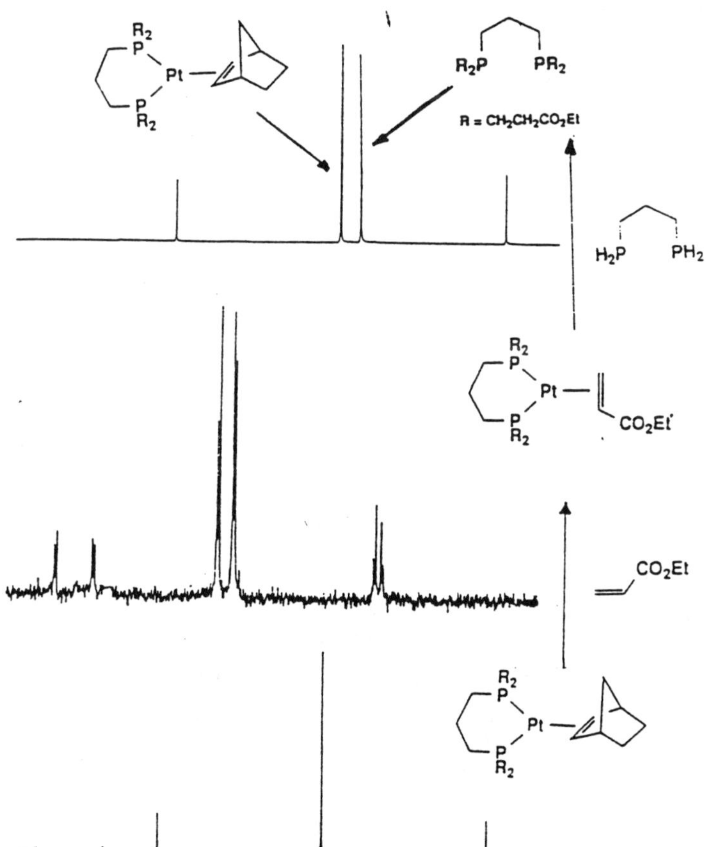

Figure 5 ^{31}P NMR spectra for the sequence of reactions shown

The spectra (Figure 5) show that, upon addition of the CH_2=$CHCO_2Et$, the norbornene is displaced from the platinum(0). The reaction with $H_2PCH_2CH_2CH_2PH_2$ is then very rapid (complete in the time taken to accumulate the NMR spectrum, *ca.* 5 min). Furthermore the final spectrum shows sharp signals for the regenerated, unsaturated platinum(0) complex and free diphosphine. Thus the 16-electron platinum(0) complex is stable even in the presence of a chelating diphosphine. Once again it appears that the β-substituents on the alkyl chains stabilise this unsaturation.

Preliminary results show that the tetraester diphosphines react cleanly with NaOH to give the corresponding tetraacetates which are very water soluble. A study of the coordination chemistry of these potentially hexadentate ligands is in progress.

4. Conclusion

The addition of P-H bonds to formaldehyde, allyl alcohol, and ethyl acrylate is a route to water-soluble, alkyl phosphines. Hydrogen-bonding makes the chemistry of hydroxymethyl phosphines very distinct from their ethyl- or methyl-phosphine analogues. Platinum(0) catalysts for the hydrophosphination of Michael-activated alkenes are more efficient and selective than radical-initiators and provide general access to a potentially large new class of diphosphine analogues of edta.

The work described in this article was carried out in collaboration with Albright and Wilson who have provided substantial financial support and, more importantly, the technical expertise of the personnel at the ITC in Warley, particularly Peter Hoye, James Ellis, Ranbir Padda and Gary Woodward.

5. References

1. Herrmann, W.A. and Kohlpainter, C.W. (1993) Water-Soluble Ligands, Metal Complexes, and Catalysis: Synergism of Homogeneous and Heterogeneous Catalysis, *Angew. Chemi., Int. Ed. (Engl.)*, **32**, 1524-1544.
2. Harrison, K.N., Hoye, P.A.T., Orpen, A.G., Pringle, P.G., and Smith, M.B. (1989) Water-Soluble, Zero-valent, Platinum-, Palladium-, and Nickel-$P(CH_2OH)_3$ Complexes: Catalysts for the Addition of PH_3 to CH_2O, *J. Chem. Soc., Chem. Comm.*, 1096-1097.
3. Pringle, P.G. and Smith, M.B. (1990) Tris(hydroxymethyl)phosphine Platinum Metal Complexes, *Platinum Metals Rev.*, **34**, 74-76.
4. Pringle, P.G. and Smith, M.B. (1990) Platinum(0)-catalysed Hydrophosphination of Acrylonitrile, *J. Chem. Soc., Chem. Comm.*, 1701-1702.
5. Ellis, J.W., Harrison, K.N., Hoye, P.A.T., Orpen, A.G., Pringle, P.G., and Smith, M.B. (1992) Water-Soluble Tris(hydroxymethyl)phosphine Complexes with Nickel, Palladium, and Platinum. Crystal Structure of $[Pd\{P(CH_2OH)_3\}_4].CH_3OH$, *Inorg. Chem.*, **31**, 3026-3033.

6. Hoye, P.A.T, Pringle, P.G., Smith, M.B., and Worboys, K. (1993) Hydrophosphination of Formaldehyde catalysed by Tris(hydroxymethylphosphine Complexes of Platinum, Palladium or Nickel, *J. Chem. Soc., Dalton Trans.*, 269-274.

7. Costa, E., Murray, M., Pringle, P.G., and Smith, M.B. (1993) NMR Study of Mononuclear, Binuclear and Trinuclear Tris(hydroxymethyl)phosphine-Platinum(II) Complexes with Aquo or Hydroxo Ligands, *Inorg. Chim. Acta*, **213**, 25-28.

8. Reuter, M. and Orthner, L. (1960) Trihydroxymethylphosphine, (German Patent (1958) 1,035,135), *Chem. Abs.*, **54**, 14124i.

9. Chatt, J., Leigh, G.J., and Slade, R.M. (1976) Rhodium(I), Rhodium(III), Palladium(II) and Platinum(II) Complexes Containing Ligands of the Type PR_mQ_{3-m} (m = 0, 1, or 2; R = Me, Et, Bu^t, or Ph; Q = CH_2OCOMe or CH_2OH), *J. Chem. Soc., Dalton Trans.*, 2021-2028.

10. English, A.D., Meakin, P., and Jesson, J.P. (1976) Mechanisms of Ligand Association in Complexes of the Type $[RML_3]^+ X^-$ (M = Ni, Pd, Pt) II. The Mechanisms of the Reactions $[HML_3]$ + L = $[HML_4]^+$ (L = PEt_3), *J. Am.Chem. Soc.*, **98**, 422-436.

11. Brüggeller, P. (1987) The Platinum(II) Hydride of Tris(2-(diphenylphosphino)ethyl)phosphine: A Novel Five Coordinate Monohydride, *Inorg. Chim. Acta*, **129**, L27-L28.

12. Drucker, A. and Grayson, M. (1970) Odorless Organic Phosphines, (U.S. Patent (1967) 3,489,811) *Chem. Abs.*, **72**, 67098t.

13. Alcock, N.W., Platt, A.W.G., and Pringle, P.G. (1989) Chelate-Stabilised Primary, Secondary, and Tertiary Alcohol and Alkoxo Platinum(II) Complexes. X-ray Structure of $[Pt(Ph_2PCH_2CH_2O)_2].H_2O$, *J. Chem. Soc., Dalton Trans.*, 139-143.

14. Bryars, K.H. (1988) Synthesis, Characterisation and Conformational Studies of Six-membered Platinacycles, PhD Thesis, University of Warwick, UK.

15. Werner, J. (1989) Synthesis of Water-Soluble Tetrairidium Clusters Suitable for Heavy Atom Labelling of Proteins, *Z. Naturforsch.*, **44b**, 79-82.

16. Podlahová, J., Kratochvíl, B., and Hasek, J. (1985) Trigonal Bipyramidal Penta-aquazinc(II) Bis(3,3'-phosphintriyltripropionato)dizinczte(II, II) Heptahydrate, *J. Chem. Soc., Dalton Trans.*, 2393-2396.

17. Wolf, E. and Reuter, M. (1961) Tris(cyanoalkyl)phosphines, (German Patent (1960) 1,082,910) *Chem. Abs.*, **55**, 16422e.

18. Podlahová, J. and Podlaha, J. (1980) Synthesis and Characterization of Tetrasodium Salt of Ethylenediphosphinetetraacetic acid, The Phosphorus Analogue of EDTA, *Coll. Czech. Chem. Commun.*, **45**, 2049-2053.

19. Podlahová, J., Kavan, J., Silha, J., and Podlaha, J. (1984) Compounds Structurally Related to Complexones 24. Nickel(II) Complexes of Ethylenediphosphine-tetraacetic acid, *Polyhedron*, **3**, 963-968.

20. Podlaha and J., Podlahová, J. (1994) Crystal Structure of Bis(ethylenediphosphine-tetraacetic acid)palladium(II) Bis(dihydrogenethylenediphosphinetetraacetato)-palladate(II) Tetrahydrate Hydrobromide(1/1) - An Unusual Case of Bromide Uncoordinated to palladium(II), *Coll. Czech. Chem. Comm.*, **59**, 359-366.

AQUEOUS MEDIA IN THE INTRAMOLECULAR HECK REACTION

DOUGLAS B. GROTJAHN and XIAOWEI ZHANG
Department of Chemistry and Biochemistry
Box 871604
Arizona State University
Tempe, AZ 85287-1604 USA

In the last several years intramolecular variants of the Heck reaction [1] have been developed to produce complex polycyclic structures, as schematically illustrated by conversion of **A** to **D**. However, the successful formation of more than one ring in intramolecular Heck reactions of **A** has depended on substrate *structure* to

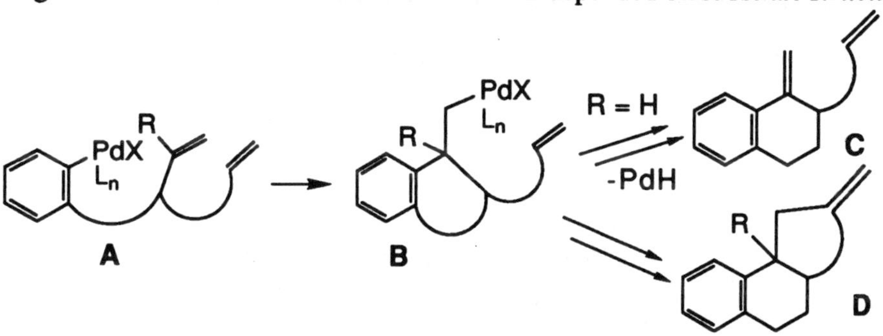

prevent β-hydride elimination from intermediate(s) (e.g., **B**, R ≠ H). For **B** in which R = H, however, we wondered if aqueous media would favor further alkene insertion over β-hydride elimination in **B**, either because the alkene-bearing chain would be poorly solvated by water (hydrophobic effect [2]) or because water would help to ionize the Pd-X bond. *Intriguingly, our preliminary experiments suggest that changing from* CH_3CN *to* H_2O-CH_3CN *or* H_2O-*ethanol mixtures, with additives such as LiCl, indeed causes dramatic changes in product distribution for some substrates but not others.*

Cyclization precursors **7** (next page) were chosen to examine the effects of varying the tether length and the hydrophobicity (and the steric properties) of the distal alkene unit. Known 1,2-addition of Grignard reagents **1** to acrolein gave **2**. The alcohol function was converted to an amino group in **4** [3] by Mitsunobu reaction with phthalimide, followed by hydrazinolysis and salt formation. Introduction of the *N*-methyl substituent, deemed wise based on earlier difficulties with cyclizations of secondary amides, was achieved by carbamate formation, giving **5**, and subsequent reduction to the amine, isolated as its salt **6**. Finally, acylation provided cyclization substrates **7**.

I. T. Horváth and F. Joó (eds.), Aqueous Organometallic Chemistry and Catalysis, 123–125.

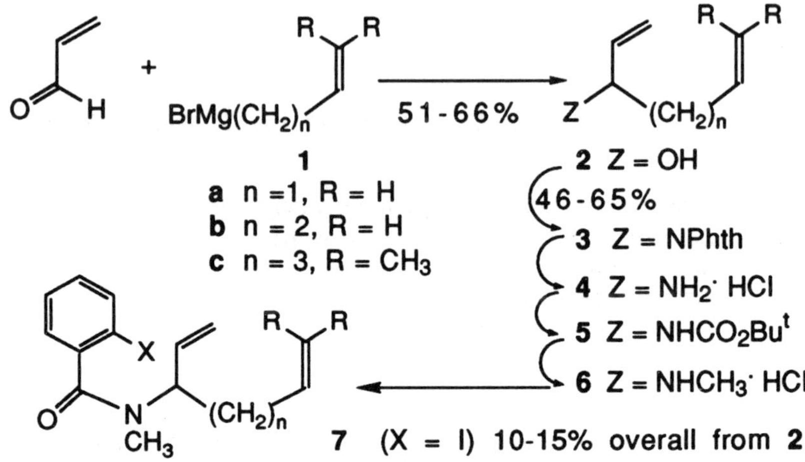

In pure acetonitrile, **7b** and **7c** underwent clean Heck reaction (for typical conditions, see text below and Table 1) to give isoquinolone derivatives **8** and **9** (total yields 51-64%, ratio varied but was near 1:1). NMR analysis of crude reaction mixtures provided no evidence whatever for tricyclic products of type **D**. Exocyclic alkene **8** presumably is the initial product of β-hydride elimination

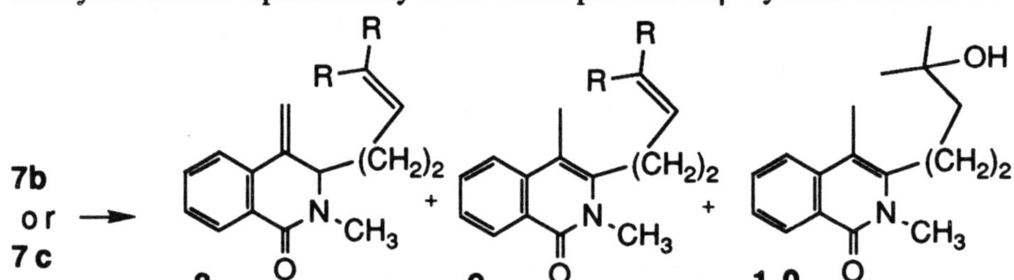

from **B**, and serves as the precursor to **9**. In ethanol-water (1:1), **7b** followed the same course as in acetonitrile, although the trisubstituted double bond in **7c** apparently suffered some hydration as evidenced by **10** (tentatively identified). Even the addition of the salting-out additive LiCl [2] at 4M concentration did not alter the cyclization of **7c**, other than to suppress formation of **10**. Exploration of **7b** under these conditions is underway.

However, initial evidence (Table 1) shows that solvent and ionic additives profoundly influence the cyclization of **7a** (X = I). In acetonitrile, major products are the bicycles **12** (rationalized as arising from isomerization of **C**) and **11**. The latter could arise from (Pd-catalyzed?) Cope rearrangement of **C**, driven by aromatization, or from **7a** by initial Cope rearrangement to an enamide, followed by Heck cyclization and β-hydride elimination. Further, traces of tricyclic product **13** [3] and **16** were accompanied by greater amounts of a compound whose spectral properties are similar, but not identical to those of **13** and **14**, suggesting **15**. Experiments are underway to determine the viability of initial Cope rearrangement of **7a** (X = H or I) followed by tandem Heck cyclization and exocyclic alkene isomerization, which could give **15**. The relative

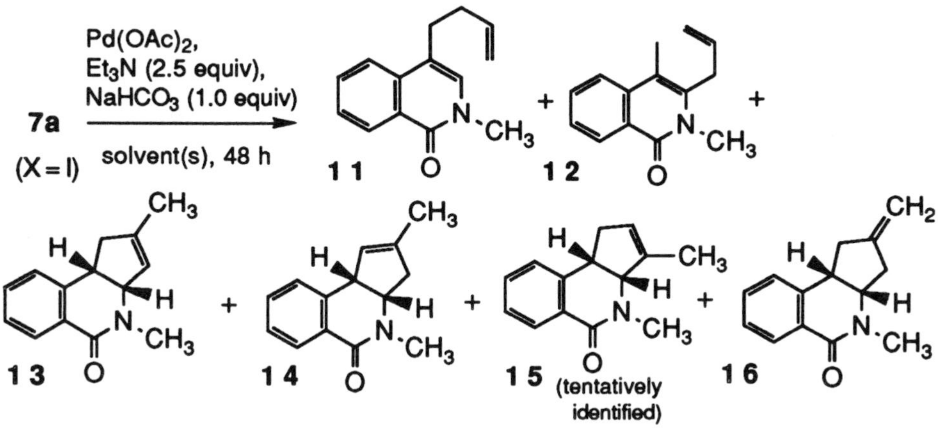

TABLE 1. Effects of medium on the cyclization of **7a** (X = I).

solvent(s)	temp.(ºC)	$Pd(OAc)_2$	additive	% yield of products[a] **11**	**12**	**13**	**14**	**15**	**16**
CH_3CN	80	0.2 equiv	none	12	17	2	0	21	5
				(5)	(9)	(2)	--	(4)	(5)
EtOH-H_2O (1:1)	60	0.2 equiv	Bu_4NCl	0	23	11	8	10	0
			(1 equiv)	-	(10)	(5)	(7)	(5)	-
EtOH-H_2O (1:1)	60	0.1 equiv	LiCl (1M)	0	≤2	16	16	8	0
EtOH-H_2O (1:1)	60	0.1 equiv	LiCl (4M)	0	6	18	26	12	0

[a]Yields determined in crude product mixture by NMR integration against an internal standard. Yields in parentheses are for product isolated by HPLC.

stereochemistry of **16** is not yet determined unequivocally, but mechanistically it would make sense that the isomer shown is a precursor to **13** and **14**. Significantly, reactions run in EtOH-H_2O do not produce detectable quantities of **11** and less of **15**, and the addition of significant quantities of LiCl suppresses the formation of **12**, results we tentatively ascribe to the combined effects of water and salting-out additive LiCl, but clearly, much remains to be done. Present efforts focus on yield optimization, determining product stability under the reaction conditions, and on unravelling the mechanistic puzzle presented by variations in product distribution. Nonetheless, the significance of this work is that the scope of tandem Heck cyclizations and indeed other organometallic processes may be expanded by the use of aqueous media.

References

1. Overman, L.E., Abelman, M.M., Kucera, D.J., Tran, V.D., and Ricca, D.J. (1992) Palladium-catalyzed polyene cyclizations, *Pure Appl. Chem.* **64**, 1813-1819 and references therein to work of Grigg, de Meijere, Negishi, Oppolzer, Trost, and others.
2. Breslow, R. (1991) Hydrophobic effects on simple organic reactions in water, *Acc. Chem. Res.* **24**, 159-164.
3. Unless described as tentatively identified, all structures were secured by a combination of ^1H, ^{13}C NMR and IR spectroscopy. **13** and **14** were further characterized by COSY, NOESY, HMBC and HMQC , **16** by COSY and NOESY experiments.

NEW EFFICIENT WATER-SOLUBLE CATALYSTS FOR TWO-PHASE OLEFIN HYDROFORMYLATION

W.A. HERRMANN *, G. ALBANESE, R. MANETSBERGER,
R. SCHMID, C. SCHWER†
Anorganisch-chemisches Institut der Technischen Universität München
Lichtenbergstrasse 4, D-85747 Garching (Germany)
†Max-Planck-Institut für Biochemie, Am Klopferspitz 18, D-82152 Martinsried (Germany)

The diphosphine 2,2′-Bis(diphenylphosphinomethylen)-1,1′-binaphthalene (NAPHOS) and the monophosphine 3,4-dimethyl-2,5,6-triphenyl-1-phosphanorborna-2,5-diene were sulfonated by means of oleum. The sulfonation conditions were optimized by in situ NMR-techniques to minimize the oxidation of the phosphorus. The resulting sulfonation products (BINAS: n = 4 to 8; NORBOS: n = 3) were separated by gel chromatography (NORBOS) and capillary electrophoresis (BINAS). Although we could obtain the 8-fold sulfonated BINAS-8, we were interested in analytical samples of lower sulfonation grades. Compared with TPPTS, the separation of the different isomers of BINAS proved to be more complicated and could not be achieved by our standard gel permeation chromatography. Similary the separation by HPLC gave only unsatisfactory results. Even ion exchange chromatography which at first sight seemed adapted to the highly ionic nature of the ligands resulted in no additional advantage. Finally we chose capillary electrophoresis connected to a UV/VIS detector as an adapted and rapid method. We could obtain a sharp separation of the various sulfonated species and with the help of our NMR-Data verify the purity of BINAS-8 species. The various sulfonated species were detected by a CE-MS-coupling using electrospray as ionisation mode. As pure compounds, the threefold sulfonated NORBOS (**1**) and the eightfold sulfonated BINAS-8 (**2**) were isolated and characterized by 2D-NMR-methods.

I. T. Horváth and F. Joó (eds.), Aqueous Organometallic Chemistry and Catalysis, 127–129.

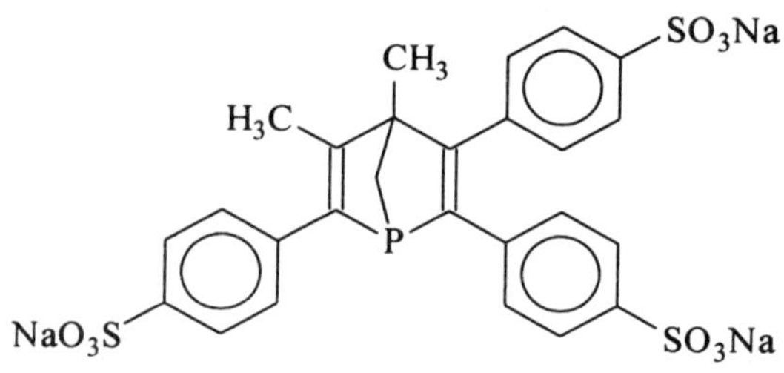

NORBOS **1**

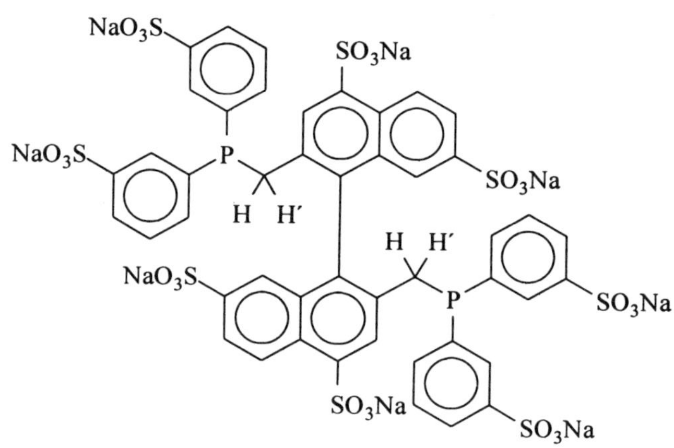

BINAS-8 **2**

In addition, the position of the sulfonate substituents in **2** were confirmed theoretically by semiempirical quantummechanical calculations (MNDO/PM3). In agreement with theory the substituents are at the 4- and 7-position of the naphthalene.

The new ligands were tested for catalytic activity toward hydroformylation of propene. The apparatus for the catalytic tests was appropriate for continuous hydroformylation. Reaction conditions can easily be varied in order to investigate dependence of catalyst properties to reaction parameters such as temperature, pressure and amount of olefin. The results of the catalytic tests can be summarized as follows, keeping in mind that the phosphorus/rhodium ratio (P/Rh) for both BINAS and NORBOS were significant lower than those typically used with TPPTS. For BINAS, the catalytic activity and productivity are as high as 178.5 and 1.0, respectively. A twelve- (five-)fold increase in activity and productivity in comparison with commonly used TPPTS is achieved. Interestingly the n/iso-isomer ratio of the butyric aldehyde increased to 98:2.

For NORBOS, the catalytic activity was 117.8. The productivity was 0.37. This performance is lower than that for BINAS. Nonetheless, these values represent an eightfold increase in activity and a productivity twice as high as for TPPTS. The reported data of the new water-soluble phosphines represent the best results so far recorded in the rhodium-catalyzed hydroformylation of propylene.

RHODIUM AND PALLADIUM WATER-SOLUBLE COMPLEXES IN THE CARBONYLATION REACTIONS. IMPROVEMENT IN THE TRANSFER BETWEEN THE ORGANIC AND AQUEOUS PHASES

Fanny MONTEIL, Laurent MIQUEL, René QUEAU and Philippe KALCK*
Laboratoire de Chimie des Procédés, Ecole Nationale Supérieure de Chimie, 118, route de Narbonne, 31077 TOULOUSE cédex (FRANCE)

1. INTRODUCTION

Homogeneous catalysis provides mild and selective routes to achieve a lot of syntheses starting from basic organic raw products. An important attractiveness over heterogeneous systems is that selectivity is obtained by a fine tuning of the coordination sphere of the metal. However, usually a major drawback is the separation of the organic products from the catalysts. A very elegant and attractive approach has been made using water-soluble ligands that are poorly soluble in organic media so that catalysis is performed in a two-phase system. [1-5] A very easy separation of the catalyst from the organic reactants and products is thus achieved by decantation and separation of the two phases. In addition, catalysis in aqueous media could decrease the toxicity inherent to organic solvents or modify the kinetics and the stereoselectivity of a given reaction. Even when the reactants and the product are gaseous, as for the hydroformylation of propene, the biphasic system decreases the contact between the product and the catalyst so that the reaction can be more selective due to the reduced formation of by-products. Organic reactions catalyzed by water-soluble coordination complexes have attracted considerable attention during the past few years owing to the above-mentionned reasons and also to the industrial use of this technology for the hydroformylation of propene. Indeed, a process using rhodium-tris(m-sulfophenyl)phosphine was patented by Kuntz for Rhône-Poulenc Industries[6,7] and developed by Ruhr Chemie (now Hoechst)[5] which produces around 300 000 tons per year of butanal.

It has been demonstrated that the reaction occurs in the bulk of the water phase instead of at the interface. Thus, the kinetic decreases as the solubility of the alkene in water. For heavy alkenes improvements in the transfers between the organic and aqueous phases are required in order to ensure correct reaction rates. This contribution deals with the use of a cosolvent for the hydroformylation of 1-octene catalyzed by rhodium-TPPTS dinuclear precursors and of a co-reactant playing also the role of cosolvent for the hydroxycarbonylation of bromobenzene catalyzed by palladium-TPPTS systems.

I. T. Horváth and F. Joó (eds.), Aqueous Organometallic Chemistry and Catalysis, 131–147.

2.THE TPPTS LIGAND

The tris(m-sulfophenyl)phosphine (or TPPTS) was prepared[6] by the direct sufonation of triphenylphosphine with oleum and subsequent neutralization with NaOH. Thus TPPTS is available as its sodium salt trihydrated with the three sulfonate groups occupying each a *meta* position on the phenyl ring with regard to the phosphorus atom (Fig. 1). This ligand can be

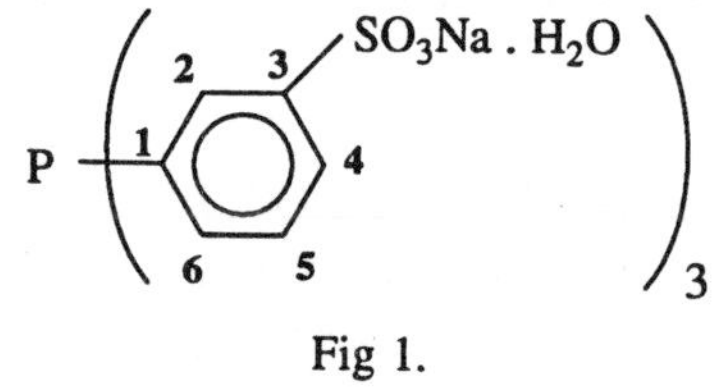

Fig 1.

purified by precipitation from an aqueous solution by slow addition of ethanol. The resulting white solid is washed successively with ethanol and ether dried under vacuum. TPPTS can be stored under nitrogen for long periods. It is highly soluble in water, until 1100 $g.l^{-1}$.[7] The elemental analysis is consistent with a $P(C_6H_4SO_3Na.H_2O)_3$ formula. ^1H, ^{31}P and ^{13}C NMR spectra are fully assigned especially when the samples are dissolved in CD_3OD. The ^{13}C NMR spectra (see fig. 2) are particularly informative : C_1 $\delta = 139.28$ (dd, $^1JC_1P = 12.0$; $^3JC_1H_5 = 7.8$ Hz) ; C_2 $\delta = 139.19$ (ddt, $^1JC_2H_2 = 162.9$; $^2JCP = 21.7$; $^3JC_2H_6 \approx {}^3JC_2H_4 = 7.2$ Hz) ; C_3 $\delta = 146.08$ (dd, $^3JC_3P = 6.6$; $^3JC_3H_5 = 6.6$ Hz); C_4 $\delta = 129.44$ (dt, $^1JC_4H_4 = 165.5$; $^3JC_4H_2 \approx {}^3JC_4H_6 = 5.6$ Hz).; C_5 $\delta = 132.34$ (dd, $^1JC_5H_5 = 164.9$; $^3JC_5P = 7.6$ Hz) ; C_6 $\delta = 132.58$ (ddt, $^1JC_6H_6 = 165.1$; $^2JC_6P = 18.2$; $^3JC_6H_2 \approx {}^3JC_6H_4 = 6.2$ Hz).

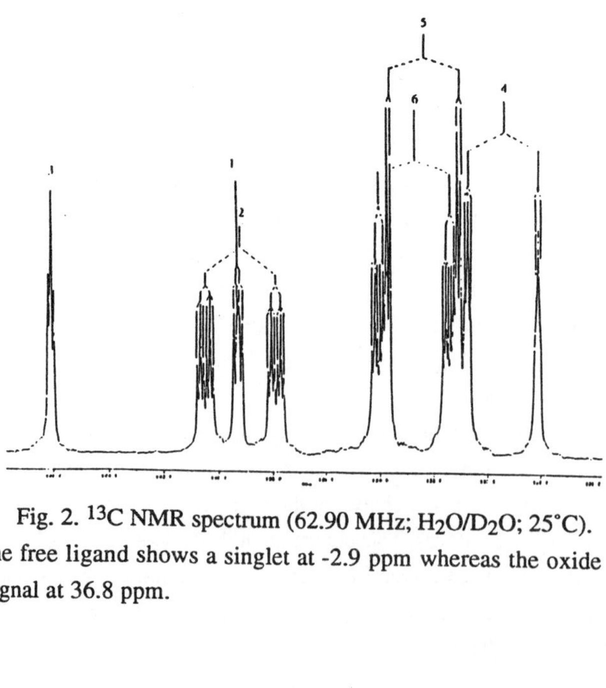

Fig. 2. ^{13}C NMR spectrum (62.90 MHz; H_2O/D_2O; 25°C).

In ^{31}P NMR, the free ligand shows a singlet at -2.9 ppm whereas the oxide OTPPTS, when present, has a signal at 36.8 ppm.

3. HYDROFORMYLATION OF 1-OCTENE[8]

Previous studies on dinuclear rhodium (I) complexes containing thiolato bridging ligands $[Rh_2(\mu\text{-}S^tBu)_2(CO)_2L_2]$, where L = $P(OMe)_3$ or $P(OPh)_3$, have shown that the hydroformylation of 1-hexene can be carried out at 5 bar and 80°C with turnover frequencies higher than 1 min^{-1}. The reaction is completely selective in aldehyde. The dinuclear complex is recovered untouched after catalysis.[9] Recently we observed that the corresponding $[Rh_2(\mu\text{-}S^tBu)_2(CO)_2(PPh_3)_2]$ complex catalyzes the reaction under the same experimental conditions with higher rates.[10] For instance, with a P/Rh molar ratio of 10, 72% conversion was achieved in 1h (TOF = 21 min^{-1}) giving a selectivity in linear (n with regard to iso) aldehyde of n/(n+i) = 78%. A higher excess of triphenyphosphine (P/Rh = 25) allows to reach better selectivities (88%), the conversion decreasing to 51% (TOF = 11.5 min^{-1}).

Similarly, we have shown that $[Rh_2(\mu\text{-}S^tBu)_2(CO)_2(TPPTS)_2]$ **1** also avoids the hydrogenation of alkenes, providing for the hydroformylation reaction a complete selectivity in aldehydes.[11,12] The behavior of this complex is quite different of that of $[HRh(CO)(TPPTS)_3]$ **2** as shown on Fig. 3. Indeed, **1** requires an induction period whereas the reaction starts immediately with **2**. Under the same conditions **1** becomes more active than **2**.

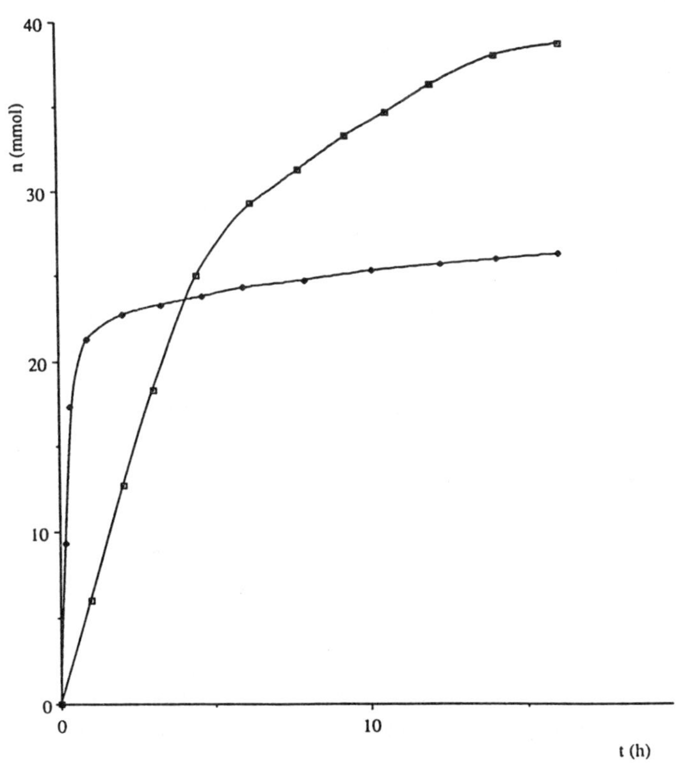

Fig. 3.

In addition, linearities higher than 97% have been achieved. Such promising results prompted us to investigate the reactivity of **1** for the hydroformylation of heavy alkenes for which we selected 1-octene as representative of this class of compounds. Of course the biphasic system is convenient for catalyst recovery but to overcome the low solubility of the reagent in water that reduces the reaction rate dramatically, it is necessary to improve the transfers between the organic and aqueous phases. We chose to explore the addition of suitable cosolvents in order to increase the solubility of the organic substrate in the aqueous phase.

3.1 SELECTION OF AN ADAPTED COSOLVENT

Neat 1-octene is very slowly coverted by **1** as shown in table 1. The selectivity in aldehydes is quite complete and the linearity is higher than 96% provided an excess of phosphine ligand is introduced. This addition of extra TPPTS allows the two phases to be easily separated by decantation.

TABLE 1. Hydroformylation of 1-octene (runs 1-3) and 1-hexene (run 4) catalyzed by **1** in a biphasic medium (**1**, 0.16 mmol ; alkene, 64 mmol ; H_2O, 150 ml ; P = 5 bar (CO/H_2 = 1/1) ; T = 80°C ; t = 15 h ; stirring rate, 2500 rev.min^{-1}).

run	P/Rh	conversion (%)	sel. in aldehydes (%)	n/(n+i)	separation
1	6/1	24	100	90	difficult
2	10/1	18	100	96	easy
3	50/1	6	100	> 98	easy
4	10/1	81	100	97	easy

The comparison of runs 2 and 4 shows that under the same experimental conditions the reaction rate is higher with 1-hexene than with 1-octene. This difference originates in the solubility of the two alkenes since for 1-hexene it is 50 10^{-6} kg/kg water whereas for 1-octene it is only 2.07 10^{-6}.

Previous studies on the hydrogenation of heavy alkenes in biphasic systems had shown that many suitable cosolvents increase the concentration of 1-octene in water and simultaneously avoid the presence of water in the organic phase in the hypothesis where the water-soluble complex might also leach into the organic phase[13]. Ethanol, methanol, acetone and acetonitrile were selected owing to their inertness toward the products, except the possible formation of acetals; tertiary-amines could also be used. The reaction rates are greatly increased, especially for 22% w/w cosolvent as shown on Fig. 4, with regard to neat 1-octene. However 5-10% of internal octenes are formed and the regioselectivity decreases significantly from 96% to about 83%.

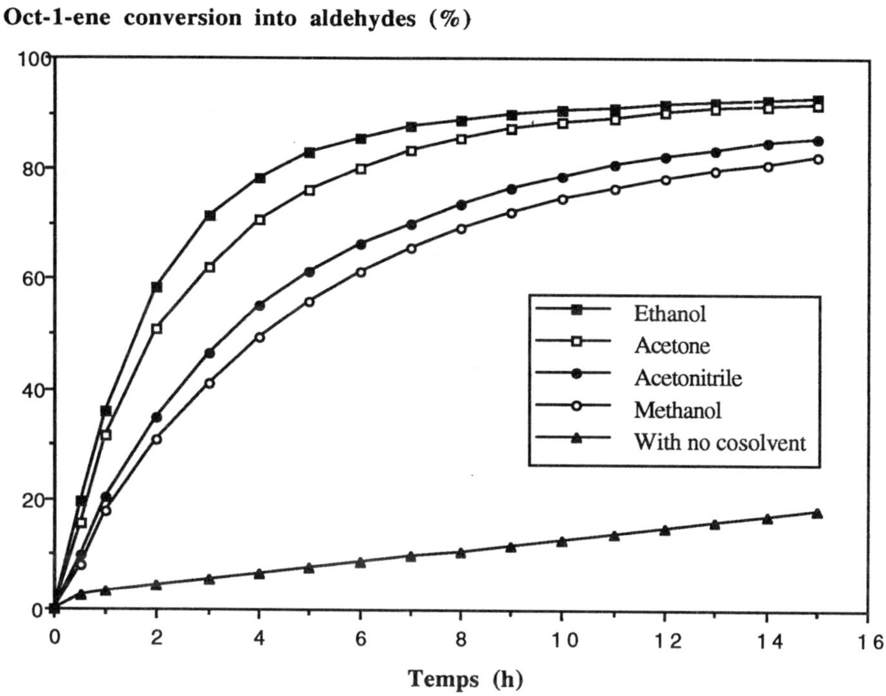

Fig. 4.

It is here interesting to consider the solvophobicity parameter Sp rather than the polarity of the medium[14]. The Sp values are established from the difference between free energies for the transfer of a given solute from water to another solvent; they have been expressed at 25°C relative to two fixed solvents and the range is between 1 for water and 0 for the most hydrophobic solvent, hexadecane. For the various cosolvents investigated, we observed a linear correlation (Fig. 5) of log (n/i) with Sp (ethanol 0.74, methanol 0.84, acetone 0.81) the mixture being 22% w/w cosolvent.

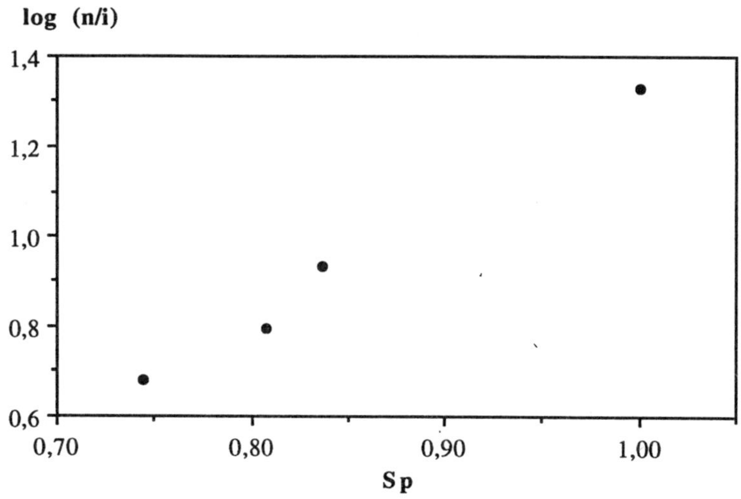

Fig. 5

The formation of the two linear and branched aldehydes obey presumably similar mechanisms, so the n/i ratio is related directly to the rate constant ratio kn/ki.

3.2 FATE OF THE DINUCLEAR RHODIUM COMPLEX 1

After a first experiment carried out with 16% w/w ethanol, the reaction mixture was transfered into a separatory funnel under argon for decantation. The conversion of 1-octene was 87%, the selectivity in C_9 aldehydes 89% and the linearity 90%. The aqueous phase which had turned from yellow to brown was reintroduced into the autoclave for a second run to give 81% conversion, then for a third run to afford 73% conversion, the selectivities being unchanged.

Analysis of the yellow organic phase by infrared revealed the presence of the tetracarbonyl complex $[Rh_2(\mu\text{-}S^tBu)_2(CO)_4]$ **3** ; around 20% of **1** was transformed into **3** as evaluated by UV. This organic phase contains also large amounts of the ^tBuSH mercaptan. In the aqueous phase complex **1** was fully transformed into **2** as shown by IR (ν_{CO} = 1922 cm^{-1}, ν_{RhH} = 2003 cm^{-1}) and ^{31}P NMR (δ = 45.1 ppm, d : $^2J_{RhP}$ = 155.1 Hz) with the presence in the medium of free TPPTS and the oxide OTPPTS (-2.9 and 35.8 ppm respectively). The hydride ligand was unambiguously characterized by a quadruplet in ^{1}H NMR at -9.6 ppm ($^3J_{PH}$ = 13.4 Hz).

Further investigation of the role played by water in the transformation of **1** into **3** showed that at room temperature or even at 80°C **1** remains unchanged in water under a dinitrogen atmospher but that as soon as CO is introduced in the medium the formation of **3** is observed. For instance at 80°C the reaction was complete in 4h. The simultaneous presence of CO and H_2O is necessary since **1** cannot be transformed when heated under reflux in methanol under CO. Thus, the water gas shift reaction occurs to generate the hydride ligand.

The mononuclear complex [HRh(CO)(TPPTS)$_3$] **2**[15] leads to a first order kinetics, as expected. However the order of the hydroformylation reaction with respect to **1** was found to be 0.7. This order could be due to the simultaneous catalytic activity of both complexes **1** and **2**. More presumably this value results from the leaching of [Rh$_2$(μ-S^{t}Bu)$_2$(CO)$_4$] in the organic phase, decreasing the concentration of **2** for catalysis. Complex [HRh(CO)(PPh$_3$)$_3$] is known to form brown solutions containing carbonyl μ-phosphido clusters which are inactive in hydroformylation.[16] We observed such brown colours and as for the triphenylphosphine analogue[16], we noted that significant amounts of internal alkene are produced. To prevent the deactivation of **1**, a large excess of TPPTS needs to be added,[15] at the expense of the rate.

Whereas the addition of ^{t}BuSH to [HRh(CO)(PPh$_3$)$_3$] was shown to afford irreversibly the complex [Rh$_2$(μ-S^{t}Bu)$_2$(CO)$_2$(PPh$_3$)$_2$],[17] here the low solubility of ^{t}BuSH and [Rh$_2$(μ-S^{t}Bu)$_2$(CO)$_4$] in water as well as the reactivity of CO and H_2O on **1** induce the formation of **2** in the aqueous phase.

3.3 WATER-SOLUBLE BRIDGING THIOLATO LIGANDS CONTAINING COMPLEXES

[Rh$_2$(μ-SR)$_2$(COD)$_2$] complexes, in wich the dimethylaminopropylthiolato ligand can be protonated at the nitrogen atom to produce water solubility, has been published recently, the addition of two phosphite or triphenylphosphine ligands leading to active species in hydroformylation.[18] According to the high yield procedure developed in our laboratory,[19] the tetracarbonyl complex [Rh$_2$(μ-SCH$_2$CH$_2$CH$_2$NHMe$_2$)(CO)$_4$]Cl$_2$ was prepared by direct addition of Me$_2$N(CH$_2$)$_3$SH to [RhCl$_2$(CO)$_2$]$^-$ in methanol. Addition of two equivalents of TPPTS to a water solution of this material led to [Rh$_2$(μ-SCH$_2$CH$_2$CH$_2$NHMe$_2$)(CO)$_2$(TPPTS)$_2$]Cl$_2$ **4** which was fully characterized by elemental analysis, infrared and ^{1}H, ^{13}C and ^{31}P NMR.[8]

Hydroformylation of 1-octene carried out at 80°C, 5 bar with ethanol as cosolvent gave 10 % conversion into C$_9$ aldehydes after 15 h. It is worth to mention that no octane nor internal octene was detected as previously observed for the [Rh$_2$(μ-SR)$_2$(CO)$_2$L$_2$] complexes in classical organic solvents.[20] As for **1**, a 84 % selectivity in linear aldehyde was determined showing that the presence of a cosolvent decreases significantly the linearity. After removing water under reduced pressure until a brown powder was obtained, complex **2** was not detected either by infrared or ^{1}H and ^{31}P NMR. Particularly no hydride signal was detected. All the NMR data are the same than those of **4** stirred in water under a CO/H$_2$ atmosphere and are consistent with a dinuclear structure containing the fragment $Rh(CO)(SCH_2CH_2CH_2NHMe_2)(TPPTS)_x$.

3.4 SUPPORTED AQUEOUS-PHASE CATALYSIS

Our attention was drawn to a new biphasic system designed to facilitate the reactions to occur at the interface of the two phases.[21-23] These catalytic systems consist of a thin film that resides on a high surface-area hydrophilic support and is composed of water-soluble organometallic complexes and water. The reaction of the liquide-phase organic reactants occurs at the film-organic interphase so that the rate is no more governed by the solubility of the reactants in water. It has been proposed that hydrogen bonding of the hydrated sodium-sulfonate groups to the OH groups present at the surface of the support (generally a silica) is responsible for the immobilization of TPPTS.[24]

For the hydrophilic support, we used Sipernat 22 silica from Degussa characterized by a mean granulometry of 100 μm and a BET surface area of 150 $m^2.g^{-1}$. This silica was contacted under argon for 2h with an aqueous solution containing **1** and an excess of TPPTS. Water was evaporated under vacuum at room temperature until a yellow powder was obtained and the vacuum was maintained for 1 day. Thermogravimetric analyses showed that the water content of this solid was 2.8 w%.

Hydroformylation of 1-octene at 5 bar and 80°C was carried out with toluene as solvent to provide a good dispersion. Direct use of the catalyst gave a yield in aldehydes of 5.5% for 15h with a selectivity of 100% and a linearity of 93%. Similar results can be obtained by introducing the support, **1**, TPPTS, the required amount of water and toluene. It is necessary to increase the water content until 8 w% to have the maximum activity (yield 87%, linearity 86%); this value corresponds to a 6.19Å thickness for the aqueous film on the surface. Beyond 10 w% the reactivity decreases (yield 72% for 12 w%) slowly. For 6-10 w% water contents of the solid support the thickness of the aqueous film corresponds to 1 to 2 molecules of water : complex **1** would be immobilized through the sulfonate groups only, as proposed previously for **2** by Horváth,[24] the hydrophobic part of the molecule being present in the organic phase and thus accessible to 1-octene. Larger amounts of water allow to dissolve fully the complex in the water layer so the biphasic situation is restored. Moreover the pores of the solid can be plugged by water thus reducing the amount of catalyst.

Although in batch experiments it is dificult to collect all the solid after a run, the reactivity in a second run can be estimated to be the same. Addition of large quantities of water to the yellow solid allow to extract the complex from the support.^{31}P NMR analyses have shown that complex **1** is still present after catalysis.

These prelimary results on supported aqueous-phase catalysis show that the reaction rate is slightly lower than that observed in the presence of a cosolvent (22 w% ethanol) since after 15h the conversion was 80%. The linearity of 86% is a little bit better than with ethanol. No isomerization of 1-octene nor hydrogenation was noted. The dinuclear structure of complex **1** is preserved.

4. HYDROXYCARBONYLATION OF BROMOBENZENE

Aryl or alkyl halides represent an important class of molecules from which it is possible to proceed to multiple functionalizations. The carbonylation reaction carried out in the presence of nucleophilic species bearing a hydrogen atom allows to prepare various organic carbonyl compounds such as aldehydes, esters, acids, amides, cetoesters, etc... The general equation is the following :

$$\text{Ar X} + \text{CO} + \text{Nu H} + \text{base} \longrightarrow \text{Ar CO Nu} + [\text{base H}]\text{X}$$

$$\text{NuH} = H_2, \text{ROH}, H_2O, R_2NH...$$

This reaction is catalyzed by palladium complexes containing triphenylphosphine. Many studies have shown that the nature and the strength of the base are important in removing the HX coproduced.[28] The triphenylphosphine ligands can be substituted with more basic phosphines in order to improve the rate-determining oxidative addition of ArX to the palladium (0) active species.[29] Specialy designed ligands,[30,31] or very basic phosphines [32] can allow to activate also the C-Cl bond.

In the general conteset of our investigations on the carbonylation reactions, we were interested in the hydroxycarbonylation of aromatic halides. As this reaction requires water as reactant and owing to the role of cosolvent that can be played by the coreactant NEt_3, we used water-soluble palladium complexes in order to carry out catalysis in the water phase. Bromobenzene was used in a first step as a representative compound of this class of molecules.[33]

4.1 GENERATION OF [Pd(TPPTS)x]

Usually a palladium precursor such as $[PdCl_2(PPh_3)_2]$ or a $Pd(OAc_2/PPh_3$ mixture is used for the carbonylation of aryl halides.[25] From these starting materials a palladium(0) active species is formed $[Pd(PPh_3)_x]$.[25] Thus, we synthesized the complex $[Pd(TPPTS)_3]$ by exchange with $[Pd(PPh_3)_4]$ according to the procedure of Herrmann et al.[34] We observed in ^{31}P NMR that this complex, characterized by a broad singlet at 24.0 ppm, is in a dynamical equilibrium with $[Pd(TPPTS)_2]$ and free ligand. Simple addition of small amounts of TPPTS shifts the signal towards high fields; for a molar TPPTS/Pd ratio of 12.5 the mean phosphine signal appears at 3.2 ppm. This phenomenon was already described for the PPh_3 analogue.[35]

Recently, it was unambigously demonstrated that addition of triphenylphosphine to $Pd(OAc)_2$ affords $[Pd(PPh_3)_x]$ and $OPPh_3$,[35,36] and that palladium (II) complexes are reduced to a palladium(0) species by their own coordinated triphenylphosphine ligands which are oxidized to phosphine oxide in the presence of OH^- ions.[37] The addition of TPPTS to $Pd(OAc)_2$ was monitored by ^{31}P NMR spectroscopy. For a P/Pd ratio of 12.5 a broad singlet was observed at 4 ppm due to the mixture of $[Pd(TPPTS)_3]$, $[Pd(TPPTS)_2]$ and TPPTS in dynamic equilibrium beside a sharp singlet at 36.3 ppm assigned to the oxide OTPPTS. As OTPPTS, like $OPPh_3$, does not play any role in catalysis under rigorously homogeneous conditions it is more convenient to introduce the non-air sensitive $Pd(OAc)_2$ salt and an excess of phosphorus ligand into the autoclave since the expected zero-valent palladium active species is very quickly produced in the medium.

4.2 HYDROXYCARBONYLATION OF BROMOBENZENE

Carbonylation of bromobenzene was carried out in the presence of triethylamine and produced ammonium benzoate which was mainly found in the aqueous phase :

$$PhBr + CO + H_2O + 2\ NEt_3 \longrightarrow [NHEt_3][PhCOO] + [NHEt_3]Br$$

A study of the sensitivity of the reaction to the main parameters has shown that a temperature of 150°C, a CO pressure of 15 bar and a molar TPPTS/Pd ratio of at least 12.5 were necessary. Particulary an excess of phosphine is required to maintain the $[Pd(TPPTS)_x]$ complex in solution as shown after a catalytic run by ^{31}P NMR. For lower ratios extensive decomposition occurs giving rise to black palladium (P/Pd=5) or small black particules in suspension, providing a green color to the solution, which coalesce and form a deposit progressively. The reaction is not sensitive to the presence or not of dihydrogen : even with a H_2 pressure of 30 bar 1% only of benzaldehyde was produced. The hydroxycarbonylation is considerably faster than hydrocarbonylation. Even introducing one equivalent of sodium formate per bromobenzene did not produce more than 1% benzaldehyde; however the formation of black palladium is responsible for significant amounts of benzene (16% beside 75% benzoic acid) arising from debromohydrogenation. In a separate experiment where all the parameters were maintained except the presence of CO and where palladium was introduced as Pd/C, it was checked that bromobenzene was slowly hydrogenolyzed into benzene.

Thus at 150°C, with P_{co} = 15 bar, PhBr/Pd = 50, TPPTS/Pd = 12.5, NEt_3/PhBr = 2.2, the two solvents being toluene and water (10 ml/10 ml) an induction period of 1h was observed ; after 3h the carbonylation was complete, giving a turnover frequency of 17 h^{-1}. Decantation of the reaction mixture allowed a rapid separation of the uncolored organic and the yellow aqueous phases. A second run was carried out with this aqueous phase and a new charge of bromobenzene leading to ammonium benzoate. However palladium was recovered as a black precipitate, affording an almost colorless aqueous phase. A ^{31}P NMR analysis of this aqueous solution revealed the presence of mainly OTPPTS at 32.9 ppm, some $OPPh(m\text{-}C_6H_4SO_3Na)_2$ (or OTPPDS) at 33.5 ppm, small amounts of TPPTS and presumably two phosphonium salts at 25.3 and 25.0 ppm. We have checked that in the absence of palladium the oxidation of TPPTS into OTPPTS and its transformation in OTPPDS does not occur. Fig. 6 shows the pathway we propose for the formation of these two oxides catalyzed by palladium.

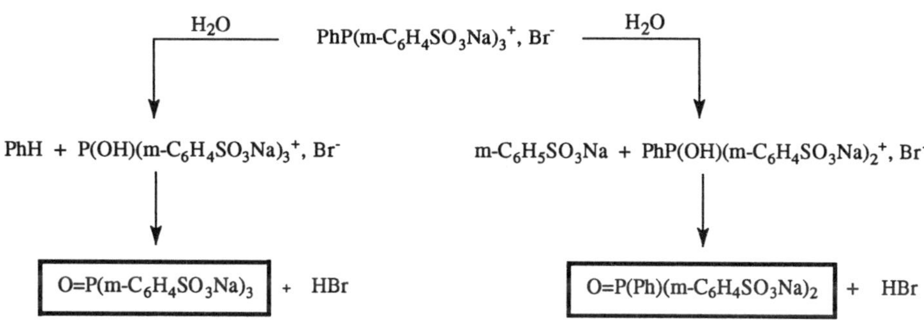

Fig. 6. Proposed pathway for the formation of OTPPTS and OTPPDS

Thus starting from or generating *in situ* $[Pd(TPPTS)_3]$ as catalyst affords a highly selective hydroxycarbonylation reaction of bromobenzene. However, as phosphine oxide is produced during catalysis, it is necessary to complement the aqueous phase with fresh TPPTS from one run to the next one.

4.3 MECHANISTIC APPROACH

In the literature it is admitted, according to the more recent investigations on the triphenylphosphine containing palladium catalysts,[38] that the active species is $[Pd(PPh_3)_2]$ and the rate-determining step is the oxidative addition of bromobenzene to afford $[Pd(Ph)Br(PPh_3)_2]$. Carbonylation gives the benzoyl species $[Pd(COPh)Br(PPh_3)_2]$ which undergoes a triethylamine-assisted attack of methanol leading to methylbenzoate, triethylammonium bromide and to the regeneration of the active species.[38]

We have observed that bromobenzene reacts slowly at 35°C in methanol with [Pd(TPPTS)$_3$] to afford quantitatively [Pd(Ph)Br(TPPTS)$_2$] as a white precipitate. Deep-garnet D_2O solutions resulting of dissolution of this complex present ^{1}H NMR signals consistent with a [Pd(Ph)(TPPTS)$_2$] fragment and in ^{31}P NMR a singlet at 27.8 ppm. We assign to this garnet complex a formula [Pd(Ph)OH(TPPTS)$_2$] or [Pd(Ph)(H_2O)(TPPTS)$_2$]Br. Any attempt to prepare [Pd(Ph)Br(TPPTS)$_2$] by a direct exchange reaction from [Pd(Ph)Br(PPh_3)$_2$] in water-chloroform led invariably to [Pd(TPPTS)$_3$], OTPPTS through the likely formation of a phosphonium salt (δ = 26.3 ppm) while benzene is produced simultaneously. Fig. 7 shows a proposed pathway which accounts for our observations.

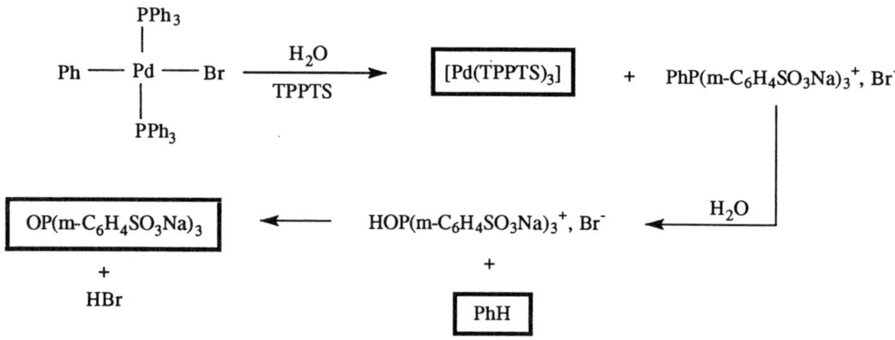

Fig. 7. Proposed pathway for the formation of [Pd(TPPTS)$_3$], OTPPTS and benzene in an attempt to replace the PPh_3 ligands by TPPTS in [Pd(Ph)Br(PPh_3)$_2$].

Direct addition of PhBr to [Pd(TPPTS)$_3$] in water leads to [Pd(Ph)OH(TPPTS)$_2$] or [Pd(Ph)(H_2O)(TPPTS)$_2$]Br. Carbonylation of this garnet complex at room temperature afforded [Pd(COPh)OH(TPPTS)$_2$] or [Pd(COPh)(H_2O)(TPPTS)$_2$]Br quite easily.

On the contrary, the addition of an aqueous phase containing TPPTS to a chloroform solution of [Pd(COPh)Br(PPh_3)$_2$] led at 50°C after 1h to [Pd(COPh)Br(TPPTS)$_2$]. This complex shows in ^{31}P NMR a singlet at 24.7 ppm. It reacts in the presence of triethylamine to give triethylammonium benzoate and [Pd(TPPTS)$_x$]. As we also observed in catalysis, reductive elimination of PhCOBr, which could be hydrolyzed, does not occur; the reaction needs to be assisted by a basic reagent.

As three compounds have been characterized, i.e. a palladium (0) species, a Pd-Ph and a Pd-COPh compounds, a catalytic cycle can be proposed which recalls that published in the literature for the PPh_3 analogue.[38] Here, a nucleophilic attack of a hydroxide ion assisted by triethylamine is considered to occur on the benzoyl species. This proposed mechanism is supported by the fact that sodium hydroxide affords the same result. Fig. 8 displays the four main steps of the simplified catalytic cycle showing a pentacoordinated carbonyl species which has not been observed.

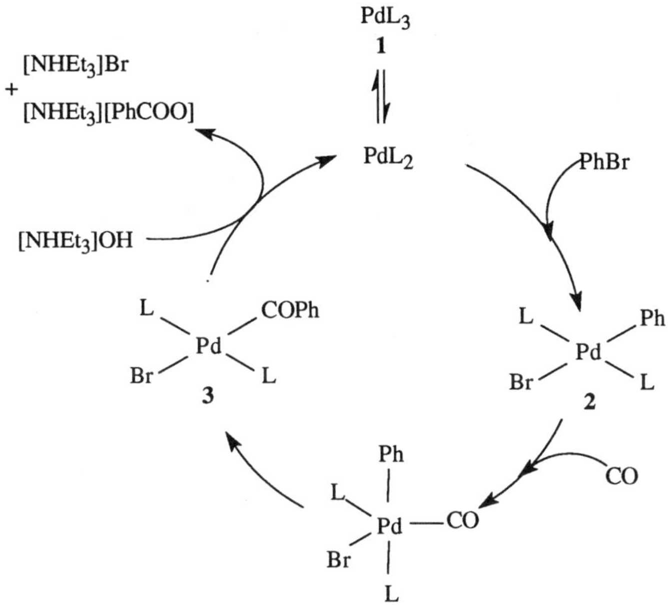

Fig. 8. Proposed catalytic cycle for the hydroxycarbonylation of bromobenzene in water

5. FUTURE PROSPECTS

Carbonylation reactions of heavy substrates, especially monoterpenes, are presently investigated, including diastereoselective reactions. The supported-aqueous phase catalysis looks a promising method to combine a good reactivity in catalysis with an easy recycling of the metal complex. In addition, catalysis in microemulsions seems an interesting field to be explored.

6. ACKNOWLEDGMENTS

We thank Rhône-Poulenc Chimie for a gift of TPPTS. Comptoir Lyon-Alemand-Louyot is gratefully acknowledged for a loan of rhodium and palladium salts. This project was mainly supported by the Ministère de la Recherche et de l'Enseignement Supérieur through a "Jeune Equipe" recognition.

7. REFERENCES

1. Joó, F. and Tóth, Z. (1980) Catalysis by water-soluble phosphine complexes of transition metal ions in aqueous and two-phase media, *J. Mol. Cat.* **8**, 369-383.

2. Sinou, D. (1987) Phosphines hydrosolubles. Synthèses et Applications en catalyse, *Bull. Soc. Chim. Fr.*, 480-486.

3. Barton, M. and Atwood, J.D. (1991) Aqueous soluble organometallic complexes, *J. Coord. Chem.*, **24**, 43-67.

4. Kalck, Ph. and Monteil, F.(1992) Use of water-soluble ligands in homogeneous catalysis, *Adv. Organometal. Chem.*, **34**, 219-283.

5. Herrmann, W.A. and Kohlpainter, C.W. (1993) Water-soluble ligands, metal complexes and catalysts : synergism of homogeneous and heterogeneous catalysis, *Angew. Chem. Int. Ed. Engl.*, **32**, 1524-1544.

6. Kuntz, E. (1975) Procédé d'hydroformylation des oléfines, 20 june 1975 to Rhône-Poulenc Industries, *Fr. Dem. 75 19407.*

7. Kuntz, E. (1987) Homogeneous catalysis.... in water, *Chemtech.*, **17**, 570-575.

8. Monteil, F. Quéau, R. and Kalck, Ph. (1994) Behaviour of water-soluble dinuclear rhodium complexes in the hydroformylation reaction of 1-octene, *J. Organometal. Chem.*, in press.

9. Kalck, Ph. (1987). Use of dinuclear templates to perform the hydroformylation reaction, in A. de Meijere and H. tom Dieck (eds), *Organometallics in Organic Synthesis*, Springer-Verlag, Berlin, Heidelberg, 297-320.

10. Kalck. Ph., Pérès, Y., Quéau, R., Molinier, J., Escaffre, P., de Oliveira, E.L. and Peyrille, B. (1992) Novel active dirhodium complexes for the low pressure hydroformylation reaction. Unexpected inhibiting effect of high partial pressures of hydrogen, *J. Organometal. Chem.*, **426**, C16-C20.

11. Besson, B., Kalck, Ph. and Thorez, A. (1988) Dinuclear and water-soluble rhodium complexes and hydroformylation catalysis therewith, *U.S. Pat. 4,778,905*, Oct. 18, 1988 (priority Oct. 16, 1984) to Rhône-Poulenc Chimie de Base.

12. Kalck, Ph., Escaffre, P., Serein-Spirau, F. Thorez, A., Besson, B., Colleuille, Y. and Perron, R. (1988) Water-soluble dinuclear rhodium complexes as catalytic precursors for the hydroformylation of alkenes in a two phases system, *New J. Chem.*, **12**, 687-690.

13. Hablot, I. (1991) Reaction liquide-liquide-gaz : études thermodynamique et cinétique d'une hydrogénation d'oléfine catalysée par un complexe de coordination solubilisé en phase aqueuse, *Thèse de docteur de l'Institut National Polytechnique de Toulouse*, Jan. 17, 1991.

14. Abraham, M.H., Grellier, P.L. and McGill, R.A. (1988) A quantitative measure of solvent solvophobic effect, *J. Chem. Soc. Perkin Trans II*, 339-345.

15. Arhancet, J.P., Davis, M.E., Merola, J.S. and Hanson, B.E. (1990) Supported aqueous-phase catalysts, *J. Catal.* **121**, 327-339.

16. Jardine, F.H. (1982) Carbonylhydridotris (triphenylphosphine) rhodium(I) *Polyhedron*, **1**, 569-605.

17. Escaffre, P., Kalck, Ph. and Thorez, A. (1986) Catalytic activity of $RhH(CO)(PPh_3)_3$ in the presence of a thiol : a new way to obtain very active species, *J. Organometal. Chem.,* **302**, C17-C20.

18. Polo, A., Claver, C., Castillon, S., Ruiz, A., Bayon, J.C., Real, J., Mealli, C. and Masi, D. (1992) Regioselective hydroformylation of cyclic vinyl and allyl ethers with rhodium catalysts. Crucial influence of the size of the phosphorus cocatalyst, *Organometallics*, **11**, 3525-3533.

19. Kalck, Ph., Senocq, F., Siani, M. and Thorez, A. (1988) A novel and easy way to prepare $Rh_2(\mu\text{-}SR)_2(CO)_4$ and $Rh_2(\mu\text{-}SR)_2(CO)_2(PR'_3)_2]$ complexes on a large scale, *J. Organometal. Chem.*, **350**, 77-80.

20. Dedieu, A., Escaffre, P., Frances, J.M., Kalck, Ph. and Thorez, A. (1986) Hydroformylation occuring on two rhodium centers during the full catalytic cycle, *New. J. Chim.*, **10**, 631-634.

21. Arhancet, J.P., Davis, M.E., Merola, J.S. and Hanson, B.E. (1989) Hydroformylation by supported aqueous-phase catalysis : a new class of heterogeneous catalysts, *Nature* , **339**, 454-455.

22. Arhancet, J.P., Davis, M.E. and Hanson, B.E. (1989) Supported aqueous-phase, rhodium hydroformylation catalysts. I. New methods of preparation, *J. Catal.,* **129,** 94-99.

23. Davis, M.E. (1992) Supported aqueous-phase catalysis, *Chetmech* , **22**, 498-502.

24. Horváth, I.T. (1990) Hydroformylation of olefins with the water-soluble $HRh(CO)[P(m-C_6H_4SO_3Na)_3]_3$ in supported aqueous-phase. Is it really aqueous?, *Catal. Lett.*, **6**, 43-48.

25. Heck, R.F (1985) Palladium reagents in organic syntheses, in Katritszky, A.R., Meth-Cohn, O. and Rees, C.W. (eds), *Best synthetic methods*, Academic Press, London.

26. Gauthier-Lafaye, J. and Perron, R. (1988) Carbon monoxide and fine chemicals synthesis, in Mortreux, A. and Petit, F. (eds), *industrial applications of homogeneous catalysis*, D. Reidel Publishnig Co, Dordrecht, pp. 19-64.

27. Colquhoun,H.M., Thompson,D.J. and Twiss, M.V. (1991), *Carbonylation. Direct synthesis of carbonyl compounds*, Plenum Press, New York and London.

28. Pri-Bar, I. and Buchman, O. (1988) Hydroxycarbonylation of aryl halides with formate salts catalyzed by palladium complexes, *J. Org. Chem.* **53**, 624-626.

29. Milstein, D. (1986) The product-forming step in palladium-catalysed methoxycarbonylation of organic halides, *J. Chem. Soc., Chem. Comm. 817-818.*

30. Ben-David, Y., Portnoy, M. and Milstein, D. (1989) Formylation of aryl chlorides catalysed by a palladium complex, *J. Chem. Soc., Chem. Comm.* 1816-1817.

31. Ben-David, Y., Portnoy, M. and Milstein, D. (1989) Chelate-assisted, Pd-catalyzed efficient carbonylation of aryl chlorides, *J. Am. Chem. Soc.* **111**, 8742-8744.

32. Huser, M., Youinou, M.T. and Osborn, J.A. (1989) Chlorocarbon activation : catalytic carbonylation of dichlorometane and chlorobenzene, *Angew. Chem., Int. Ed. Engl.* **28**, 1386-1387.

33. Monteil, F. and Kalck, Ph. (1994) Carbonylation of bromobenzene in a biphasic medium catalysed by water-soluble palladium complexes derived from tris(3-sulphophenyl)phosphine, *J. Organometal. Chem.* in press.

34. Herrmann, W.A., Kellner, J. and Riesel, H. (1990) III*. Neve wasserlösliche Metallkomplexe des sulfonierten Triphenyl-phosphans(TPPTS) : Mn, Fe, Ru, Co, Rh, Ir, Ni, Pd, Pt, Ag, Au *J. Organometal. Chem.* **389**, 103.

35. Amatore, C., Jutand, A. and M'Barki, M.A. (1992) Evidence of the formation of zerovalent palladium from $Pd(OAc)_2$ and triphenylphosphine, *Organometallics* **11**, 3009-3013.

36. Ozawa, F. Kubo, A. and Hayashi, T. (1992) Generation of tertiary phophine - coordinated Pd(0) species from $Pd(OAc)_2$ in the catalytic Heck reaction. *Chem. Lett.* 2177.

37. Grushin, V.V and Alper, H. (1993) Alkali-induced disproportionation of palladium (II) tertiary phosphine complexes, $[L_2PdCl_2]$, to LO and palladium (0). Key intermediates in the biphasic carbonylation of ArX catalyzed by $[L_2PdCl_2]$, *Organometallics,* **12**, 1890-1901.

38. Moser, W.R., Wang, A.W. and Kildahl, N.K. (1988) Mechanistic studies of the palladium-catalyzed reaction of methanol with bromobenzene and CO to produce methyl benzoate. 1. Stoichiometry study, *J. Am. Chem. Soc.* **110**, 2816-2820.

NEW WATER SOLUBLE PHOSPHINES FOR ORGANOMETALLIC CHEMISTRY AND CATALYSIS IN THE AQUEOUS PHASE

Brian E. Hanson, Hao Ding, Tamas Bartik, and Berit Bartik

Department of Chemistry
Virginia Polytechnic Institute and State University
Blacksburg, VA 24061-0212
USA

1. Introduction

The fact that water soluble phosphines may also be surface active was first noted by Wilkinson et al. in their study of the hydroformylation of 1-hexene with rhodium complexes of $P(C_6H_5)_2(C_6H_4SO_3Na)$, TPPMS.[1] Specifically it was stated that aqueous solutions of TPPMS have a tendency to form emulsions that separated slowly. While TPPMS has definite hydrophobic and hydrophilic regions it does not have the general shape of micelle forming surfactants. Specifically it is more disk shaped than rod-like.

Of the water soluble phosphine ligands reported in the literature [2] many, in fact, should demonstrate surface activity. Only a few, however, have been designed specifically to be surface active; notable among these are the zwitterionic tris-pyridyl phosphines, **1**, shown below.[3]

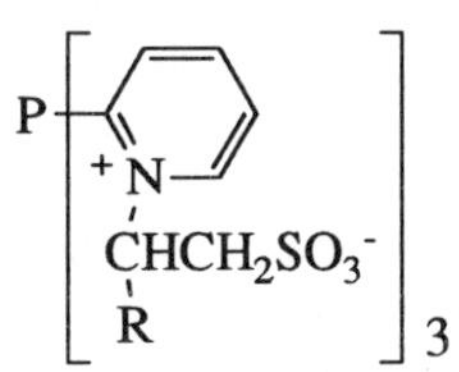

$R = C_nH_{2n+1}$; n = 6 or 12

1

With these phosphines it was reported that when n = 12, stable emulsions were generated under two-phase hydroformylation conditions. While the concentrations of phosphine used for catalysis were expected to be above the critical micelle concentration direct measurements to demonstrate aggregation of the phosphines in water were not done.

The chiral phosphine, DIOP, has been prepared with a pendant polyethylene ether group, **2**.[4] This compound should behave as a nonionic surfactant, although its surface activity was not investigated. Other water soluble phosphines that should exhibit

I. T. Horváth and F. Joó (eds.), Aqueous Organometallic Chemistry and Catalysis, 149–158.

surface active character include, $Ph_2P(CH_2)_4SO_3K$;[5] $P[(CH_2)_xC_6H_4\text{-p-}SO_3Na]_3$, x = 1, 2, 3, 6;[6,7] and diphenylphosphine substituted sulfonated polystyrene.[8] Of these it is noted here that the ligands, $P[(CH_2)_xC_6H_4\text{-p-}SO_3Na]_3$, show enhanced reaction rates for the two phase hydroformylation of 1-octene at low ligand to rhodium ratios.[7]

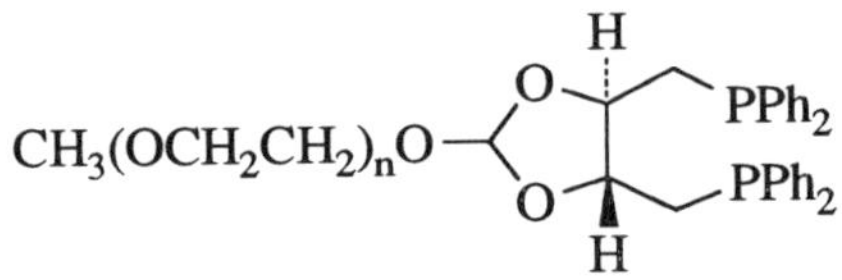

2

The possibility that potentially surface active phosphines aggregate in solution has only recently been investigated.[9] The potential for aggregation in aqueous solution as noted above with **1** should impact catalytic activity under two-phase reaction conditions. In general, the degree of aggregation of a ligand in water may also affect reaction selectivity under two-phase reaction conditions, although reaction selectivity was not affected by the ability of **1** to form micelles,

The new phosphines, $P(Menthyl)((CH_2)_8C_6H_4\text{-p-}SO_3Na)_2$, **3**,[9] and $P[C_6H_4\text{-p-}(CH_2)_xC_6H_4\text{-p-}SO_3Na]_3$,, **4**, [10] are demonstrated to aggregate under certain conditions in water.[10] The catalytic activity of rhodium complexes of the new phosphines compared to $P(C_6H_4SO_3Na)_3$, TPPTS, in light of their tendency to aggregate, gives insight to the mechanism of water soluble hydroformylation catalysts based on sulfonated phosphines.

2. Preparation of the Surface Active Phosphines

The new phosphines, **3** and **4**, were prepared as described previously.[9, 10] The disulfonated phosphine **3** was synthesized by direct sulfonation of bis(8-phenyloctyl)menthyl phosphine which in turn was prepared by the reaction of $C_6H_5(CH_2)_8MgBr$ and $(Menthyl)PCl_2$ in ether.[11] The trisulfonated triaryl phosphines, **4**, were synthesized by sulfonation of the phosphine generated from the reaction of $C_6H_5(CH_2)_mC_6H_4Li$ with PCl_3. In both cases the sulfonation procedure follows the method previously described by us.[6] Sulfonation requires concentrated sulfuric acid and a total reaction time from 1 to 5 hours. Schematic representations of the ligands are shown below. Analytical data for the phosphines have been previously reported.[9, 10]

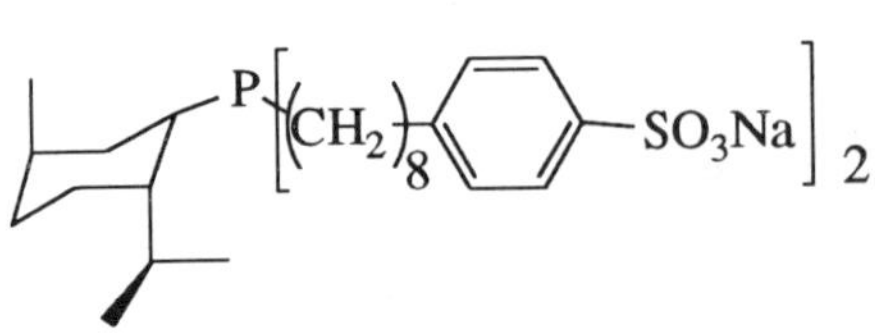

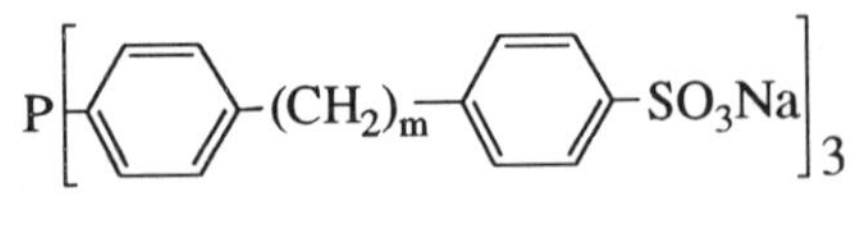

3

4a, m = 3; **4b**, m = 6

3. Aggregation of Sulfonated Phosphines in Water

Dynamic light scattering experiments were done on a Biotage dp-801 molecular size detector using aqueous solutions of the phosphines. The solutions were prepared in degassed distilled deionized water and filtered twice through a 0.1 μm syringe filter. The measurements were done at ambient temperature, 22°C, and atmospheric pressure. For the ligand **4a** at 0.05 M concentration the average hydrodynamic radius was consistently measured to be 7 Å. This demonstrates that no aggregation of the ligand occurs under these conditions. However at a NaCl concentration of 0.25 M the average hydrodynamic radius increases to 20 Å. The corresponding value for a sample of **4b** that contained 5 % by weight Na_2SO_4 is 20Å. In 0.25M aqueous NaCl the hydrodynamic radius increased to 32 Å.

In contrast to **4a** and **4b** light scattering experiments on aqueous solutions of TPPTS showed no evidence for aggregation at all concentrations of TPPTS and NaCl investigated (up to 0.1 M TPPTS and 0.5M NaCl).

The hydrodynamic radius of 7Å for **4a** in water is consistent with isolated molecules of the ligand. The phosphine is comprised of three linear pendant groups terminated with a sulfonate group and linked at the hydrocarbon end by phosphorus. In the absence of added salt electrostatic repulsion's between the sulfonate groups are expected to define the solution conformation. It is speculated that this leads to a nearly trigonal planar arrangement of sulfonate groups with respect to the phosphorus atom. Aggregation apparently takes place as the ionic strength of the solution is increased by the addition of salt. If the relationship of the sulfonate groups to the phosphorus atom is pyramidal then

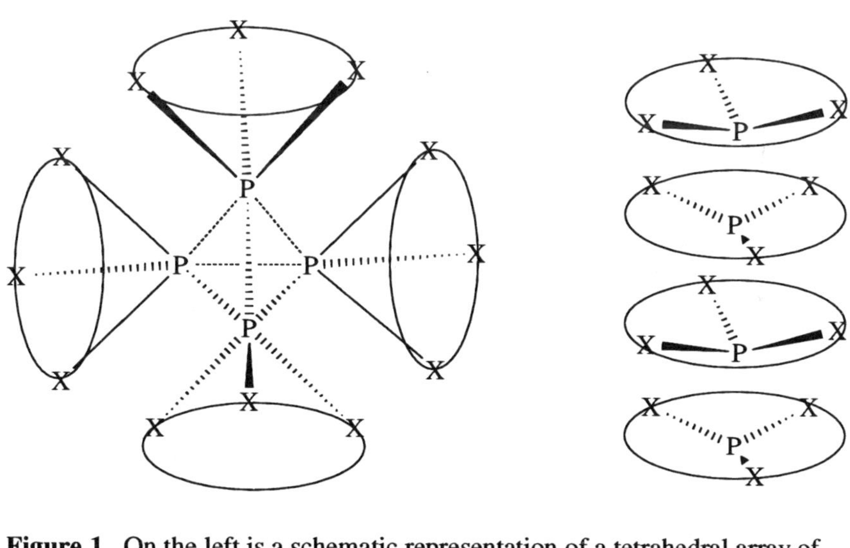

Figure 1. On the left is a schematic representation of a tetrahedral array of sulfonated phosphines, e.g. **4a.**, which generates a nearly spherical shape. On the right is a stacked arrangement of phosphines to give a rod-like shaped micelle. (X = p-$C_6H_4SO_3Na$)

a tetrahedral array of four phosphines, with the phosphorus atoms comprising the hydrophobic core of a small micelle, fits the experimental hydrodynamic radius of 20 Å. Alternatively the phosphines can stack with either a staggered or gauche arrangement of "arms" with pendant sulfonate groups. The stacked phosphines give the general shape of a rod-shaped micelle. These two possibilities are shown schematically in Figure 1.

The sulfonated menthyl phosphine, **3** behaves similarly. That is, aggregates of radius 30 Å are observed in the absence of added salt and dynamic light scattering experiments on aqueous solutions that are 0.01 M in **3** and 0.25 M in NaCl show the presence of aggregates of r = 40 Å. All available samples of the menthyl phosphine contained some sodium sulfate thus it is impossible to determine the inherent ability of **3** to aggregate in the absence of salt. The important feature is that the phosphine has the ability to form micelles in aqueous solution.

Aqueous solutions of the surface active phosphines have many of the characteristics expected of solutions that contain surface active compounds. For example water solutions of **3** dissolve the water insoluble dye Orange OT. This is consistent with the formation of micelles. Also salt solutions of all the surface active phosphines, **3**, **4**, and **4b** foam when shaken. This behavior is associated with the presence of detergents. Presence of the structures indicated in Figure 1 in water could account for the micellar character of the solutions.

4. Size and Basicity of Sulfonated Phosphines

As reported previously the electron donating ability and size of water soluble phosphines can be determined spectroscopically.[6] This is accomplished by preparation of the appropriate nickel and palladium complexes and recording, respectively, the infrared spectrum of $Ni(CO)_3L$[12] and the ^{31}P NMR spectrum of *trans*-$PdCl_2L_2$,[13] where L = the water soluble phosphine. Comparison with the infrared spectrum of $Ni(CO)_3(PnBu_3)$ is accomplished by a choice of solvent that dissolves both the n-butyl phosphine derivative and the sulfonated phosphine derivative. The values are then corrected to methylene chloride since literature values are reported in this solvent. The ^{31}P NMR chemical shifts are not very solvent dependent and are not corrected for solvent.

It was noted that sulfonated phosphines that have the sulfonate group removed from the phosphorus atom by three or more methylene groups are sterically and electronically similar to their unsulfonated analogs. Thus for the phosphines shown here the steric and electronic parameters χ and θ are given for the unsulfonated analogs in Table 1.

Table 1.

Phosphine	θ	χ (nonsulfonated analog)
3	138°	3.5
4a	145° ^{a}	12.0
4b	145° ^{a}	12.0

a. Estimated from models.

5. Rate Enhancement with Surface Active Phosphines

In Figure 2 the styrene hydroformylation activity of **3** compared to TPPTS in methanol/water as a function of time is summarized. These data were generated at a constant ligand to rhodium ratio of 3.[9] Clearly the rate of styrene hydroformylation increases in the presence of the more surface active phosphine, **3**.

In a separate set of experiments to compare rate as a function of ligand to rhodium ratio; when L = TPPTS, the yield of aldehyde reaches its maximum value at L/Rh = 3 and drops slightly with increasing L/Rh thereafter. The sulfonated phosphine, **3**, shows a different trend in conversion as a function of ligand concentration. The maximum yield of aldehydes, about 90%, is achieved at a L/Rh ratio of approximately two; the yield drops to about 50% at L/Rh of 5 and remains virtually unchanged with further L/Rh variation.[9] The diminished activity at high ligand-to-rhodium ratios is attributed

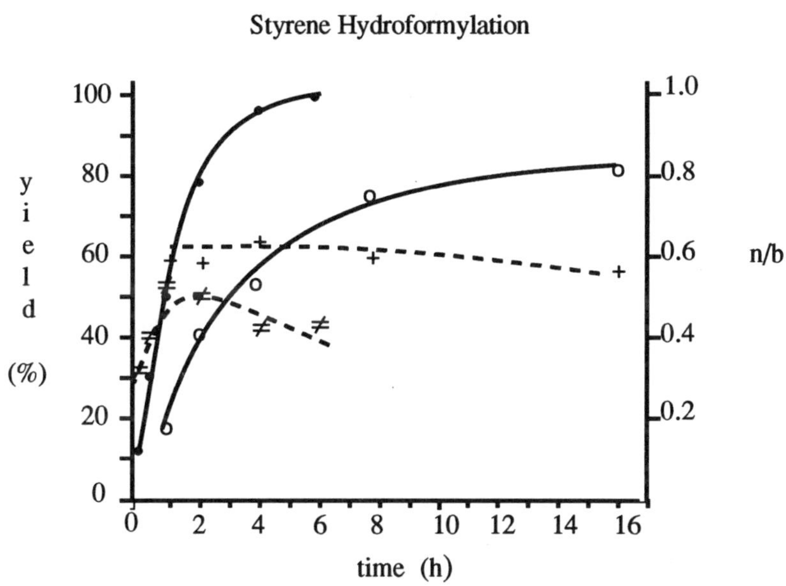

yield		n/b
•	$PR[(CH_2)_8C_6H_4\text{-p-}SO_3Na]_2/Rh(acac)(CO)_2 = 3/1$	≠
o	$TPPTS/Rh(acac)(CO)_2 = 3/1$	+

Figure 2. Styrene hydroformylation activity of TPPTS and **3** as a function of time at a ligand to rhodium ratio of 3

to the low probability of dissociation of the relatively basic ligand **3** from $HRh(CO)L_3$ complexes.[9] Loss of phosphine is generally considered to be prerequisite for catalytic activity.

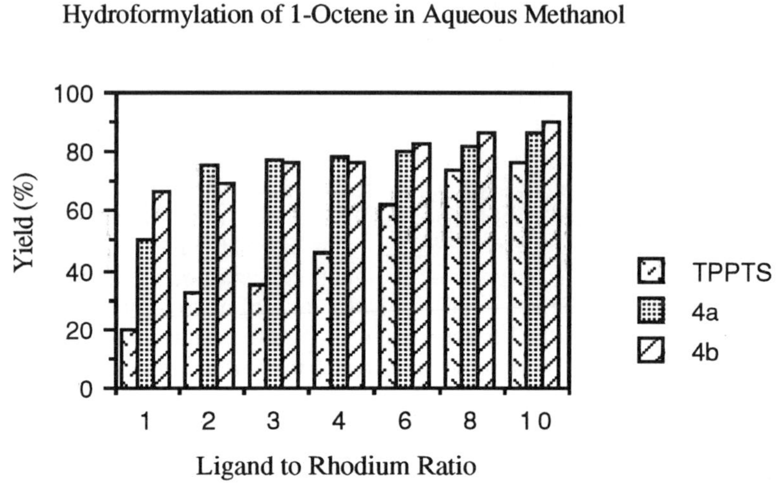

Figure 3. Hydroformylation yield (nonanals and nonanols) as a function of L/Rh ratio. All reactions were run for 5 hours at 120°C and 19 atm. Octene/Rh = 500; [Rh] = 0.005 M.

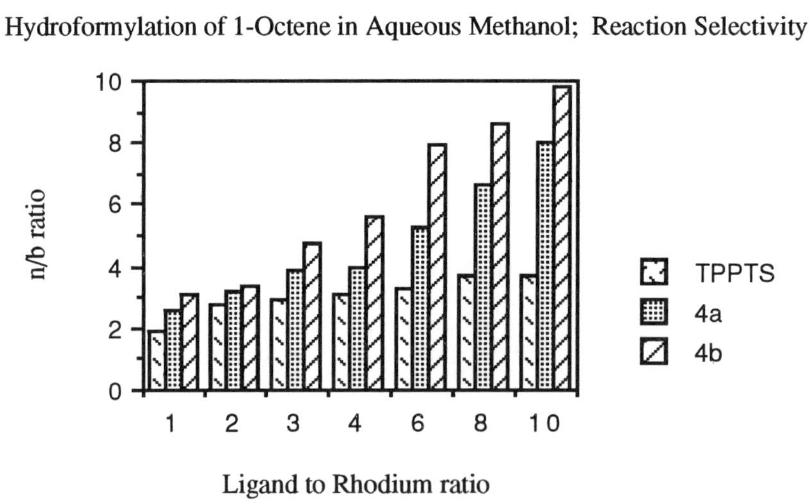

Figure 4. Corresponding reaction selectivities for the hydroformylation results presented in Figure 3.

In Figure 3 the activity of **4a** and **4b** compared to TPPTS for the hydroformylation of 1-octene in methanol/water is summarized.[10] The corresponding data for reaction selectivity are summarized in Figure 4. [10]

The results with **4a** and **4b** compared to TPPTS for 1-octene hydroformylation are qualitatively similar to those noted above for styrene hydroformylation with **3** and TPPTS. That is the reaction rates are faster with the surface active phosphines. Significantly in the case of hydroformylation of 1-octene the n/b ratios are consistently higher with the surface active phosphines compared to TPPTS (see Figure 4). This result is unique to the aqueous methanol as the solvent. A possible mechanism for the observed selectivities is outlined in Section 6.

When the 1-octene hydroformylation reaction is done in water alone as the catalyst containing phase different trends are observed with respect to activity and selectivity. The results in water are summarized in Figures 5 and 6.[14]

Reaction Yields for the Hydroformylation of Octene in Water

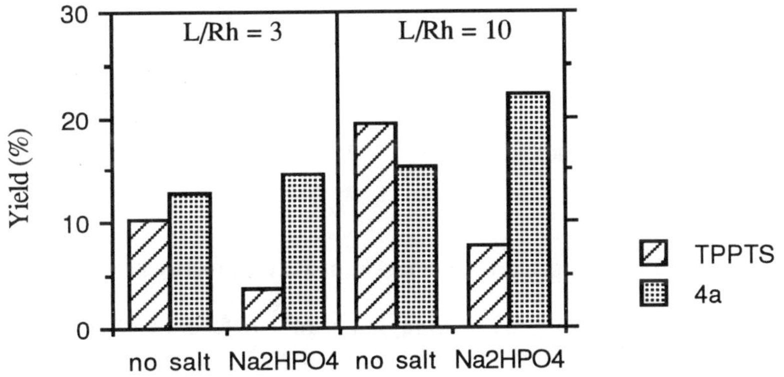

Figure 5. Hydroformylation yield (nonanals and nonanols) as a function of L/Rh ratio and salt concentration. All reactions were run for 24 hours at 120°C and 19 atm. Octene/Rh = 500; [Rh] = 0.005 M. In those reactions with Na_2HPO_4 the salt concentration was 0.25 M

The reaction rates are clearly diminished for both **4a** and TPPTS in water alone as the catalyst bearing phase compare to aqueous methanol. This is expected due to the low solubility of octene in water. The addition of salt has a significant effect on the rate of catalysis. With catalysts derived from TPPTS the rate drops significantly upon the addition of salt, while with **4a** as the ligand reaction rates increase.

In Figure 6 it is seen that reaction selectivity is also strongly dependent on salt concentration. First for TPPTS the reaction selectivity in water is greater than in aqueous methanol; and the addition of Na_2HPO_4 to water phase TPPTS catalysts gives a dramatic further increase in n/b ratio. The same trend is observed for **4a** but the magnitude of selectivity enhancement upon addition of Na_2HPO_4 is not as great.

Importantly in both cases, TPPTS and **4a**, n/b ratios increase as the solution ionic strength increases.

Rection selectivity in Hydroformylation of Octene in Water

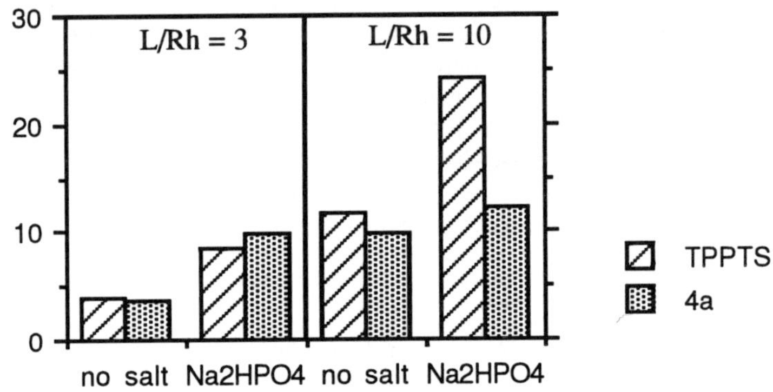

Figure 6. Corresponding reaction selectivities for the hydroformylation results presented in Figure 5.

6. The Effect of Ionic Strength on Aqueous-Phase Hydroformylation

The results in Section 5 on the hydroformylation of 1-octene can be summarized as follows. (i) Catalysts prepared from surface active phosphines give better reaction rates for the hydroformylation of water insoluble substrates than those prepared from TPPTS. (ii) Reaction selectivity, as measured by n/b ratio, is solvent and ligand dependent. For a given ligand reaction selectivity increases with increasing solution ionic strength, i.e. selectivity increases in the order $MeOH/H_2O < H_2O <$ 0.25 M aqueous Na_2HPO_4. (iii) Reaction activity goes in the following order for the surface active phosphine, **4a**: $MeOH/H_2O >> H_2O <$ 0.25 M aq Na_2HPO_4; the corresponding trend of TPPTS is $MeOH/H_2O >> H_2O >$0.25 M Na_2HPO_4.

As ionic strength increases the solubility of octene in waster is further diminished thus the expected effect of added salt is to decrease the rate of a catalytic reaction that takes place in water. This is the observed trend for the TPPTS catalysts. A different trend is noted for the **4a** catalysts, namely the reaction rate increases upon going from H_2O to 0.25 M Na_2HPO_4. The increase in rate is consistent with formation of micelles under catalytic conditions. The presence of micelles is expected to increase the solubility of octene in the aqueous phase and thus increase the rate.

Ionic strength also is expected to play a role in determining the stability of metal complexes of sulfonated phosphines. For example, the complex (nonanoyl)Rh(CO)L_2

is an expected intermediate in the hydroformylation of 1-octene. When L = **4a**, this intermediate can be schematically represented as shown in Figure 7. The orientation of the sulfonate groups around a hydrophobic core is expected to be determined by ionic repulsions; the molecule is thus expected to have the general shape of a micelle. The

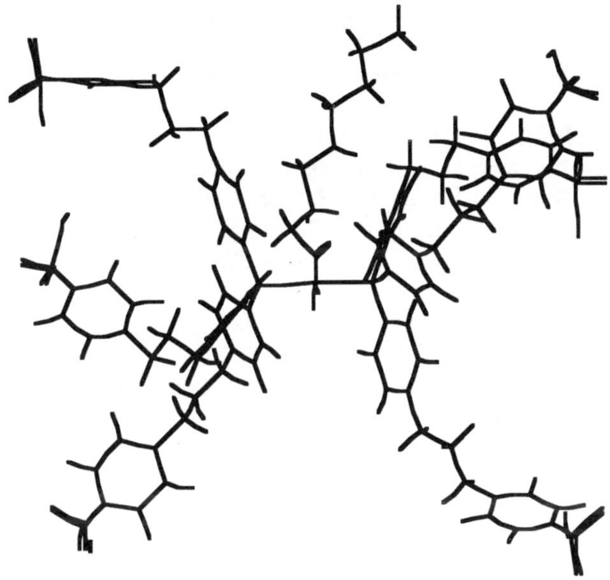

Figure 7. Schematic representation of the potential intermediate (nonanoyl)Rh(CO)L_2 drawn so that the sulfonate groups are far apart.

stability of this "micellar complex" to a certain extent is dependent on the minimization of repulsions of the sulfonate groups. Intramolecular repulsions expected to be minimized in solutions of high ionic strength. Thus high salt concentrations favor complexes of two or more sulfonated phosphine ligands. These are exactly the complexes needed for improved reaction selectivity. Furthermore if this model is correct the effect of solution ionic strength on selectivity should be more pronounced in TPPTS complexes than in complexes of **4a** for the simple reason that the sulfonate groups in (nonanoyl)Rh(CO)$(TPPTS)_2$ lie in closer proximity than those in (nonanoyl)Rh(CO)$(\mathbf{4a})_2$. In fact it has been estimated that the sulfonate groups in the TPPTS complex, HRh(CO)$(TPPTS)_3$ lie on the surface of a sphere of radius 8 Å.[15]

7. Acknowledgments

Work at Virginia Tech on water soluble phosphines is supported by NSF (CTS-9101846 and CHE-9319881) and the Exxon Education Foundation. Berit Bartik gratefully acknowledges a fellowship from the Deutsche Forschungsgemeinschaft (1991-93).

8. References

1. Borowski, A. F.; Cole-Hamilton, D. J.; Wilkinson, G.; *Nouv, J. Chem.* (1978) **2,** 137.
2. Herrmann, W. A.; Kohlpaintner, C. W.; *Angew. Chem. Int. Ed. Engl.* (1993) **32**, 1524.
3. Fell,B.; Pagadogianakis, G.; *J. Mol. Catal.* (1991) **66**, 143.
4. Amrani, Y.; Ph.D. Thesis, Univ. Lyon, **1986**,
5. Paetzold, E.; Kinting,A.; Oehme, G. *J. Prakt. Chem.* (1987) **329**, 725.
6. Bartik, T; Bartik, B.; Hanson, B. E.; Guo, I.; Tóth, I., *Organometallics* (1993) *12*, 164.
7. Bartik, T; Bartik, B.; Hanson, B. E.; *J. Mol. Catal.* (1994) **88**, 43.
8. Manassen, J.; Dror, Y.; U. S. Patent (1983) 4,415,500.
9. Bartik, T; Ding, H.; Bartik, B.; Hanson, B. E.; *J. Mol. Catal.*, submitted.
10. Ding, H.; Hanson, B. E.; Bartik, T.; Bartik, B.; *Organometallics,* in press.
11. Kosolapoff, G. M.; Maier, L.; *Organic Phosphorus Compounds*, vol 1, Wiley-Interscience, New York (1972).
12. Tolman, C. A.; *Chem. Rev.* (1977) **77**, 313.
13. Bartik, T.; Himmler, T.; *J. Organomet. Chem.,* (1986) **293**, 343.
14. Ding, H.; Hanson, B. E.; *J. C. S. Chem. Commun.*, submitted.
14. Ding, H.; Hanson, B. E.; T. E. Glass, *Inorg. Chim. Acta*, submitted.

HEXAAQUARUTHENATE(II) AS CATALYST PRECURSOR IN AQUEOUS ORGANOMETALLIC CATALYSIS

ATTILA C. BÉNYEI
Institute of Physical Chemistry, Lajos Kossuth University
P.O. Box 7, Debrecen, H-4010 (Hungary)

1. Introduction

It is equally important from theoretical and practical point of view to study aqua complexes of transition metals. Among them, $Ru(H_2O)_6(tos)_2$ (**I**, tos=*p*-toluenesulfonate) can be a versatile starting material in aqueous organometallic chemistry. It offers a genuine *chlorine free* entry into several classes of the ruthenium chemistry. *In situ* preparation of catalysts starting from **I** and the appropriate ligand is also possible. In this work earlier studies on $Ru(H_2O)_6{}^{2+}$ has reviewed and new results reported to explore the applicability of $Ru(H_2O)_6{}^{2+}$ in synthetic chemistry and catalysis.

2. Preparation of $Ru(H_2O)_6(tos)_2$

Although the first preparation of hexaaquaruthenium(II) by means of electrochemical reduction of ruthenium(III) was reported[1] as early as 1965, only chemical methods developed by Kallen and Early[2] and later modified by Ludi and Bernhard[3,4] made larger quantities of pure $Ru(H_2O)_6{}^{2+}$ salts available. The physical and chemical characteristics of $Ru(H_2O)_6{}^{2+}$ was investigated by this research group[5] in detail including normal coordinate analysis of the IR and Raman spectra to estimate the self exchange rate of the ruthenium(II)/ruthenium(III) couple[6]. Mechanism of this electron exchange[7] was studied with direct ^{17}O and ^{99}Ru NMR methods. Rate of water exchange was also measured by NMR[8].
The synthesis of hexaaquaruthenium(II) is outlined in Scheme 1. Starting from chlorine free ruthenium(III) or ruthenium(IV) salts RuO_4 is prepared and then reduced to ruthenium(II) by means of lead or tin. Because of the

I. T. Horváth and F. Joó (eds.), Aqueous Organometallic Chemistry and Catalysis, 159–171.

reactivity of either Ru(VIII) or Ru(II), the medium of choice for the reduction can be HBF_4 or H_2SiF_6. Ion exchange on strong acidic ion exchange resin, elution with *p*-toluenesulfonic acid or trifluoromethanesulfonic acid, evaporation and recrystalization yields the appropriate ruthenium(II) hexaaqua salt.

SCHEME 1. Preparation of $Ru(H_2O)_6(tos)_2$

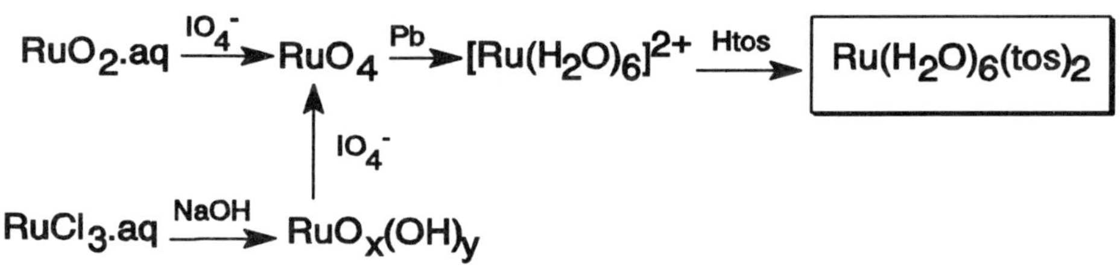

This preparation is tedious and cumbersome: very toxic RuO_4 is an intermediate, the reaction takes several days and argon atmosphere should be maintained throughout the reaction as well as the ion exchange. However, the product hexaaquaruthenium(II) salt is stable and well characterized including X-ray structure. The *p*-toluenesulfonate salt, in the solid state, is stable even under ambient conditions, in the air. The triflate salt should be kept in the refrigerator, under argon.

Hexaaquaruthenium(II) salts are convenient to handle and contain only easy to substitute water molecules in the coordination sphere of ruthenium. These characteristics makes aqua complexes of ruthenium(II) superior to $RuCl_3.3H_2O$ which is the traditional starting material in ruthenium chemistry but its polymeric and ill defined nature[9] can cause reproducibility problems. Altogether, hexaaquaruthenium(II) can serve as suitable starting material in the preparation of other complexes or can be the source of ruthenium for *in situ* preparation of catalysts especially in water solutions and aqueous/organic two phase systems. A further advantage is that this complex does not contain chloride ligand so *chlorine free* entry into several classes of the ruthenium chemistry is possible.

3. Reaction of $Ru(H_2O)_6(tos)_2$ with phosphorous, nitrogen or diene ligands

Either **I** or $Ru(H_2O)_2(THF)_2(tos)_2$, which is formed in the reaction of **I** with THF, reacts[9] with tertiary monodentate or chelating phosphorous ligands giving the appropriate phosphine complexes (Scheme 2, dppe=1,2-bis(diphenylphosphino)ethane). In the presence of $NaAlH_4$ hydride complexes such as *cis*-$RuH_2(dppe)_2$ or $RuH_2(PPh_3)(C_6H_6)$ could be prepared. Adding PTA (PTA=1,3,5-triaza-7-phosphaadamantane) to the THF solution of **I** under reflux gives a very reactive complex, presumably $Ru(H_2O)_2(PTA)_3(tos)_2$, in more than 80% yield. This complex can be one of the intermediates in the hydrogen transfer reduction of aldehydes catalyzed by $RuCl_2(PTA)_4$. In this[11] catalytic system a ruthenium species containing three PTA ligand proved to be the most active catalyst. Starting from **I** such complex can be prepared without the application of extraction techniques.

SCHEME 2. Reaction of $Ru(H_2O)_6(tos)_2$ with phosphine ligands

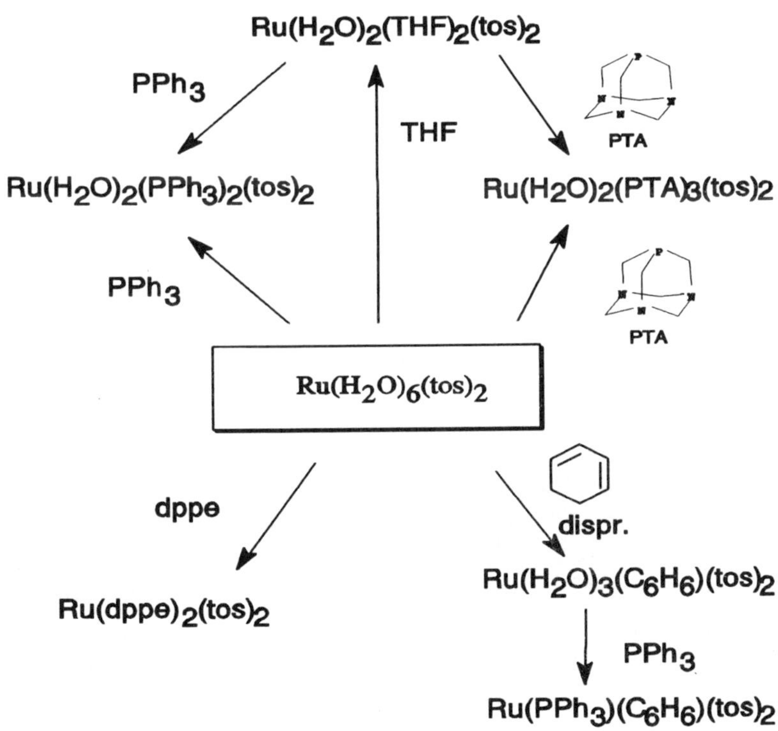

Reaction of $Ru(H_2O)_6(tos)_2$ with some heterocyclic nitrogen donors have also been studied[12] (Scheme 3). Catalytic activities for the pyridazin or pyrazine complexes weren't reported. According to our studies the reaction between **I** and neocuproine hydrate (neocuproine=2,9- dimethyl-1,10-phenanthroline) in refluxing THF gives $Ru(H_2O)(neocuproine)_2(tos)_2$.

SCHEME 3. Reaction of $Ru(H_2O)_6(tos)_2$ with nitrogen donor ligands

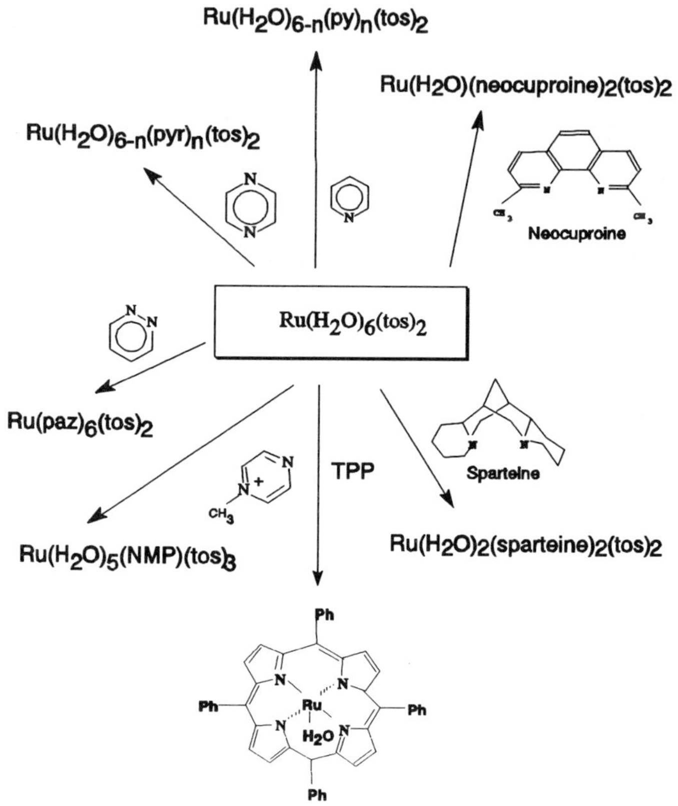

The X-ray structure of this neocuproine complex (Figure 1) shows that one of the axial positions around the ruthenium center is occupied by a tosylate ion. This structure is in sharp contrast to that of other known ruthenium aqua complexes[3,14,15,16] where the tosylate group is only a counter ion and fixed in the lattice via hydrogen bonded to coordinated water molecules. According to the X-ray structure, the neocuproine complex is better to formulate as $[Ru(H_2O)(neocuproine)_2(tos)]^+[tos]^-$. Hexaaquaruthenium(II) was reacted with TPP (TPP=5,10,15,20-tetraphenyl-21H,23H-porphine) in THF, too. The resulting $Ru(H_2O)(TPP)$ complex was characterized by IR and NMR methods.

FIGURE 1. X-ray structure of $[Ru(H_2O)(neocuproine)_2(tos)]^+[tos]^-$

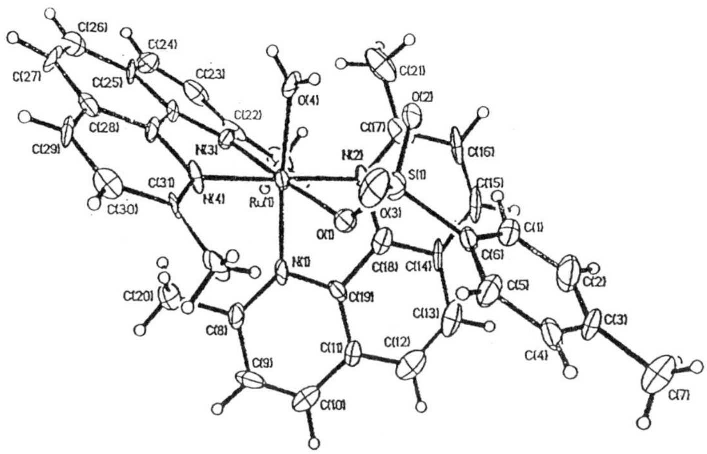

Based on these studies $Ru(H_2O)_6(tos)_2$ offers a convenient entry into the vast chemistry of ruthenium phenantroline and porphine complexes. According to our preliminary results[13] nitric acid oxidation of $Ru(H_2O)(neocuproine)_2(tos)_2$ shows interesting kinetic behavior *i.e.* autocatalysis, similarly to the analogous iron complex. Reaction of **I** with the chiral nitrogen donor sparteine gives a black precipitate, presumably $Ru(H_2O)_2(sparteine)_2(tos)_2$. This compound or the *in situ* prepared complex catalytically active in hydrogenation and hydrogen transfer reactions.
Ruthenium hexaaqua(II) reacts with dienes as well. In the presence of excess cyclohexadiene the benzene complex $Ru(\eta^6\text{-}C_6H_6)(H_2O)_3(tos)_2$ is formed[14] while the diene disproportionates. In the presence of cyclooctadiene $Ru(H_2O)_4(cod)(tos)_2$ is synthesized[14]. Ruthenium complexes of functionalized olefins can also be prepared[16]. Similar olefin complexes are intermediates or products in olefin isomerization[17] or ring opening polymerization[18] reactions catalyzed by $Ru(H_2O)_6(tos)_2$.

4. Catalytic activity of $Ru(H_2O)_6(tos)_2$ and its derivatives

We studied the catalytic activity of $Ru(H_2O)_6(tos)_2$ and some of its derivatives in the hydrogenation of carbon-carbon double bonds as well as hydrogen transfer reduction of carbonyl compounds. The catalytic reduction of unsaturated carboxylic acids is quite slow at 333K and under 1 atm of H_2 (Table 1). The reaction was studied in wide pH range (pH = 3-12). In acidic medium the reaction did not go while at pH above 10 the system turned to be heterogeneous according to the mercury test.

TABLE 1. Catalytic hydrogenation of C=C bonds with $Ru(H_2O)_6(tos)_2$

Substrate	10^5 initial rate / Ms^{-1}
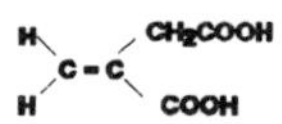	2.2[a]
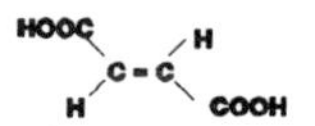	2.5[a]
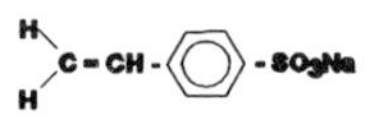	1.2[a]
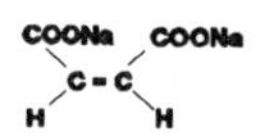	0.16[b]
	0.49[b]
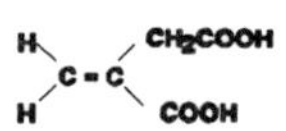	1.18[b]
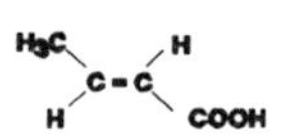	3.68[b]

Conditions: 333K, 1 atm, H_2, [substrate]=50mM, [Ru]=1.8mM, a: in water, b: in 0.067M Na_2SO_4.

The initial rate of the hydrogenation dropped significantly when 0.067M Na_2SO_4 was applied as medium instead of pure water. In the presence of added ligands **I** shows catalytic activity in hydrogen transfer reduction of benzaldehyde when water solution of sodium formate is the hydrogen donor (Eq. 1 and Figure 2). In the presence of TPPMS (TPPMS=*meta*-sulphonato-phenyl-diphenylphosphine) the reaction is very fast similarly to the $RuCl_2(TPPMS)_3$ containing system[19] while in the presence of PTA or sparteine the catalytic reduction is slow.

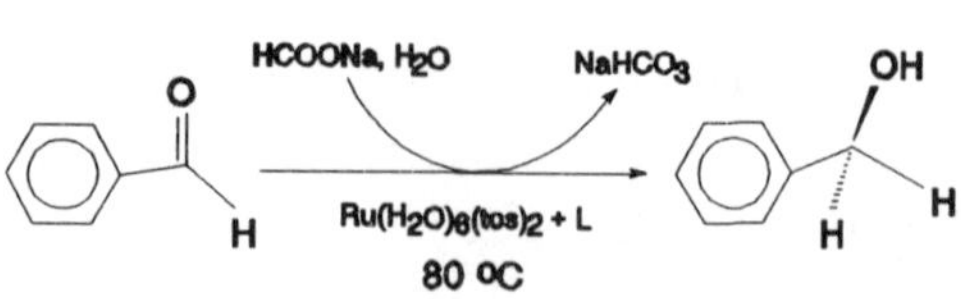

(1)

FIGURE 2. Catalytic activity of $Ru(H_2O)_6(tos)_2$ in the presence of added ligands

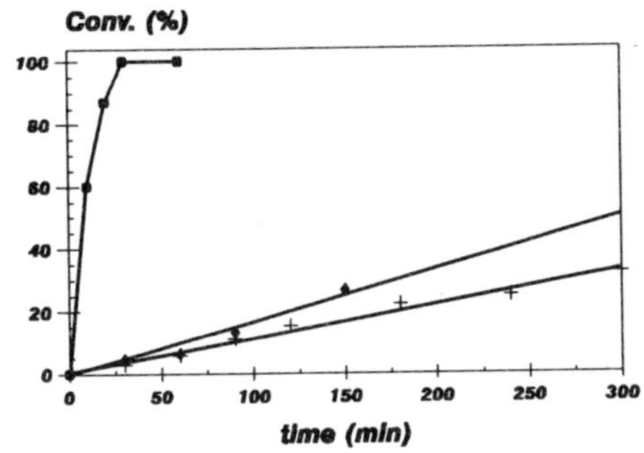

Conditions: 5mL chlorobenzene, 3ml of 5M HCOONa, 0.02mmol $Ru(H_2O)_6(tos)_2$, 2.5mmol benzaldehyde, 353K, GC conversion; [Ru]:[L]=1:10 for TPPMS (■), 1:3 for PTA (+) and 1:2 for sparteine (♦).

According to the results summarized in Table 2, the sparteine complex is quite a good catalyst for the hydrogen transfer reduction of acetophenone when *i*-propanol is the reducing agent and KOH is the promoter (Eq. 2). Addition of water or excess KOH inhibited the reduction.

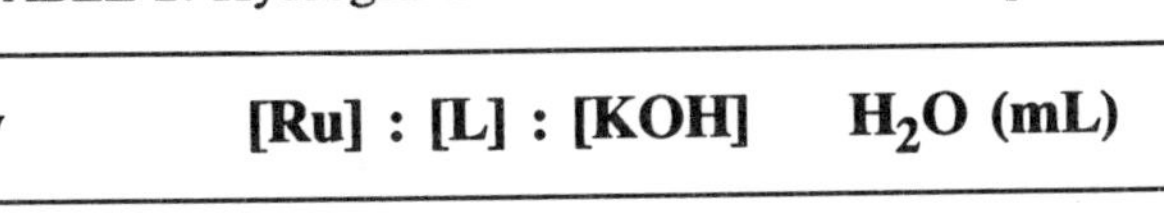

(2)

TABLE 2. Hydrogen transfer reduction of acetophenone.

Entry	[Ru] : [L] : [KOH]			H_2O (mL)	Yield(%)
1.	1	-	11		8
2.	1	3	11	-	44
3.	**1**	**4**	**11**	-	**72**
4.	1	3	22	0.5	7.3
5.	1	3	22	-	7.8
6.	1	3	11	3	0
7.	1	3	22	3	2
8.	3	9	44	-	18.7

Conditions: 13mL *i*-PrOH, [Ru]=1.2mM, [Ru]:[acetophenone]=1:140, reflux, 4 hours, HPLC yield.

The same complex is active in hydrogenation of itaconic acid with molecular hydrogen. However, in this case the catalytic reaction is also slow and induction periods of bad reproducibility were observed.

5. Reaction of $Ru(H_2O)_6(tos)_2$ with various ligands

Kinetics of substitution reaction of $Ru(H_2O)_6(tos)_2$ in the presence of various ligands was studied[20] by NMR and UV-VIS methods (Scheme 4). To prevent the hydrolysis of hexaaquaruthenium(II) the medium was 0.1 M *p*-toluenesulfonic acid solution in most cases. The ionic strength was approximately 0.5M, in some studies even 5M[7]. It is interesting to note, that $Ru(H_2O)_6(tos)_2$ reacts with dinitrogen[21], too and the reaction with carbon monoxide[22] or ethylene[23] is surprisingly slower then expected for soft ruthenium(II) in such a hard environment what water molecules can offer.

SCHEME 4. Reaction of $Ru(H_2O)_6(tos)_2$ with different ligand

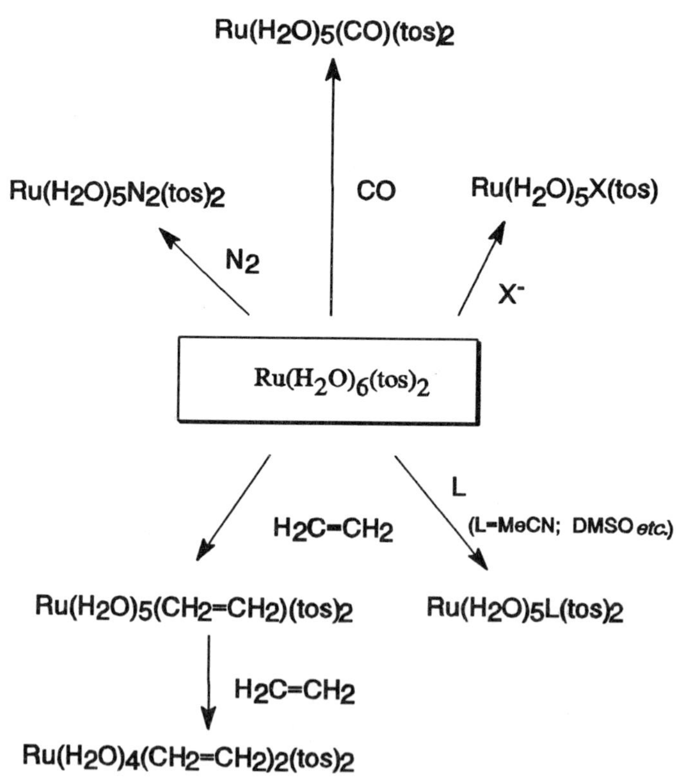

6. Stability of $Ru(H_2O)_6(tos)_2$ in water

It is well known[1,3], that $Ru(H_2O)_6(tos)_2$ reacts with various counter ions *e.g.*, ClO_4^- is reduced into Cl^-. It is one of the reasons why ligand substitution or water exchange reaction of hexaaquaruthenium(II) was studied in the presence of large excess of *p*-toluenesulfonic acid or trifluoromethanesulfonic acid. On the other hand, without these additives, ruthenium(II) is readily oxidized into ruthenium(III). A sudden pH drop can be observed when $Ru(H_2O)_6(tos)_2$ is dissolved in pure water (Figure 3). There is a significant increase in the absorbance at 300 nm, too.

FIGURE 3. Change of pH and A_{300} when $Ru(H_2O)_6(tos)_2$ is dissolved

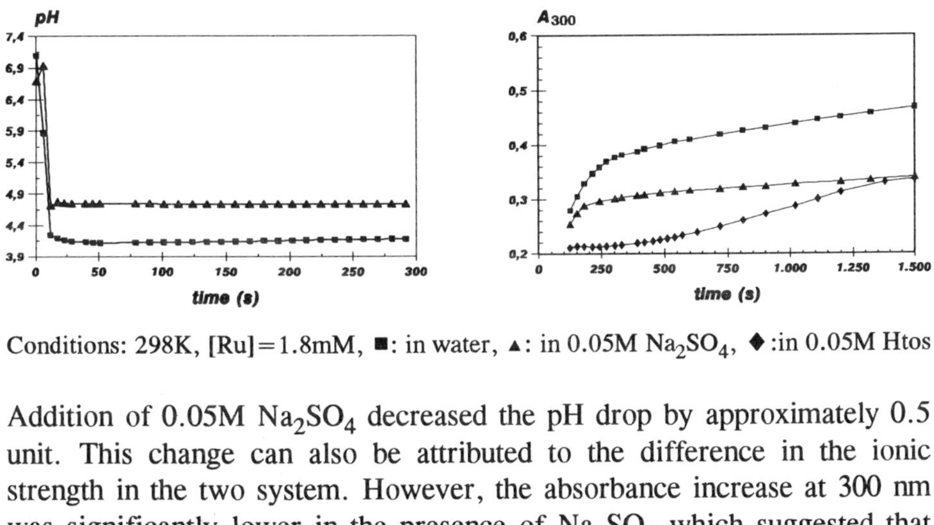

Conditions: 298K, [Ru]=1.8mM, ■: in water, ▲: in 0.05M Na_2SO_4, ◆:in 0.05M Htos

Addition of 0.05M Na_2SO_4 decreased the pH drop by approximately 0.5 unit. This change can also be attributed to the difference in the ionic strength in the two system. However, the absorbance increase at 300 nm was significantly lower in the presence of Na_2SO_4 which suggested that SO_4^{2-} ions can prevent at least some of the hydrolytic processes of $Ru(H_2O)_6^{2+}$.

TABLE 3. Reduction of itaconic acid in various medium

Medium	10^5 initial rate / Ms^{-1}
H_2O	3.7
0.067M Na_2SO_4	1.7
0.2M Na-tos	2.4
0.2M H-tos	1.8
0.067M H_2SO_4	1.4

Conditions: 333K, 1atm, H_2, [Ru]=3.6mM, [S]=50mM.

We tried to prove and utilize this possible stabilizing effect of Na_2SO_4 in catalytic reactions. However, in the reduction of itaconic acid the initial rate of hydrogenation decreased significantly when Na_2SO_4, H_2SO_4, Natos or Htos was applied as added electrolyte (Table 3).

Possible reactions when $Ru(H_2O)_6(tos)_2$ is dissolved in water is shown on Scheme 5. Traces of oxygen can oxidize ruthenium(II) into ruthenium(III). Under the conditions of catalytic reactions even contaminants of the substrate olefin can oxidize the catalyst quantitatively. $Ru(H_2O)_6{}^{2+}$ easily disproportionates and precipitation of metallic ruthenium can shift this equilibrium. If minute amount of base (KOH) was added to the water solution of hexaaquaruthenium(II) a black precipitate is formed which can be dissolved in dilute HCl. Other deprotonated aquo complexes of ruthenium also tend to aggregate[24]. Formation of μ-oxo complexes and/or reduction of water by $Ru(H_2O)_6{}^{2+}$ can't be excluded either.

SCHEME 5. Possible hydrolytic reactions of $Ru(H_2O)_6(tos)_2$

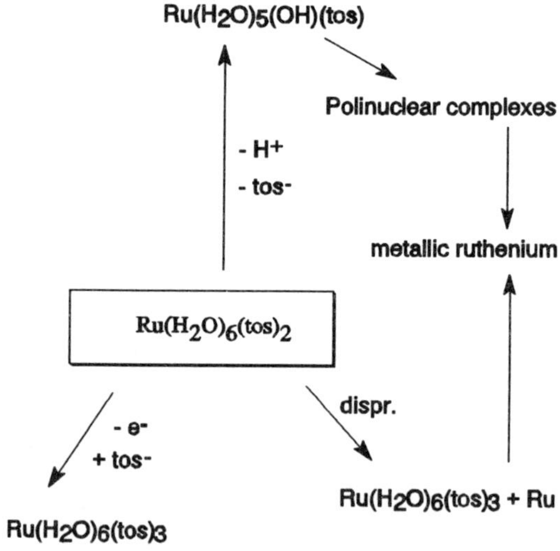

7. Summary

Reactions of $Ru(H_2O)_6(tos)_2$ with the small water soluble phosphine, PTA (PTA=1,3,5-triaza-7-phosphaadamantane) as well as various nitrogen donor ligands were studied. Catalytic activity of **I** and complexes with PTA or nitrogen donor ligands found to be low or moderate in hydrogenation or hydrogen transfer reactions. Stability of **I** in water solution was also investigated. According to these results the pH drop or absorbance increase at 300 nm, which were observed upon dissolution of **I** in water, can be decreased by addition of Na_2SO_4 to the solution. However, initial rate of catalytic hydrogenation reactions was significantly decreased when Na_2SO_4 was added to the reaction medium.

Further studies to check the catalytic activity of $Ru(H_2O)_6(tos)_2$ and its derivatives in other reactions as well as to clear the role of different hydrolytic pathways are under progress.

8. Acknowledgement

The author is grateful to Prof. F. Joó for his continuous interest in the research project and the helpful discussions. Generous loan of ruthenium from Johnson Matthey is greatly appreciated. Financial support provided by the Hungarian Research Foundation (OTKA Grant No. F4021) is acknowledged. Special thanks to Prof D.J. Darensbourg and J.H. Reibenspies for the X-ray measurements and S. Lehel for technical assistance.

9. References

1. Mercer, E.E. and Buckley, R.R. (1965) Hexaaquoruthenium(II) *Inorganic Chemistry* **4**, 1692-1695.
2. Kallen, T.W. and Earley, J.E. (1971) Aquoruthenium(II) catalysis of substitution on aquoruthenium(III) *Inorganic Chemistry* **10**, 1149-1151.
3. Bernhard, P., Bürgi, H.B., Hauser, J., Lehmann, H. and Ludi, A. (1982) Synthesis and crystal and molecular structures of hexaaquaruthenium(II) *p*-toluenesulfonate and hexaaquaruthenium(III) *p*-toluenesulfonate, $[Ru(H_2O)_6](C_7H_7SO_3)_2$ and $[Ru(H_2O)_6](C_7H_7SO_3)_3$.3H2O *Inorganic Chemistry,* **21**, 3936-3941.
4. Bernhard, P., Biner, M. and Ludi, A. (1990) A modified synthesis of ruthenium(II) hexaaqua salts *Polyhedron,* **9**, 1095-1097.
5. Bernhard, P., Lehmann, H. and Ludi, A. (1983) Ruthenium aqua ions - their comparison to ruthenium ammines *Comments Inorganic Chemistry*, **2**, 145-156.
6. Bernhard, P. and Ludi, A. (1984) Infrared and raman spectra of the hexaaqua-ruthenium ions: Normal-coordinate analysis for $Ru(H_2O)_6{}^{2+}$ and $Ru(H_2O)_6{}^{3+}$ *Inorganic Chemistry* **23**, 870-872.
7. Bernhard, P., Helm, L., Rapaport, I., Ludi, A. and Merbach, A.E. (1984) The rate of the hexa-aquaruthenium(II)/(III) self exchange reaction: ^{17}O and ^{99}Ru N.m.r. spectroscopic studies *J.C.S. Chemical Communications*, 302-303.
   Bernhard, P., Helm, L., Ludi, A. and Merbach, A.E. (1985) Direct measurement of a prominent outer-sphere electron self-exchange: kinetic parameters for the

hexaaquaruthenium(II)/(III) couple determined by oxygen-17 and ruthenium-99 NMR *Journal of the American Chemical Society* **107**, 312-317.

8. Rapaport, I., Helm, L., Merbach, A.E., Bernhard, P. and Ludi, A. (1988) Variable-temperature and variable-pressure NMR kinetic study of solvent exchange on $Ru(H_2O)_6^{3+}$, $Ru(H_2O)_6^{2+}$, and $Ru(CH_3CN)_6^{2+}$ *Inorganic Chemistry* **27**, 873-879.
9. Cotton, F.A. and Wilkinson, G. (1988) Advanced inorganic chemistry (5th edition), John Wiley & Sons, New York, pp. 884-885.
10. Bailey, O.H. and Ludi, A. (1985) Preparative uses of hexaaquaruthenium(II): Synthesis of phosphine complexes *Inorganic Chemistry*, **24**, 2582 -2585.
11. Darensbourg, D.J., Joó, F., Kannisto, M., Kathó, Á., Reibenspies, J.H. and Daigle, D.J. (1994) Water-soluble organometallic compounds. 4. Catalytic hydrogenation of aldehydes in an aqueous two-phase solvent system using a 1,3,5-triaza-7-phosphaadamantane complex of ruthenium *Inorganic Chemistry*, **33**, 200-208.
12. Bernhard, P., Lehmann, H. and Ludi, A. (1981) Synthesis and properties of substituted ruthenium aqua complexes *J.C.S. Chemical Communication*, 1216-1217.
13. Nagy, I.P., Szilágyi, E. and Bényei, A.C.,(1994) unpublished results.
14. Stebler-Röthlisberger, M., Hummel, W., Pittet, P.A., Bürgi, H.B., Ludi, A. and Merbach, A. E. (1988) Triaqua(benzene)ruthenium(II) and triaqua(benzene)osmium(II): synthesis, molecular structure and water-exchange kinetics *Inorganic Chemistry*, **27**, 1358-1363.
15. Kölle, U., Flunkert, G., Görissen, R, Schmidt, M.U. and Englert, U. (1992) Preparation, ligand exchange reactions and structure of a diene(aqua)ruthenium complex $[(cod)Ru(H_2O)_4](OTs)_2$ *Angewandte Chemie International Edition English*, **31**, 440-442.
16. McGrath, D.V. and Grubbs, R.H., (1991), Aqueous ruthenium(II) complexes of functionalized olefins: The X-ray structure of $Ru(H_2O)_2(\eta^1(O):\eta^2(C,C')$-$OCOCH_2CH{=}CHCH_3)_2$ *Journal of the American Chemical Society* **113**, 3611-3613.
17. Karlen, T. and Ludi, A. (1992) Isomerization of olefins catalyzed by the hexaaquaruthenium(2+) ion *Helvetica Chimica Acta*, **75**, 1604-1606.
    McGrath, D.V. and Grubbs, R.H., (1994), The mechanism of aqueous ruthenium(II) catalyzed olefin isomerization *Organometallics* **13**, 224-235.
18. Novak, B.M. and Grubbs, R.H., (1988) Catalytic organometallic chemistry in water: The aqueous ring-opening metathesis polimerization of 7-oxanorbornene derivatives *Journal of the American Chemical Society* **110**, 7542-7543.
    Mühlebach, A., Bernhard, P., Bühler, N., Karlen T. and Ludi, A. (1994) Ring opening metathesis polymerization of bicyclo[2.2.1]hept-2-ene (2-norbornene) and *exo,exo*-5,6-bis(methoxycarbonyl)-7-oxabicyclo [2.2.1]hept-2-ene using Ru(II) and Ru(III) complexes: Polymerization kinetics and ruthenium content in polymers *Journal of Molecular Catalysis*, **90**, 143-156.
19. Bényei, A.C. and Joó, F., (1990) Organometallic catalysis in aqueous solutions: The biphasic transfer hydrogenation of aldehydes catalyzed by water soluble phosphine complexes of ruthenium, rhodium, and iridium *Journal of Molecular Catalysis,* **58**, 151-163.
20. Aebischer, N., Laurenczy, G., Ludi, A. and Merbach, A.E. (1993) Monocomplex formation reactions of hexaaquaruthenium(II): A mechanistic study *Inorganic Chemistry*, **32**, 2810-2814.
    Patel, A., Leitch, P. and Richens, D.T. (1991) Kinetic studies of the complexation

of oxalate to the t_{2g}^6 hexaaqua ions of ruthenium(II) and ruthenium(III) *J.C.S. Dalton Transaction*, 1029-1036.

21. Laurenczy, G., Helm, L. and Merbach, A.E. (1991) The binding of dinitrogen to ruthenium(II) in aqueous solution *Inorganica Chimica Acta*, **189**, 131-133.
22. Laurenczy, G., Helm, L., Ludi, A. and Merbach, A.E. (1991) Quantitative formation in water of $[Ru(CO)(H_2O)_5]^{2+}$ from hexaaquaruthenium(II) and carbon monoxide *Helvetica Chimica Acta*, **74**, 1236-1238.
23. Laurenczy, G. and Merbach, A.E. (1993) Aqueous catalytic dimerisation of ethylene: Characterization of the reaction intermediates $Ru(CH_2{=}CH_2)(H_2O)_5(tos)_2$ and $Ru(CH_2{=}CH_2)_2(H_2O)_4(tos)_2$ (tos=toluene-p-sulfonate) *J.C.S. Chemical Communication*, 187-189.
24. Hung, Y., Kung, W.J. and Taube, H. (1981) Aquo chemistry of monoarene complexes of osmium(II) and ruthenium(II) *Inorganic Chemistry*, **20**, 457-463.

RUTHENIUM-CATALYZED OXIDATION OF CHLORINATED HYDROCARBONS IN AQUEOUS MEDIUM

M. BRESSAN
Università G.D'Annunzio, Facoltà di Scienze
via dei Vestini, 66100 Chieti Scalo, Italy

Ruthenium (II) complexes are effective catalysts for the deep oxidation of a variety of chlorinated hydrocarbons at ambient temperature and in the liquid phase. Factors such as solvent, substrate, oxidant and concentration affect the reactions. The substrates examined are a representative series of chloroaliphatics, of α-chlorinated olefins and of chlorinated benzenes. When the reactions are carried out in strongly polar media, such as water, in the presence of a surfactant agent and of persulfate degradations of the substrates are definitely faster: for many α-chlorinated olefins rates in excess of 100 turnover/min are observed.

The request of suitable methods for the quantitative mineralization of chlorinated organics continues to be a major question of environmental chemistry research. Liquid-phase catalysts are close models to the monooxygenase enzymes used by the biological systems for the oxidation of a wide variety of xenobiotics, including chlorinated organics [1], and could be a gentle alternative to the conventional methods of destruction based on gas-phase oxidation.

As a continuation of our previous studies on ruthenium catalysts for the oxidation of organic substrates [2], we report here on the effective deep oxidation of a number of chlorinated hydrocarbons by $KHSO_5$ in the presence of the dimethylsulfoxide-'solvated' $[Ru(H_2O)_2(Me_2SO)_4](BF_4)_2$ ion, a convenient chlorine-free and water-soluble precursors of ruthenium tetroxide. Ruthenium tetroxide is a multipurpose oxidizing agent and probably the reagent of choice for the oxidative fission of aromatic nuclei, which surprisingly proceeds as rapidly as the oxidation of olefinic substances. Other secondary oxidants were tested, but the best catalytic performances were obtained with mono-persulfate. Hypochlorite also showed good reactivity, but it is a source of hypochlorous acid, a powerful chlorinating agent.

Chlorobenzene is only slightly affected by the oxidation with the conventional ruthenium trichloride-hypochlorite system in water-carbon tetrachloride (or dichloromethane) double-phase, with less than 10% of the substrate being oxidized after 100 h reaction at room temperature. The initial oxidation experiments with aqueous oxone (≤500 mM in mono-persulfate), in the presence of 10 mM solutions of the ruthenium catalysts utilized

I. T. Horváth and F. Joó (eds.), Aqueous Organometallic Chemistry and Catalysis, 173–175.

dichloromethane as the organic solvent and showed complete disappearance of the substrate (100 mM) within hours. Carbon dioxide and inorganic chlorine (hydrochloric acid and/or dichlorine) were formed, together with minor amounts of chloroform and other oxidation products, among which we were able to identify formic acid. The evolution of inorganic chlorine was found to exceed by ca. two-fold the amount expected on the basis of the quantity of substrate destroyed, a clear indication that extensive oxidation of dichloromethane takes place in the conditions used.

To avoid the above undesirable side-reactions, we replaced dichloromethane with nitromethane, a non-chlorinated and non-easily oxidizable solvents alternative to acetonitrile, which causes massive salt precipitation from relatively concentrated oxone solutions. In nitromethane the oxidative fission proceeds definitely faster than in the chlorinated solvents by at least one order of magnitude. The reaction can be also performed in the water phase only, by simply adding chlorobenzene to aqueous solutions of mono-persulfate in the presence of the water soluble catalysts and of suitable amounts of a surfactant agent, like cetyltrimethylammoniun hydrosulfate [3], which effectively 'dissolves' the organic substrate. The oxidation rates became comparable to those of the reactions conducted in nitromethane. The investigations were typically conducted with $KHSO_5$ 50 mM in aqueous medium (pH ca. 3) in the presence of the catalyst (0.1 mM), of cetyltrimethylammonium hydrosulfate (usually 2.5% w/w) and the substrates (10 mM).

The measured amounts of inorganic chlorine formed during the oxidation of chlorobenzene in the presence of chlorine-free catalysts and solvents indicate quantitative dechlorination of the substrate (>95%). The amount of carbon dioxide produced was constantly smaller than the chlorobenzene destroyed (ca 85%) and indeed examinations of the acidic aqueous phase gave clear indication for the presence of formic acid (ca. 2%) and other unidentified organic products. No evidence of carbon monoxide was obtained. Highly disactivated (and highly hindered) polychlorobenzenes are unaffected by the oxidation and only 1,3-dichlorobenzene, the only one to possess a relatively reactive and accessible ring position (i.e. position 5), exhibited moderate sensitivity to the oxidative fission. Oxidation completely failed for 1,2- and 1,4-dichlorobenzene and for other hyperchlorinated benzenes.

Among simple benzene-derivatives, nitrobenzene is the most resistant to the oxidation: yet, to our knowledge, no oxidizing system is reported to be able to effectively degrade this recalcitrant substrate. Instead, when the reaction is conducted in nitromethane or in water, a slow but significant degradation took place with mono-persulfate, leading to the complete disappearance of nitrobenzene within ca 24 h. The organic nitro-group is quantitatively (ca. 95%) converted into nitrate ions and the organic moiety into CO_2 (ca. 85%).

Our results on the oxidation of tetra-, tri- and cis-dichloroethylene, of 1,1'-dichloropropene and of 2-chloro(bromo)-2-butene by the ruthenium-persulfate system show that the resistance to the oxidation roughly depends on the degree of chlorination, by being the conversion rate of tetrachloroethylene to carbon dioxide and hydrochloric acid smaller by almost two order of magnitude with respect to trichloroethylene (the latter and the other examined substrates also yield substantial amounts of formic or acetic acid and other minor products).

In the same conditions the reactions also proceed at satisfactory, although lower, rates when oxone is replaced by sodium hypochlorite. To the contrary the presence of an apolar organic phase like carbon tetrachloride or dichloromethane results in a dramatic decrease of the rates of the oxidation of the α-chloro-olefins up to three orders of magnitude. In the reaction conditions also these chlorinated solvents are oxidized, although very slowly. Notice that in the aqueous emulsion conditions described above oxidations of the substrates proceed at exceedingly high rates even in the absence of the ruthenium catalyst, contrary to what observed with the double phase systems, where the presence of the catalyst is imperative.

During the oxidation, i.e. with excess oxidant, the dominant species for chlorine is Cl_2, in agreement with the large E° value (1.8 V) reported for the HSO_5^-/HSO_4^- couple [4]. Cl_2 is rapidly replaced by HCl when the persulfate is consumed: at this point the solutions are intensely violet-colored and contain a strongly reducing ruthenium(IV) derivative, which can be considered the 'resting' state of the catalyst.

The positive role played by the surfactant agents may depend on several factors, among which the orientation of the reactants induced by the formation of the micelles could be decisive: one can speculate that the anionic oxidant can be strongly bound to the cationic head groups of the surfactant, whereas the hydrophobic substrates may undergo comicellization.

Finally, the fact that mono-persulfate is more active than hypochlorite may suggest that an oxidizing species other than ruthenium tetroxide is formed, tentatively a peroxometal species [M-O-O-SO_3H], as recently proposed by Meunier for related metalporphyrinate-catalyzed oxidations [5]. Inspection of the reaction mixtures by UVV spectroscopy gave however no evidence of detectable quantities of chromophores other than ruthenium tetroxide.

The method described in this paper presents several advantages because it requires easily available reagents and it readily occurs under extremely simple experimental conditions, in high yield and at very low substrate concentration.

References

1. Sigel H. and Sigel, A. (1992) *Metal ions in Biological Systems (Vol.. 28)*, Marcel Dekker, New York.
2. Bressan M. and Morvillo A. (1988) *J. Chem. Soc. Chem. Commun..*, 650; (1989) *Ibid.*, 421; (1989) *Inorg. Chem.*, **28**, 950; Bressan M., Morvillo A. and Romanello G. (1990) *Ibid.*, **29**, 2976; Bressan M., Forti L. and Morvillo A. (1994) *J. Chem. Soc. Chem. Commun.* , 253 .
3. Fringuelli F., Germani R., Pizzo F., Santinelli F. and Savelli G. (1992) *J. Org. Chem.* **57**, 1198 and ref. therein.
4. Meunier B. (1992) *New J. Chem.* **16**, 203.
5. Hoffmann P., Robert A. and Meunier B. (1992) *Bull. Soc. Chim. Fr.*, **129**, 85.

HYDROSOLUBLE LIGANDS FOR A NEW TECHNOLOGY

E.G. KUNTZ

C.P.E. LYON

Bd du 11 novembre 1918

69100 VILLEURBANNE FRANCE

1 - Introduction

Since 1974 the most widely studied water soluble phosphines are those with sulfonated phosphines , the sodium salts of :

* metasulfophenyl-diphenylphosphine (TPPMS)
* tri-(metasulfophenyl) phosphine (TPPTS)

In 1984 Ruhrchemie, using TPPTS and Rhone Poulenc laboratory results as a base, developed an industrial process for propene hydroformylation.

In 1988 Rhone Poulenc using TPPTS developed a new industrial process for the synthesis of vitamin E.

En 1992 Kuraray annonced the first commercialization of butadiene hydrodimerisation using a palladium catalyst modified by a phosphonium salt of TPPMS.

2 - Catalyst separation

Many different laboratory reactions are catalysed by transition metals. It's easy to separate the catalyst in heterogeneous systems, and even in homogeneous systems it is often economic to "lose" the catalyst. However, when the catalyst is expensive or toxic, its separation from the reaction medium becomes important. One can remove products from the catalyst by distillation, by addition of a non miscible solvent, by use of polymers and membranes or by separation using classical analytical methods.

The distillation of the products from the reaction medium is used in the Wacker process - the aldehyde is distilled from a reaction medium that contains the Pd-Cu catalytic system. Other examples are the Monsanto acetic acid process or the Union Carbide butanal process. There are three major conditions : high volatibility of organic products, stability of catalyst for several months, low yields of heavy products.

I. T. Horváth and F. Joó (eds.), Aqueous Organometallic Chemistry and Catalysis, 177–181.

The addition of a nonmiscible solvent after the reaction depends on the polarity of the products. The synthesis of adiponitrile by hydrocyanation of 3-pentenenitrile in the presence of nickel and arylphosphite illustrates a polar system. The high boiling point of adiponitrile precludes distillation. The catalyst is separated by liquid extraction with cyclohexane in the Dupont adiponitrile process.

When reaction products are non -polar, the catalyst must be maintained during the reaction in a nonmiscilble solvent. For example, the oligomerisation of ethylene used to prepare linear olefins is catalysed by nickel. It can be carried out in polar solvents such as butanediol, which is a poor solvent for olefins (two phase system in reactor of Shell Higher Olefins Process).

Although the nonmiscible solvent approach is realistic it has limitations : in most cases the distribution factors of reagents and catalyst between the two phases is not high enough to allow continuous separation inside the reactor. It requires an extraction column and a distillation column. In chemical speciality products, produced in a batch process, the catalyst is separated by classical analytical methods. As far as we know the two approches which consist of using organic polymers and semipermeable membranes have not been developed in any major industrial process. This is due to : degradation of polymer, leaching of metal, or poor productivity.

The use of water which is immiscible with most organic products is an obvious step forward. But to carry out homogeneous catalysis in water requires the development of a new chemistry which was unexplored in 1970.

3 - Organometallic chemistry in water with phosphines before 1976

TPPMS was prepared as early as 1958 through J. Chatt through sulfonation of triphenylphosphine . The purpose of this preparation was to study the complexation of silver and cadmium with a phosphorus compound. With silver rather high constants of formation were found at 25°C : logK 1 = 8, logK 2 = 6, logK 3 = 5

In 1973, the first attempt to carry out a transition metal catalysed reaction in an aqueous solution in the presence of hydroxyalkylphosphine was reported by J. Chatt. These systems, however, did not catalyse the hydrogenation or the hydroformylation of olefins in water.

In March 1975, a monosulfonated triphenylphosphine was reported by F. Joo and M.T. Beck. They hydrogenated pyruvic acid to lactic acid using an aqueous solution of ruthenium or rhodium with TPPMS in a one phase system.

In 1974 in Rhone Poulenc our purpose was to take out a general patent on organometallic chemistry in water involving phosphines. The general principle of two-phase catalysis with water as solvent was to be claimed. The publication of F. Joo and M.T. Beck forced us to take out one patent for each type of reaction. The first one, in June 1975, was the hydroformylation of olefins using sulfonated water soluble phosphines, principally TPPTS.

4 - Hydroformylation of propylene in water

Propylene is only slightly soluble in water. Hydroformylation of propylene in water constitutes the first published and probably the best example of biphasic organometallic catalysis for which the catalyst is maintained in the polar solvent.

Comparative experiments carried out with water show the role of TPPTS in the n/iso selectivity and the rhodium partition of the oxo reaction , at 80° C and at 40 bar pressure in an equimolar mixture of carbone monoxide and hydrogen :

* Without ligand the selectivity of rhodium salts is 45% and the reaction takes place in the organic phase.
* With the triphenylphosphine ligand, the selectivity reaches a value of 65% and the reaction takes place in the organic phase (G. Wilkinson system at a low ratio of phosphorus/rhodium.
* We should notice that in a one phase organic system like ethanol the tris (tetrabutylamnonium) salt of TPPTS gives rise to the same selectivity as TPP (65%)

* With TPPMS the selectivity reaches a value of 91% and the reaction goes on almost entirely in the water phase.

* With TPPTS the selectivity reaches a value of 96% and the reaction goes on totally in the aqueous phase.

The reasons why TPPTS is a ligand which can prevent the rhodium from going to the organic phase are triple :

1 - extremely high solubility in water
2 - extremely low solubility in organic medium
3 - extremely high complexation ability towards rhodium as it occurs with the silver. The first equilibrium constant involving the substitution of the carbon monoxide by TPPTS in the rhodium complex is necessarely very high to maintain the rhodium in water.

THIS IS THE CENTRAL ELEMENT OF THE GENERAL PRINCIPLE OF BIPHASIC CATALYSIS USING WATER AS THE SOLVENT.

As with theWilkinson catalyst, the productivity depends on the pressure of propylene, hydrogen and carbon monoxyde. As early as 1976 it was understood that the optimum temperature for such a reaction was 125° C, a temperature at which the solubility of propylene in butanal is not too high. At 125° C the productivity decreases when the phosphine concentration is higher than 300G/L. The selectivity for n aldehyde increases with phosphine concentration up to a value of 96% and a good stability of the catalytic system is obtained.

Although the complexation free energies of the rhodium with TPPTS or TPP are the same order of magnitude, the use of water permits a better selectivity, for the linear product for two reasons :

- firstly, the " Rh (TPPTS)2" complex probably gives a selectivity of 95 - 96% compared to 90 -92% with the "Rh (TPP)2 " complex because of the steric effect of the ligands.
- secondly, the low solubilities of propylene, carbon monoxide and the organic products in water promote the appropriate rhodium species having two phosphines and one carbon monoxide molecule , thereby giving better selectivities.

5 - **Telomerisation of dienes**

Telomerization of dienes is a very versatile reaction which can work with water, alcools and various compounds containing a mobile hydrogen atom. In a laboratory with butadiene and water one can obtain primary and secondary C 8 alcohols with catalyst separation. Conversion as high as 70% can be reached with a rather good regioselectivity for the terminal alcohols

In 1992, Kuraray developed the hydrodimerisation of butadiene. They have started to hydrodimerise butadien using a palladium catalyst which employs a phosphonium of TPPMS to solubilize the catalyst in a sulfolane water phase. Triethylamine is used to stabilize the catalyst and the reaction is carried out under an atmosphere of carbone dioxide. The hydrodimer is isomerized, hydroformylated and hydrogenated to yield 1,9 nonanediol.

Telomerization of isoprene with methanol is an interesting reaction to obtain isoprenic compounds. Here the strategy must be different because TPPTS is not soluble in methanol. Therefore it is necessary to use TPPMS which is soluble in methanol. With palladium, telomerization occurs giving 70% yield with a good regioselectivity in the carbon - carbon bond formation but unfortunately a bad regioselectivity in the oxygen carbon bond formation.

In water, diethylamine reacts with isoprene to give 1 - 4 adducts. The reactions observed with palladium and platinum gave a fairly good selectivity for addition at position 1 and 4 . The reaction was studied several years later by Morel and Mignani using rhodium. In this case the functionnalization is 100% selective for the positon 4.

The high regioselectivity of rhodium in the functionalization of unsymmetrical dienes was also reported by Morel from Rhone Poulenc in 1980 using ethyl acetoacetate. With isoprene or myrcene a regioselectivity of 99% for the right isoprenic compounds is achieved. In simular conditions palladium is much less selective. In theory, with a biphasic rhodium system, the catalytic species "Rh - two TPPTS, one ethyl acetylacetate" would be prevalent having a steric hindrance which isn't possible with palladium and platinum.

In 1988 Rhone Poulenc started a new synthesis of vitamin E which constitues a dramatic improvment of this synthesis. Now it is possible to start from a cheap easily available C10 compound myrcene and ethyl acetoacetate. After hydrolysis and decarboxylation, geranylacetone is obtained. Here again the use of a biphasic system proved to be the best choice for a commercial process with a costly metal.

6 - Conclusion

Homogeneous catalysis in water constitutes a very original approach to the general problem of catalyst separation. In the field of hydroformylation and telomerization it has thus been possible to design new ligands with rhodium and new catalysts which have allowed the development of industrial processes. Which exhibit better yields and selectivities.

In the last decade the biphasic principle has given rise to new studies in water : on micellar phases, on the cyclodextrin system, on colloïdal phases and on supported aqueous phase catalysts.

COMPARATIVE STUDY OF BIPHASE REDUCTION OF CARBONYL COMPOUNDS CATALYZED BY HOMOGENEOUS AND WATER SOLUBLE Ru CATALYSTS

S. ŠABATA, J. VČELÁK, and J. HETFLEJŠ
Institute of Chemical Process Fundamentals,
Acad. Sci. of Czech Republic
Rozvojová 135, 165 02 Prague 6, Czech Rep.

1. Introduction

In recent years, the extensive study was made of transition metal complexes containing ambivalent P,O ligands (review see [1]). Although their hemilability could affect favourably transition metal-catalyzed processes, much less attention has been paid to their application as catalysts, except for methanol carbonylation (cf.[2]). Our interest in poly(oxyalkylene)phosphines as potential ligands of water-soluble transition metal catalysts led us to examine also the behaviour of parent ether-phosphines $Ph_2P(CH_2)_nOR$ (n=2, R= Me, Et, Bu; n= 1,4-dioxanyl, 2-tetrahydrofuranyl). In present work, ruthenium-ether-phosphine complexes containing the above ligands were used as catalysts in the title reaction and their efficiency was compared with that of reported homogeneous [3,4] and water- soluble [5] Ru(II) complexes.

2. Results

2.1 REDUCTION OF CYCLOHEXANONE AND HEPTANAL CATALYZED BY $RuCl_2(PPh_2CH_2CH_2OCH_3)_2$ (*I*)

The reduction under phase transfer conditions (A) is first order in *I* and both substrates (Eq.(1)), similarly to that catalyzed by $RuCl_2(PPh_3)_3$ (*II*) [3]. The k´s demonstrate the efficiency of both catalysts (TABLE 1). The ether-phosphine complex *I* is best for the aldehyde reduction, especially in the presence of the excess phosphine. The activation energies (E_{act} for $(CH_2)_5C{=}O$ = 96.3 kJ mol^{-1} (*I*) and 61.9 kJ mol^{-1} (*II*)) confirm

$$RCHO_{(org)} + HCOONa_{(aq)} \text{ --------------> } RCH_2OH_{(org)} + NaHCO_{3(aq)} \quad (A)$$

Reaction conditions: see TABLE 1

I. T. Horváth and F. Joó (eds.), Aqueous Organometallic Chemistry and Catalysis, 183–185.

$$-d[S]/dt = k\,[cat_o]\,[S] \qquad (1)$$

TABLE 1. A comparison of Ru(II) complexes *I* and *II* (1.8 mmol substrate, 0.045 mmol Ru complex, 2.5 ml C_6H_5Cl, 5 ml water, 2.5 ml 2 M-HCOONa, 0.18 mmol cetylpyridinium bromide)

Ru cat.	$k\,10^2$, $l\,mol^{-1}\,s^{-1}$	
	$(CH_2)_5C{=}O^a$	$CH_3(CH_2)_5CH{=}O^b$
I	1.4	13
	0.8^c	48^c
II	9.4	3.1
	9.8^c	13^c

[a] 50 °C, [b] 60 °C, [c] 0.85 mmol of the P-ligand added.

the chemical control of this phase transfer reaction. The ether-phosphine complexes $RuCl_2(PPh_2CH_2L)_2$ (L=tetrahydrofuranyl or 1,4-dioxanyl) were less active in both cases (k´s ranged from 0.4 to 0.8 $l\,mol^{-1}\,s^{-1}$).

2.2 EFFICIENCY OF *in situ* Ru CATALYSTS

Although our attempts at preparing well defined ruthenium(II) complexes from other ether-phosphines $PPh_2(CH_2CH_2OR)$ (R= OEt, OBu-n) and C-analogue $PPh_2(CH_2CH_2CH_2CH_3)$ by procedures similar to the syntheses of *I* and the Ru(II)-dioxanyl complex [6] have failed, $RuCl_3 \cdot H_2O$ and several phosphines (when used in excess) do form efficient catalytic systems *in situ* under reaction conditions. The reaction obeys Eq.(1), the activity sequence for the aldehyde reduction being: $MeOCH_2CH_2PPh_2$ >> PPh_3 > $PPh_2(C_6H_4SO_3Na\text{-}m)$ > $BuOCH_2CH_2PPh_2$, $EtOCH_2CH_2PPh_2$ > $BuPPh_2$ > $P(i\text{-}Pr)_3$ (TABLE 2).

3. Conclusions

The ether-phosphine Ru complex *I* seems to be the catalyst of choice for the biphase reduction of saturated aliphatic aldehydes. Of synthetic interest is the possibility of its *in situ* formation under reaction conditions. Experimental activation energies of the reduction obtained with *I* indicate that the rate determining step of this phase transfer reaction is chemically controlled, similarly to the already reported reduction [4] catalyzed by the Ru-PPh_3 complex *II*. However contrast to *II*, the aldehyde reduction with *I* proceeds to a significant extent also in the absence of the phase transfer catalyst.

TABLE 2. Activity of *in situ* prepared ruthenium catalysts in the biphase reduction of heptanal (1.8 mmol heptanal, 0.045 mmol Ru compound, 0.85 mmol free phosphine, 50°C, the other conditions as in TABLE 1)

Phosphine added	k 10^2 l mol-1 s-1,[a]	Heptanal conversion ,% [b]
none	0	0
$P(C_6H_5)_3$	0 c	0
	0.7	9.2
$P(C_6H_5)_2CH_2CH_2OCH_3$	0.2 [c]	2.8
	20.4	95.5
$P(C_6H_5)_2CH_2CH_2OC_2H_5$	0.9	11.8
$P(C_6H_5)_2CH_2CH_2OC_4H_9$-n	1.1	15.5
$P(C_6H_5)_2CH_2CH_2CH_2CH_3$	0.4	5.7
$P(i\text{-}C_3H_7)_3$	0.2	3.0
$PPh_2(C_6H_4SO_3Na$-*m*)	1.4[c]	19.4

[a] Equation (1). [b] The reaction time 1 h, determined by gas chromatography, other products than n-heptanol were not detected. [c] The reaction was carried out in the absence of the phase transfer catalyst (cetylpyridinium bromide).

4. References

1. Bader, A. and Lindner, E. (1991) Coordination chemistry and catalysis with hemilabile oxygen-phosphorus ligands, *Coord. Chem. Rev.* **108**, 27- 110
2. Lindner, E., Speidel, R., Fawzi, R., and Hiller, W. (1990) Novel basic ligands for the homogeneous catalytic carbonylation of methanol. XXVIII. Synthesis, properties and fluxional behaviour of (ether-phosphine)palladium complexes, *Chem. Ber.* **123**, 2255-2260.
3. Bar, R., Sasson, Y., and Blum, J. (1984) Transition metal-catalyzed transfer reduction of saturated aldehydes and ketones by sodium formate under phase transfer conditions, *J. Mol. Catal.* **26**, 327-332.
4. Bar, R., Bar, L.K.,and Sasson, Y. (1985) Phase transfer-catalyzed reduction of aromatic aldedehydes by aqueous sodium formate in the presence of dichlorotris(triphenylphosphine)ruthenium(II): a kinetic study, *J. Mol. Catal* **33**, 161-177.
5. Bényei, A. and Joó, F. (1990) Organometallic catalysis in aqueous solutions: the biphasic transfer hydrogenation of aldehydes catalyzed by water-soluble phosphine complexes of ruthenium, rhodium and iridium, *J. Mol. Catal.* **58**, 151-163.
6. Lindner, E., Schober, R., Fawzi, R., Englert, U., and Wegner, P. (1987) Synthesis and reactivity of ruthenium(II) complexes containing ether-phosphine ligands, Chem. Ber. **120**, 1621 - 1628.

THREE APPROACHES TO CATALYTIC AQUEOUS ORGANOMETALLIC CHEMISTRY INVOLVING WATER SOLUBLE LIGANDS, SOME MODIFIED CYCLODEXTRINS AS LIGANDS, AND REACTIONS IN AN ALUMINOPHOSPHATE CAVITY

W. ROY JACKSON, JOHN R. ANDERSON, EVA M. CAMPI, CIPTATI, QUENTIN J. McCUBBIN AND ZHENGPING YANG
Department of Chemistry, Monash University
Clayton, Victoria, Australia 3168

Abstract

Three different approaches to catalytic reactions involving organometallic catalysts in aqueous media will be described all involving hydroformylation chemistry.

1. Introduction

There is much current interest in carrying out reactions of organic compounds involving organometallic catalysts in aqueous media. The use of water as a process solvent in industry is exemplified by Ruhrchemie/Rhône-Poulenc's propylene hydroformylation technology involving triphenylphosphine trisulfonate as ligand [1] [2]. In this paper we describe hydroformylations carried out using organorhodium catalysts which have been modified so as to make them suitable for use in an aqueous medium. In particular we describe reactions of water soluble organic salts which could not readily be hydroformylated in organic solvents.

2. Results and Discussion

2.1. REACTIONS INVOLVING WATER SOLUBLE LIGANDS

We have been interested in hydroformylation reactions of unsaturated amines as novel synthetic routes to nitrogen heterocycles [3] [4]. Heterocyclic formation involves the lone pair of electrons on the nitrogen atom in all cases. We have found that the final products of these reactions can arise from either direct carbonylation [5] [6] or followed by cyclisation to an imine [3] [4] (Figure 1). In order to elucidate details of the mechanism of some of these reactions we were interested in trapping aminoaldehydes (1). In addition to mechanistic interest such compounds would be valuable intermediates in synthesis. Hydroformylation of amine salts allows for the possibility of isolating aminoaldehydes (1) as their salts which are not capable of cyclisation.

The complexities which can arise from hydroformylation reactions of unsaturated amines are typified by the reaction of N-allylaniline (2), (Figure 2) which in ethyl

I. T. Horváth and F. Joó (eds.), Aqueous Organometallic Chemistry and Catalysis, 187–194.

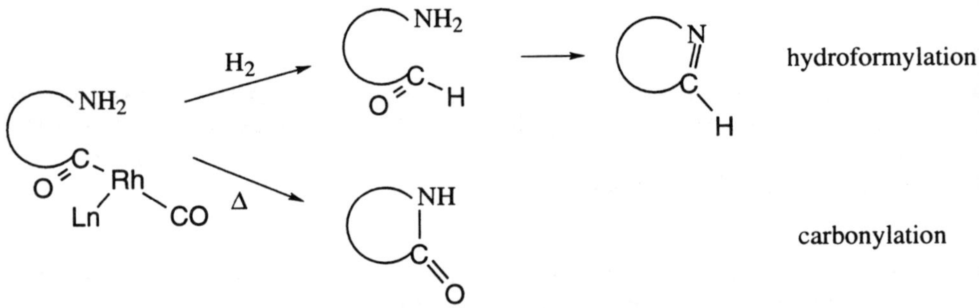

Figure 1. Heterocycle formation *via* hydroformylation or carbonylation

acetate gave the dipyrroloquinoline (5) as a mixture of diastereoisomers (80%) [3]. The formation of (5) can occur by the route outlined in Figure 2 in which the enamine (4) formed from the terminal aldehyde (3) adds to its iminium salt tautomer and the resulting

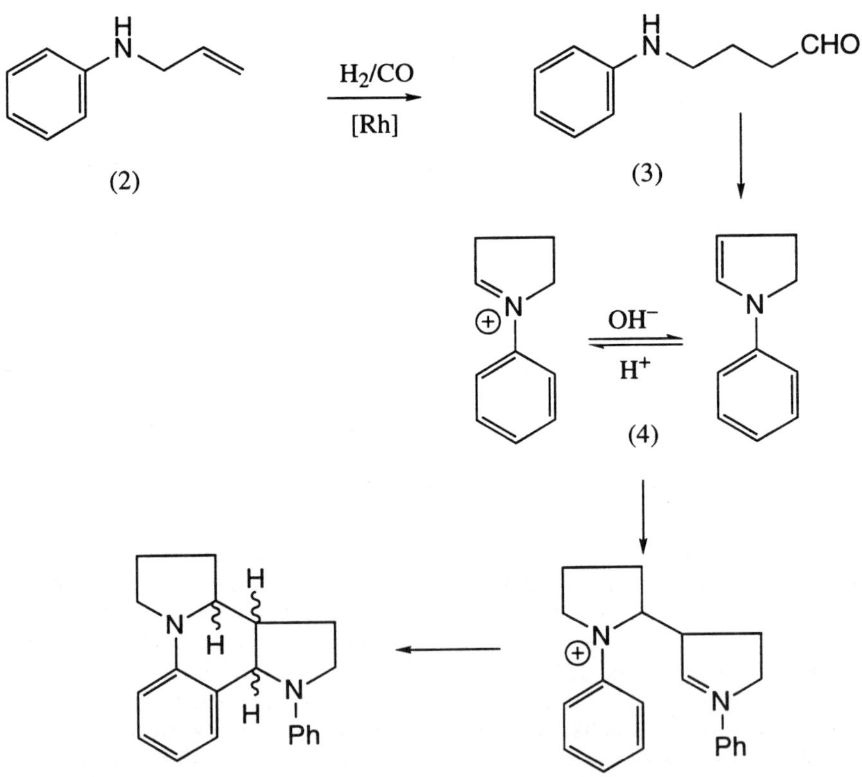

(5)

Figure 2. Hydroformylation of N-allylanine

adduct undergoes a further cyclisation. One interesting aspect of this reaction is the high regioselectivity shown in the initial hydroformylation. This high regioselectivity has also been observed in hydroformylation reactions of other N-allylbenzenamines [4].

In attempt to reduce the complexity of this reaction and to illustrate the utility of hydroformylation reactions in aqueous media we focussed our interest on the hydrochloride salt of N-allylaniline (6; R=H). Reaction of an aqueous solution of this salt with H_2/CO in the presence of $RhCl_3$ and the water-soluble phosphine ligand, tppms, surprisingly gave N-butylaniline (7; R=H) and the isomeric N-2-methylpropylaniline (8; R=H) (*ca.* 50%) as the major products together with some aniline (9; R=H) (*ca.* 35%) arising from deallylation. Similar results were obtained for the 2-methylanalogue (6; R=Me) in that the straight and branched chain isomers (7 and 8; R=Me) (*ca.* 50%) were obtained together with *o*-toluidine (9; R=Me), (30%).

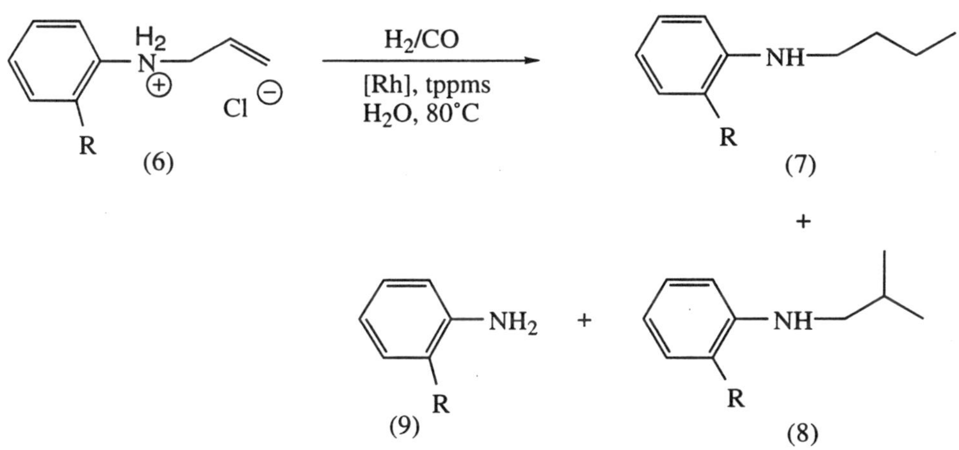

Figure 3 . Hydroformylation of N-allylanilinium salts in aqueous solution

The strategy to eliminate complex reaction sequences involving cyclisation reactions of aminoaldehydes had obviously been successful but the formationof the N-butyl (7) and N-2-methylpropylanilines (8) by a hydroformylation reduction system was most unexpected.

Reduction of the initially formed aldehyde salts to the corresponding alcohols is a not uncommon feature of hydroformylation reactions but loss of oxygen either by an elimination-reduction sequence or by direct hydrogenolysis is without direct literature precedence. However we have observed reductive loss of oxygen in the rhodium catalysed reaction of propargylamines [7]. It was of note that the product was isolated as the free amine. The point at which hydrogen chloride was lost is not clear. The ratio of linear, n to branched, i - isomers was *ca.* 80:20 not significantly different for the 70:30 value commonly obtained from rhodium catalysed hydroformylations of hex-1-ene in organic solvents. In contrast, similar reaction of the free amine, N-allylaniline in an organic solvent, ethyl acetate (as shown in Figure 2) was shown to be highly regio-selective giving products arising from the linear aldehyde only [3].

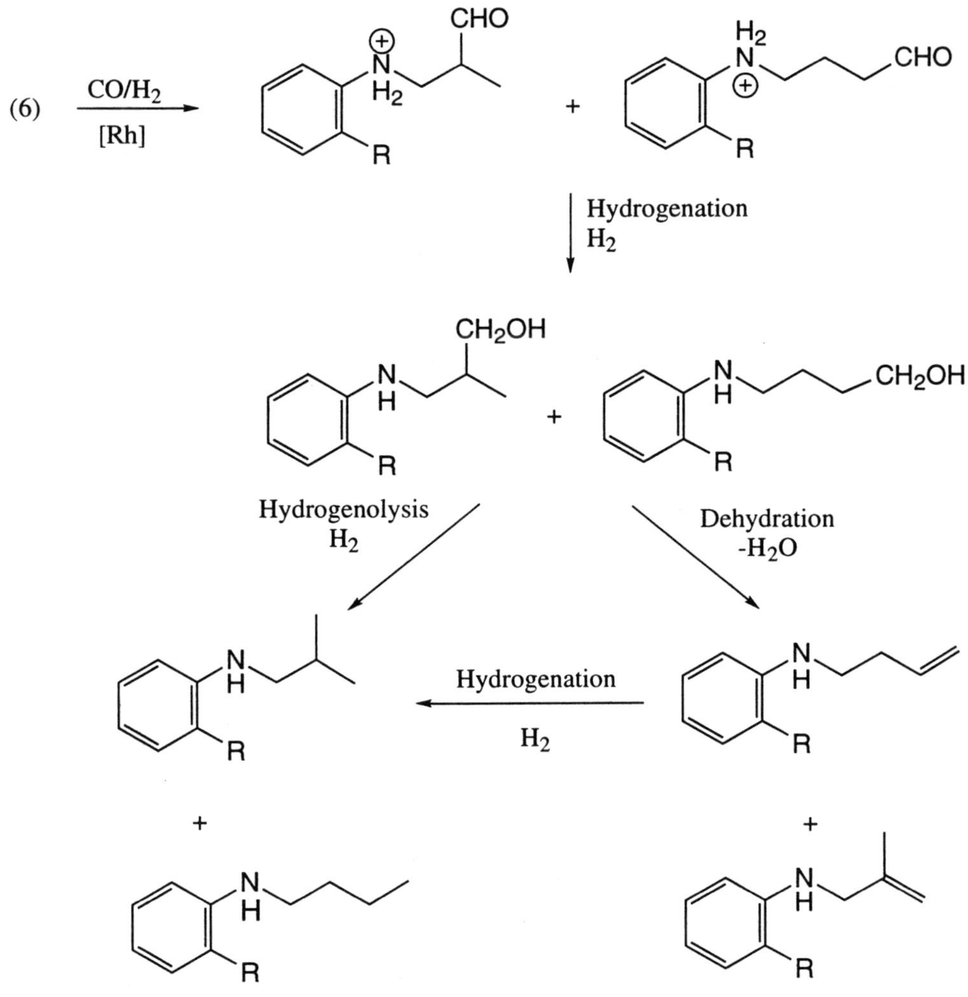

Figure 4 Possible intermediates in the formation of alkylanilines

Attempts to carry out these reactions using sodium *p*-diphenylphosphinobenzoate as the water solubilising ligands were not successful and the free phosphinoacid was found to have been precipitated during the attempted reaction.

It was of interest to see if water-soluble anionic organic substrates would also react under these conditions. Reaction of an aqueous solution of sodium 4-styrenesulfonate with H_2/CO using the $RhCl_3$ catalyst system gave what appears to be a mixture of the aldehyde (10) and alcohol (11), but these compounds have not been fully characterised.

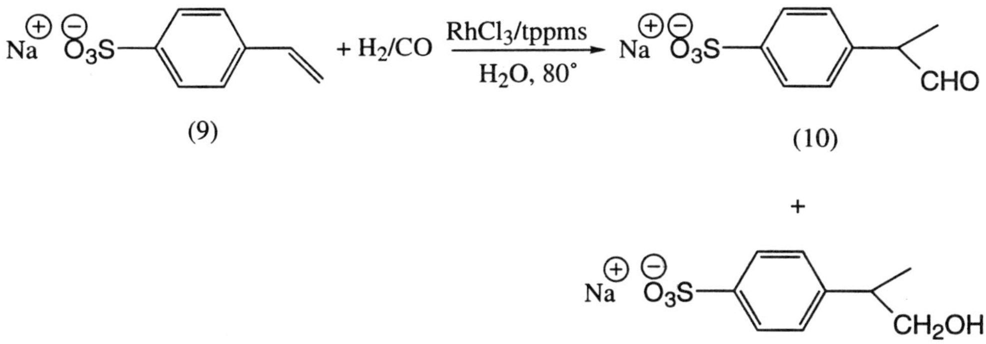

(9) (10) (11)

2.2 REACTIONS IN AN ALUMINOPHOSPHATE CAVITY

The pores of aluminosilicates and aluminophosphates have been used as sites for reactions of organic molecules [8]. The large pore structure of the recently discovered aluminophosphate, VPI-5 allows for the inclusion of large molecules e.g. buckminsterfullerene [9]. We have utilised this capacity to include organometallic compounds routinely used as homogeneous catalysts in organic synthesis and used them in reactions of water-soluble organic reactants in aqueous media. In one example we have studied the hydroformylation of N-allylaniline hydrochloride using $HRh(CO)(PPh_3)_3$. The hydrochloride (6; R=H) does not react with H_2/CO in the presence of $HRh(CO)(PPh_3)_3$ in aqueous medium either alone or when a sample of VPI-5 is also added. However, when the catalyst is included into the cavity of the VPI-5, the hydroformylation reaction proceeded to completion after 20 hours at 90°C. The only product isolated in quantitative yield was the dipyrroloquinoline (5) which we had previously isolated from reactions of the free amine in organic media (see Figure 2). The formation of (5) suggests that some free amine base must be liberated within the cavity to allow cyclisation to proceed. The origin of the basic sites are as yet unknown. The possibilities that lattice oxygen could function in this manner or that small amounts of basic impurities are present are being investigated. The contrast in the products obtained from the reactions of the anilinium salt (6; R=H) under these conditions with those described previously using $RhCl_3$/tppms as catalyst (Figure 3) highlight the sensitivity of these reactions to what appear to be relatively minor changes in conditions.

Again, it was of interest to see if the water-soluble anionic substrates, sodium 4-styrenesulfonate (9) would also react under these conditions. Preliminary studies showed that complete conversion of (9) occurred at 90°C but that the major products were polymeric in nature. Reactions at lower temperatures gave decreasing amounts of polymer together with the branched chain aldehyde (10) and related alcohol (11).

2.3 REACTIONS INVOLVING MODIFIED CYCLODEXTRINS

An alternative strategy to carrying out aqueous phase reactions of organic molecules in the pores of microporous inorganic solids involves incorporation of organic substrates into the cavities of cyclodextrins. Attempts to enhance reactivity [10] and to carry out reactions in an aqueous phase [11] have been attempted using cyclodextrins. The possibility of obtaining modified regioselectivities as a result of reaction occurring all or in part in the cyclodextrin cavity also exists.

Attempts to carry out hydroformylation of hex-1-ene using a water-soluble rhodium catalyst were retarded by the addition of α-cyclodextrin. It was envisaged that inclusion of the hexene into the cyclodextrin cavity could lead to regioselective hydroformylation at the exposed terminal end of the double bond but these expectations were not realised. A more sophisticated approach involves attachment of a rhodium species through an appropriately functionalised cyclodextrin. Copper bound in this way have been used in enantiomeric recognition of aromatic aminoacids [12]. More pertinently, two recent reports have described the preparation of modified cyclodextrins as ligands. Ito [13] has described a phosphine substituted ferrocene and Reetz [14] a phosphinoethylthio derivative which was used to prepare a cationic rhodium complex.

We have adopted a similar strategy and synthesised an isophthalamidic acid (12) and two phosphinobenzoic acid derivatives of 6-amino-6-deoxy-β-cyclodextrin (13,14).

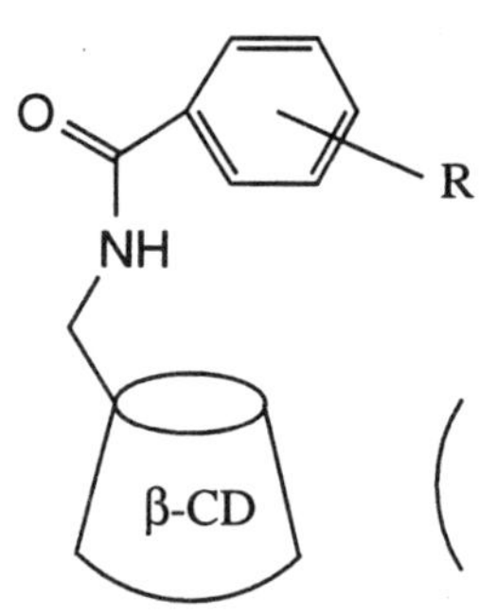

(12; R = *m*-$CONH_2$
13; R = *p*-PPh_2
14; R = *m*-PPh_2)

These ligands (13) & (14) were converted into rhodium complexes by reaction with $Rh(acac)(CO)_2$ or $Rh(nbd)_2$ and the resulting materials evaluated as catalysts for the hydroformylation of styrene in an aqueous medium. Each complex gave *ca.* 65% isolated yield of aldehydes in a ratio *ca.* 90:10;

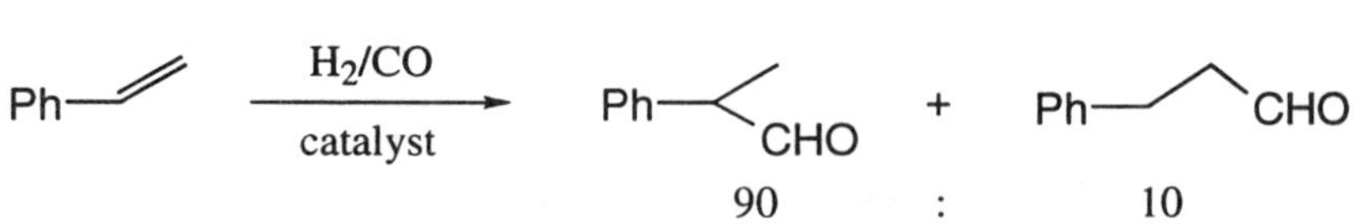

90 : 10

comparable to that obtained using rhodium complexes with phosphine ligands in an organic solvent. The branched chain aldehyde showed no significant e.e.

References

1. Haggin, J. (1994) Catalysis critical to benign process design, *Chemical and Engineering News* **72**, 22-25.

2. Hermann, W.A. and Kohlpaintner, C.W. (1993) Water-soluble ligands, metal complexes, and catalysts: synergism of homogeneous and heterogeneous catalysis, *Angew. Chem. Int. Ed. Engl.* **32**, 1524-1544.

3. Anastasiou, D., Campi, E.M., Chaouk, H., Fallon, G.D., Jackson, W.R., McCubbin, Q.J. and Trnacek, A.E. (1994) Rhodium-catalysed reactions of hydrogen and carbon monoxide with alkenylanilines, *Aust. J. Chem.* **47**, 1043-1059.

4. Campi, E.M., Jackson, W.R., McCubbin, Q.J. and Trnacek, A.E. (1994) The preparation of quinazolines and quinazolones involving rhodium-catalysed hydroformylation of alkenyl anilines, *Aust. J. Chem.* **47**, 1061-1070.

5. Campi. E.M., Chong, J.M., Jackson, W.R. and Van der Schoot, M. (1994) Synthesis of substituted α-methylene lactams by rhodium catalysed carbonylation of acetylenic amines, *Tetrahedron* **50**, 2533-2542.

6. Anastasiou, D. and Jackson, W.R. (1991) Rhodium catalysed reactions of unsaturated amines with hydrogen and carbon monoxide. A new synthesis of 2-piperidinones, *J. Organometa. Chem.* **413**, 399-410.

7. Campi, E.M., Fallon, G.D., Jackson, W.R. and Nilsson, Y. (1992) Rhodium-cAtalysed reactions of propargylamines with CO/H_2. Formation of pyrroles and butenolides, *Aust. J. Chem.* **45**, 1167-1178.

8. Pines, H. (1977) The chemistry of catalytic hydrocarbon conversions, Elsevier, Amsterdam.

9. Anderson, M.W., Shi, J., Leigh, D.A., Moody, A.E., Wade, F.A., Hamilton, B. and Carr, S.W. 91993) The confinement of Buckminsterfullerence in one-dimensional channels, *J. Chem. Soc. Chem. Commun.* 533-536.

10. Lewis, L.N. (1992) Cyclodextrin modification of the hydrosilylation reaction, Sixth International Cyclodextrin Symposium, Chicago L-36.

11. Anderson, J.R., Campi, E.M. and Jackson, W.R. (1991) Hydroformylation of olefin with water-soluble rhodium catalysts in the presence of α-cyclodextrin, *Catal. Lett.* **9**, 55-58.

12. Cucinotta, V., D'Allesandro, F., Impellizzeri, G. and Vecchio, G. (1992) The copper(II) complex with the imidazole-bound histamine derivative of β-cyclodextrin as a powerful chiral discriminating agent, *J. Chem. Soc., Chem. Commun.* 1743-1745.

13. Sawamura, M., Kitayama, K. and Ito, Y. (1993) Synthesis and properties of a new chiral diphosphine ligand bearing a cyclodextrin-based molecular recognition site and its palladium(II) complex, *Tetrahedron Asymmetry* **4**, 1829-1832.

14. Reetz, M.T. and Rudolph, J. (1993) Synthesis of a phosphine-modified cyclodextrin and its rhodium complex, *Tetrahedron Asymmetry* **4**, 2405-2406.

MODELS FOR SUPPORTED AQUEOUS-PHASE CATALYSIS

I. TÓTH #, I. GUO, and B. E. HANSON
Virginia Polytechnic Institute and State University, Blacksburg, Virginia 24061

1. Introduction

Immobilized homogeneous catalysts constitute a class of catalysts,which attempt to combine the merits of classical homogeneous and heterogeneous catalysts, i.e. high selectivities and mild reaction conditions with an ease of catalyst separation and recycling [1]. One of the most recent techniques for the immobilization of homogeneous catalysts is to adsorb a thin layer of aqueous solution of a water soluble catalyst precursor onto a hydrophyllic support such as silica [2]. It has been shown that the thus formed immobilized catalysts, termed supported aqueous-phase (SAP) catalysts, are effective for a variety of catalytic reactions in solvents which are immiscible with water [3]. It has also been proven that when a hydrophobic organic solvent is used with SAP catalysts, the adsorbed thin layer of water provides hydrogen bridges between the polar functional groups of the catalyst complex and those of the silica support strong and mobile enough to give an active immobilized catalyst which does not leach into the organic phase [2-4]. However, the nature of SAP catalysis is not yet fully explored. For example, it is ambiguous where exactly the actual catalytic process takes place with an SAP catalyst.

Supported aqueous-phase catalysts can be modeled in the following ways:

1. Homogeneous process in the organic phase (catalyst is immobilized into the substrate-product phase).

2. Homogeneous process in the supported aqueous phase (substrate diffuses into the aqueous phase which contains the active catalyst).

3. Real interfacial process at the interface of organic and supported aqueous phases.

2. Methodology

The characterization of operating SAP catalysts lacks the availability of absolute analytical methods. One possible approach to this goal is to seek similarities with appropriate analogous systems, (i) which operate in organic solvents identical to those used with SAP, (ii) which work in aqueous solutions and which (iii) are immobilized

I. T. Horváth and F. Joó (eds.), Aqueous Organometallic Chemistry and Catalysis, 195–197.

by other methods without the presence of an aqueous phase. The reaction phase (media) has a dramatic influence on the chemistry and performance of catalyst complexes. For example, it has been shown that olefin hydroformylation proceeds through different catalytic intermediates, thus with different selectivities and activities by the use of $HRh(CO)(TPPTS)_3$ [5] or $Co_2(CO)_6(TPPTS)_2$ [6] (TPPTS = trisulfonated-triphenylphosphine) in water than by the use of analogous PPh_3 complexes in organic solvents. Based on the similarities in selectivities of olefin hydroformylation by SAP-$HRh(CO)(TPPTS)_3$ to those by $HRh(CO)(PPh_3)_3$ at identical conditions, the first model has been proposed for SAP catalysis [4]. Further evidences have been provided for this model by the observed analogy in the performance of silica- (SAP) and ion-exchange resin-supported $HRh(CO)(TPPTS)_3$ hydroformylation catalysts [7] and by the observation that it is possible to generate a Pt-$SnCl_3$ bond (which is sensitive to hydrolysis) in the operating SAP-Pt-TPPTS + $SnCl_2$ system [8]. In order to investigate whether the first model can be extended to SAP applications other than hydroformylations catalyzed by Rh- [2-4], Co- [6] and Pt-TPPTS [8] complexes, some new catalyst systems have been tested.

3. Results

In addition to the above quoted literature data, SAP olefin hydroformylation results are presented by the use of $HRh(CO)(HexDPPDS)_3$, $PtCl_2(TAPTS)_2$ + $SnCl_2$ and $Co_2(CO)_6(TPrPTS)_2$. (HexDPPDS = Disulfonated-hexyldiphenyphosphine; TAPTS = trisulfonated-tris-(ω-alkylphenyl)phoshines [9] , where alkyl = CH_2, TBePTS; = C_2H_4, TEtPTS; = C_3H_6, TPrPTs.) Furthermore, some data are shown on SAP asymmetric hydroformylation of styrene by the use of $PtCl_2[(S,S)\text{-BDPP-}(p\text{-}NMe_3)_4](BF_4)_4$ + $SnCl_2$ system and on SAP asymmetric hydrogenation of dehydro-phenylalanine derivatives with $Rh(COD)[(S,S)\text{-BDPP-}(p\text{-}NMe_3)_4](BF_4)_5$ and $Rh(COD)[(S,S)\text{-Chiraphos-}(p\text{-}NMe_3)_4](BF_4)_5$ complexes. (For the synthesis of the latter funtionalized chiral ligands see ref. [10].) For the sake of comparison, the two-phase catalytic results with all of the precursors above are also given, as well as the appropriate (organic) homogeneous values, which were obtained by the use of analogous complexes containing the respective non-funtionalized ligands. The chiral catalysts have also been immobilized to ion-exchange resins and used in similar reactions.

The main findings of this study can be summarized as follows:

1. The selectivities of various SAP catalytic reactions closely resemble to those obtainable with the appropriate analogous (organic) homogeneous systems.

2. In each of the cases investigated, the activity of an SAP catalyst is a fraction of that of an analogous homogeneous catalyst at similar reaction conditions.

3. A two-phase system shows no similarities to the analogous SAP system.

4. In several instances, close analogies have been found in the performance of SAP and other analogous immobilized (ion-exchanger-supported) catalysts.

4. Conclusion

In various applications that SAP catalysts are best modeled as immobilized homogeneous catalysts which act in the organic phase. Consequently, water in SAP catalysis serves rather as an immobilizing agent than as solvent, as previously suggested [4] for the Rh-TPPTS system. Further experimental proof is provided by the observation that water as immobilizing agent in SAP can be readily replaced by other immobilizing agents such as silicon-ethers [11].

5. References

Present address: DSM Research, P. O. Box 18, 6160 MD Geleen, The Netherlands.
[1] Hartley, F. R. (1985) *Supported Metal Complexes*: Reidal, Dordrecht.
[2] a Arhancet, J. P., Davis. M. E., Merola, J. S., Hanson, B. E. (1989) Hydroformylation by supported aqueous phase catalysis, Nature **339**, 454. b Arhancet, J. P., Davis. M. E., Merola, J. S., Hanson, B.E. (1990) Supported aqueous-phase catalysts *J. Catal.*, **121**, 327-339.
[3] Davis, M. E. (1992) Supported aqueous-phase catalysis,*CHEMTECH* , August, p. 498-502 and references therein.
[4] Horváth, I. T. (1990) Hydroformylation of olefins with the water soluble $HRh(CO)[P(m-C_6H_4SO_3Na)_3]_3$ in supported aqueous phase. Is it really aqueous? *Catal. Lett.*, **6**, 43-48.
[5] Horváth, I. T., Kastrup, R. V., Oswald, A. A., Mozeleski, E. J. (1989) High-pressure nmr studies of the water soluble rhodium hydroformylation system, *Catal. Lett.*, **2**, 85-90.
[6] Guo, I., Hanson, B. E., Tóth, I., Davis, M. E. (1991) Bis[tris(m-(sodium sulfonato)phenyl)phosphine] hexacarbonyl dicobalt, $Co_2(CO)_6[P(m-C_6H_4SO_3Na)_3]_2$ in a supported aqueous phase for the hydroformylation of 1-hexene, *J. Organomet. Chem.*, **403**, 221-227.
[7] Tóth, I., Hanson, B. E., Guo, I., Davis, M. E. (1991) Immobilization of $HRh(CO)[P(m-C_6H_4SO_3Na)_3]_3$ on an anion exchange resin for the hydroformylation of higher olefins, *Catal. Lett.*, **8**, 209-214.
[8] Guo, I., Hanson, B. E., Tóth, I., Davis, M. E. (1991) Hydroformylation of 1-hexene with $Pt[P(m-C_6H_4SO_3Na)_3]Cl_2$ and its tin chloride analogue on a controlled pore glass, *J. Mol. Catal.*, **70**, 363-368.
[9] Bartik, T., Bartik, B., Hanson, B. E., Guo, I., Tóth, I. (1993) Water soluble electron-donating phosphines: Sulfonation of Tris(ω-phenylalkyl)phosphines, *Organometallics*, **12**, 164-170.
[10] Tóth, I., Hanson, B. E. (1990) Novel chiral water soluble phosphines I. Preparation of amine functionalized diop, chiraphos and BDPP derivatives and quaternization of their rhodium complexes, Tetrahedron Asymmetry, **1**, 895-912.
[11] Pugin, B.; Spindler, F.; Muller, M. (1992), Silangruppen enthaltende diphosphine, immobilisierte diopphosphine und deren verwendung als hydrierkatalysatoren, CIBA-GEIGY EP- 0 496 700.

STOICHIOMETRIC AND EXHAUSTIVE HYDROLYSIS OF *N,N*-DIALKYLCARBAMATES

DANIELA BELLI DELL'AMICO, FAUSTO CALDERAZZO, FABIO MARCHETTI AND GUIDO PAMPALONI
Dipartimento di Chimica e Chimica Industriale, Universita' di Pisa, Sezione di Chimica Inorganica, Via Risorgimento 35, I-56126 Pisa, Italy.

1. Introduction

N,N-dialkylcarbamates represent a class of compounds of relatively recent acquisition; they contain the O_2CNR_2 ligand, which is uninegative and substantially planar, due to extensive charge delocalization.

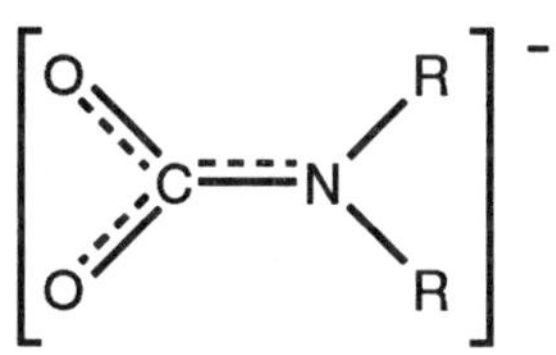

2. Homoleptic compounds

The preparative aspects of the homoleptic (*i.e.* containing a number of negative ligands O_2CNR_2 corresponding to the oxidation state of the metal) compounds, of general formula $[M(O_2CNR_2)_n]_m$, have been reviewed recently [1]; here, it will suffice to say that several preparative methods have been utilized for this class of compounds. They are outlined below. Reaction 1, namely the carbonatation of metal *N,N*-dialkylamides, is chronologically the first one to appear in the literature [2].

I. T. Horváth and F. Joó (eds.), Aqueous Organometallic Chemistry and Catalysis, 199–209.

$$M(NR_2)_n + n\,CO_2 \longrightarrow M(O_2CNR_2)_n \qquad (1)$$

$$MX_n + n\,CO_2 + 2n\,R_2NH \longrightarrow M(O_2CNR_2)_n + n\,[NH_2R_2]X \qquad (2)$$

$$[M(MeCN_4)]^{2+} + 2\,CO_2 + 6\,R_2NH \longrightarrow$$
$$trans\text{-}M(O_2CNR_2)_2(NHR_2)_2 + 2\,[NH_2R_2]^+ + 4\,MeCN\ \{M{=}Pd[3]\} \qquad (3)$$

$$M_2O + 2\,CO_2 + 2\,R_2NH \longrightarrow H_2O + 2\,M(O_2CNR_2)\ \{M{=}Ag\ [4]\} \qquad (4)$$

Conversion of a given compound into its homologues can be carried out by the exchange reaction 5 [1]. By this variety of preparative procedures,

$$M(O_2CNR_2)_n + n\,R'_2NH \xrightarrow{CO_2} M(O_2CNR'_2)_n + n\,R_2NH \qquad (5)$$

several new *N,N*-dialkylcarbamates of transition metals and Main Group elements have been prepared [1].

N,N-dialkylcarbamates $[M(O_2CNR_2)_n]_m$ of various metal complexities *m* are known. Generally, the molecular complexity decreases as the oxidation state of the metal increases and the steric bulkiness of the alkyl group increases. Mononuclear compounds are those of Group 4 metals [5] (titanium, zirconium and hafnium in their IV oxidation state), and those of palladium [3] and silicon [6]. The *N,N*-dialkylcarbamates (R=Et) of iron(II) [7], manganese(II) [8] and cobalt(II) [9] are hexanuclear. *Ceteris paribus*, with metal cations usually characterized by high coordination numbers, molecular complexity is generally high: the *N,N*-di-*iso*-propylcarbamates of lanthanide(III), i.e. ytterbium(III) [10], and neodymium(III) [11] are tetranuclear with heptacoordinated lanthanide. The *iso*-propryl derivative of aluminium(III) [12], $Al_2(O_2CNPr^i{}_2)_6$, is dinuclear with hexacoordinated aluminium, and both bridging and terminal di-*iso*-propylcarbamato groups. Molecular complexity is realized by modulation of the various bonding arrangements

(a) through (f), which have been observed experimentally for the O_2CNR_2 ligand, by means of X-ray diffraction methods [1].

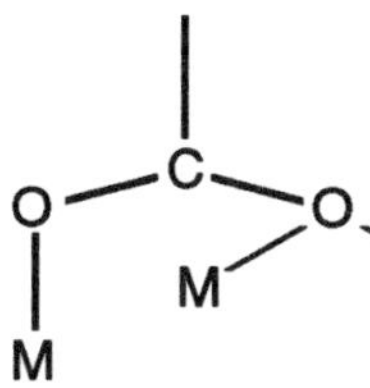

(a)

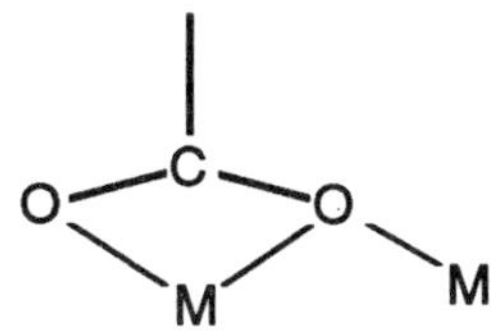

(b)

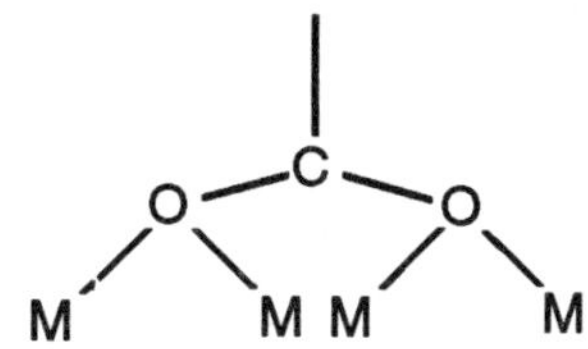

(c)

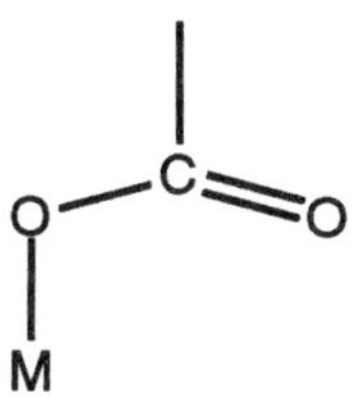

(d)

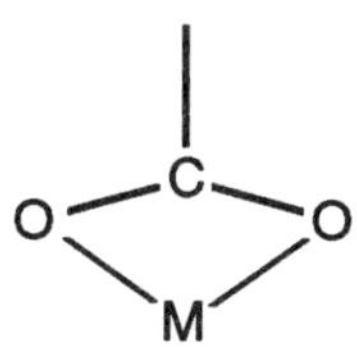

(e)

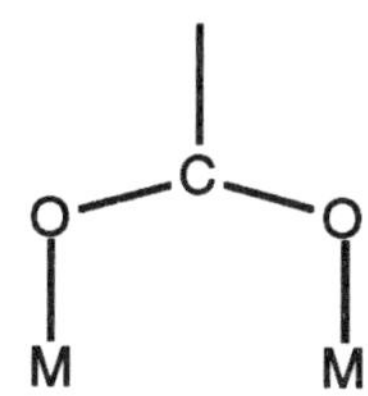

(f)

3. μ-Oxo-*N,N*-dialkylcarbamates. A special case of non-homoleptic compounds

The ligand O_2CNR_2 in $M(O_2CNR_2)_n$ is readily attacked by electrophilic reagents, including water. The inorganic products are μ-oxo complexes and carbon dioxide, while the organic product is represented by the secondary amine. The main contribution to the driving force of the reaction is the formation of carbon dioxide.

The reaction is believed to occur by electrophilic attack of the water proton on the oxygen atoms of the coordinated carbamato ligand, as shown in Scheme 1, where the reaction at a metal atom carrying two carbamato ligands is depicted. The M-OH bond formed in the first step is presumably of sufficient Brønsted acidity to bring about the further attack either intramolecularly or intermolecularly.

Scheme 1. Hydrolysis of a metal-bonded *N,N*-dialkylcarbamato ligand.

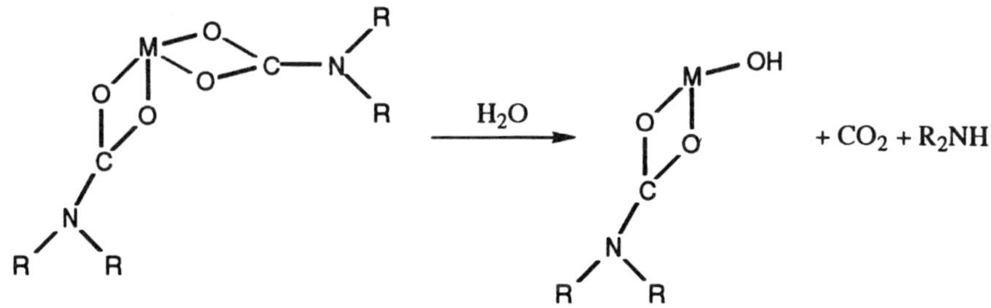

The first compound of this series to be synthesized was $U_4(\mu_3\text{-}O)_2(O_2CNEt_2)_{12}$, obtained [13] by hydrolysis of the corresponding homoleptic *N,N*-diethylcarbamato complex of uranium(IV). Figure 1 shows a simplified view of the compound , which contains two μ_3-O units in its composition.

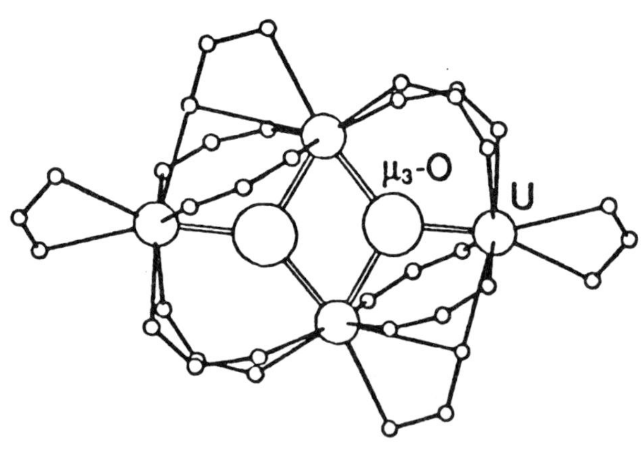

Fig. 1. A simplified view of $U_4(\mu_3\text{-}O)_2(O_2CNEt_2)_{12}$ [13]. NEt_2 fragments not shown.

Another compound which contains the μ_3-O unit is the chromium(III) derivative, $Cr_3(\mu_3\text{-}O)(O_2CNEt_2)_6Cl(NHEt_2)_2$ [14], see Figure 2, which is believed to result from an adventitious hydrolytic attack during a preparation of the homoleptic $Cr(O_2CNEt_2)_3$ derivative from the $CrCl_3$ / $HNEt_2$ / CO_2 system.

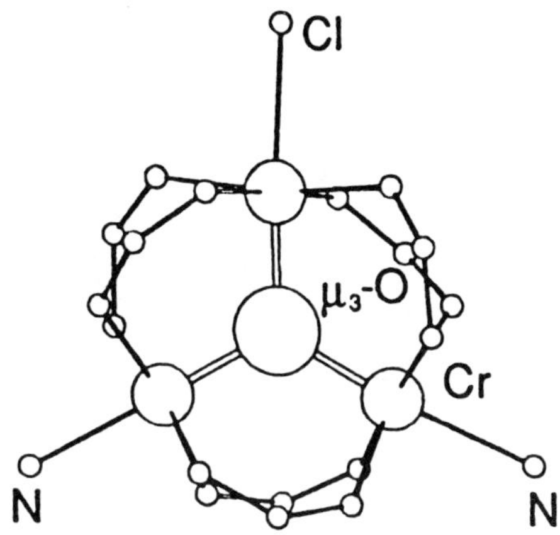

Fig. 2. Atom connectivity in $Cr_3(\mu_3\text{-}O)(O_2CNEt_2)_6Cl(NHEt_2)_2$ [14]. Alkyl chains not shown.

More complex structures arise from the hydrolysis of copper(II), and iron(II) *N,N*-dialkylcarbamates. The products resulting from the hydrolysis, of formula $M_8(\mu_4\text{-}O)_2(O_2CNPr^i_2)_{12}$, M=Fe(II) [7], Cu(II) [15], contain two μ_4-O units and are shown in figures 3 and 4, respectively.

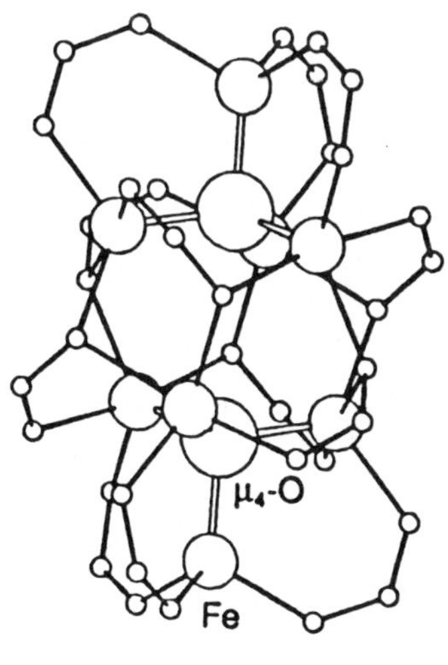

Fig. 3. Atom connectivity in $Fe_8(\mu_4\text{-}O)_2(O_2CNPr^i_2)_{12}$ [7]. NPr^i_2 fragments not shown.

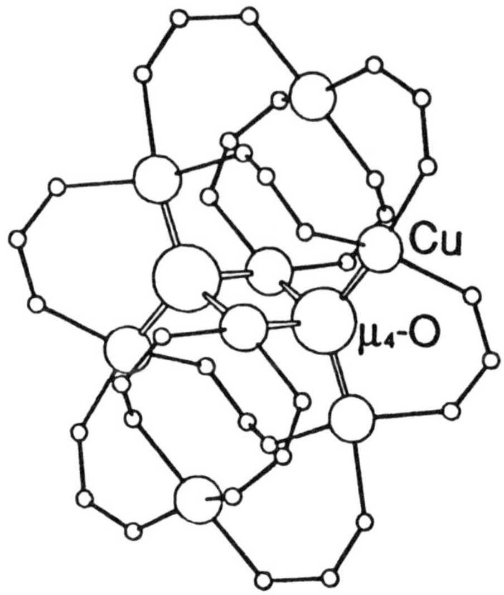

Fig. 4. Atom connectivity in $Cu_8(\mu_4\text{-}O)_2(O_2CNPr^i_2)_{12}$ [15]. NPr^i_2 fragments not shown.

It is interesting to note that in spite of the rather high nuclearity, these compounds still mantain a moderate solubility in aromatic hydrocarbons. From a hydrolytic process also originates the mixed-metal octanuclear compound, containing both zinc(II) and nickel(II) [16]. The atom connectivity is substantially the same as that shown for the iron(II) system of Figure 3, except that the labelled iron atom is substituted by zinc, the other metal positions being occupied by nickel.

It is interesting that the hydrolytic process carried out under anaerobic conditions with *N,N*-dialkylcarbamates can lead to μ-oxo derivatives otherwise very difficult to obtain. Such an example is represented by the iron(II) μ_4-oxo derivative [7] mentioned above, which is, to the best of our knowledge, the first crystallographically established neutral compound of iron(II) containing μ-oxo ligands. A mixed-valence Fe(II),(III) μ_6-oxo derivative containing both carboxylato and methoxo ligands with nuclearity m=10, has recently been reported by Lippard and coworkers [17].

4. μ-Oxo *N,N*-dialkylcarbamato complexes arising from non-hydrolytic processes

μ-Oxo derivatives within the class of *N,N*-dialkylcarbamates can be also formed by non-hydrolytic processes. One such example is the product obtained by the reaction of $ZrCp_2(CO)_2$ and $Zr(O_2CNEt_2)_4$ resulting in the formation of the ketenylidene complex of the reaction sequence 6 followed by further reaction with carbon dioxide to give deoxygenation of CO_2 and formation of the trinuclear complex of Figure 5 which contains both μ_2- and μ_3-oxides [18].

$$Cp_2Zr(CO)_2 \xrightarrow{Zr(O_2CNEt_2)_4} Zr_3Cp_2(O_2CNEt_2)_6(\mu_3\text{-}O)(\mu_2\text{-}CCO)$$

$$\xrightarrow{CO_2} Zr_3Cp_2(O_2CNEt_2)_6(\mu_3\text{-}O)(\mu_2\text{-}O) \qquad (6)$$

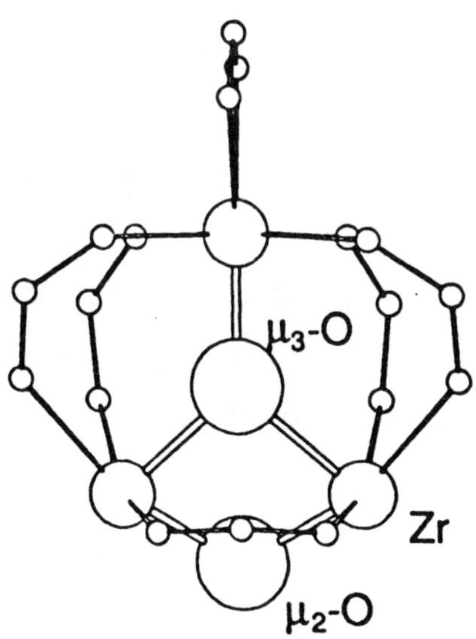

Fig. 5. Atom connectivity in $Zr_3Cp_2(O_2CNEt_2)_6(\mu_3\text{-}O)(\mu_2\text{-}O)$ [18]. NEt_2 fragments and Cp rings not shown.

Another deoxygenation process of carbon dioxide gives the μ_4-oxo derivative of zinc(II), shown in Figure 6, of formula $Zn_4(\mu_4\text{-}O)(O_2CNEt_2)_6$, with the *N,N*-diethylcarbamato groups bridging the zinc centers along the edges of the tetrahedron formed by the four zinc atoms [19]. This is an highly symmetrical structure, with a central tetrahedrally arranged oxo ligand.

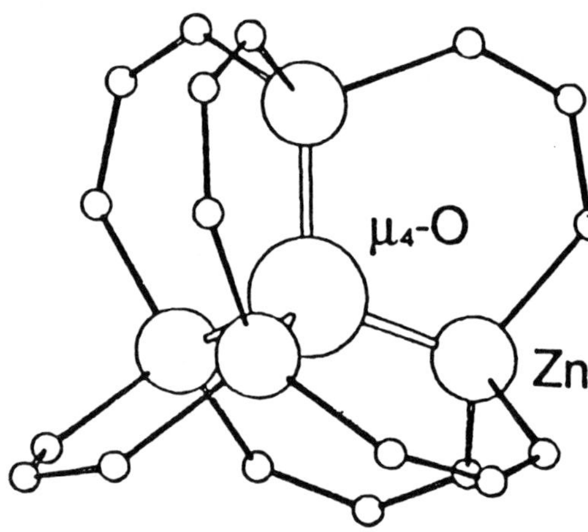

Fig. 6. Atom connectivity in $Zn_4(\mu_4\text{-}O)(O_2CNEt_2)_6$ [19]. NEt_2 fragments not shown.

The relationship between the μ_4-oxo derivative of zinc(II) and the octanuclear systems of iron(II) and Zn,Ni(II) is quite evident: the building block of the octanuclear structure is the tetranuclear μ_4-oxo-$M_4(O_2CNEt_2)_6$ unit, which is further joined by bridging tri-coordinated *N,N*-dialkylcarbamato groups, *i.e.* of type (a), to form the octanuclear compounds.

5. Exhaustive hydrolysis of *N,N*-dialkylcarbamates

If one of the homoleptic compounds of this series is treated at room temperature with an excess of water, with respect to the amount required to hydrolize the ligands, normally oxo compounds are the resulting products. The first application of this reaction was the hydrolysis of *N,N*-dialkylcarbamates of silicon [20], see eq. 7.

$$n\ Si(O_2CNR_2)_4 + (2n+1)\ H_2O \longrightarrow [NH_2R_2]_2\ Si_nO_{2n+1} + (4n-2)\ NHR_2 + 4n\ CO_2 \qquad (7)$$

Aluminium(III) *N,N*-dialkylcarbamates behave similarly.

6. Cation Implantation *via* *N,N*-dialkylcarbamates

The acidity of weak acids such as acetic acid and acetylacetone is sufficient to trigger the attack of the *N,N*-dialkylcarbamato ligand, as shown in Scheme 1. Thus, surface silanol groups of silica [21], whose pK_a is believed to be comprised between 4 and 7, should be regarded as chemically reactive functions for *N,N*-dialkylcarbamates.

In agreement with this view, several *N,N*-dialkylcarbamates have been utilized in an organic medium for cation implantation on a commercial silica [22]. Further work is in progress to assess the chemical and catalytic potentialities of the materials obtained by this new method.

7. Acknowledgements

The authors wish to thank the Consiglio Nazionale delle Ricerche (C.N.R., Roma), partly through the Progetto Finalizzato Materiali Speciali, and the Ministero della Ricerca Scientifica e Tecnologica (M.U.R.S.T.) for supporting this work.

8. References

1. D. Belli Dell' Amico, F. Calderazzo, Proceedings of the Meeeting on Syntheses and Methodologies in Inorganic Chemistry, Bressanone, Italy, December 2-7, 1991, pp. 1-17.

2. (a) G. Chandra, A. D. Jenkins, M. F. Lappert, R.C. Srivastava, *J. Chem. Soc. (A)* , 2550 (1970); (b) K.W. Bagnall, E. Yanir, *J. Inorg. Nucl. Chem.*, **36**, 777 (1974); (c) M. H. Chisholm, M. W. Extine, *J. Am. Chem. Soc.*, **99**, 782 (1977), and references therein.

3. A. Anillo, D. Belli Dell' Amico, F. Calderazzo, M. Nardelli, G. Pelizzi, L. Rocchi, *J. Chem. Soc., Dalton Trans.*, 2845 (1991).

4. R. Alessio, D. Belli Dell' Amico, F. Calderazzo, U. Englert, *Gazz. Chim. Ital.*, **123**, 719 (1993); R. Alessio, D. Belli Dell' Amico, F. Calderazzo, U. Englert, L. Labella, P. Strasser, manuscript in preparation.

5. F. Calderazzo, S. Ianelli, G. Pampaloni, G. Pelizzi, M. Sperrle, *J. Chem. Soc., Dalton Trans.*, 693 (1991).

6. D. Belli Dell' Amico, F. Calderazzo, M. Dell' Innocenti, B. Güldenpfennig, S. Ianelli, G. Pelizzi, *Gazz. Chim. Ital.*, **120**, 819 (1990).

7. D. Belli Dell'Amico, F. Calderazzo, L. Labella, C. Maichle-Mössmer, J. Strähle, *J. Chem. Soc., Chem. Commun.*, 1555 (1994).

8. A. Belforte, F. Calderazzo, P. F. Zanazzi, *J. Chem. Soc., Dalton Trans.*, 2921 (1988).

9. D. Belli Dell' Amico, F. Calderazzo, B. Giovannitti, G. Pelizzi, *J. Chem. Soc., Dalton Trans.*, 647 (1984).

10. D. Belli Dell' Amico, F. Calderazzo, F. Marchetti, G. Perego, *J. Chem. Soc., Dalton Trans.*, 483 (1983).

11. C. Della Porta, Thesis , Universita' di Pisa, 1994.

12. D. Belli Dell' Amico, F. Calderazzo, M. Dell' Innocenti, B. Güldenpfennig, S. Ianelli, G. Pelizzi, P. Robino, *Gazz. Chim. Ital.*, **123**, 283 (1993).

13. F. Calderazzo, G. Dell' Amico, M. Pasquali, G. Perego, *Inorg. Chem.*, **17**, 474 (1978).

14. D. Belli Dell' Amico, F. Calderazzo, F. Gingl, L. Labella, J. Strähle, *Gazz. Chim.* Ital., in the press.

15. E. Agostinelli, D. Belli Dell' Amico, F. Calderazzo, D. Fiorani, G. Pelizzi, *Gazz. Chim. Ital.*, **118**, 729 (1988).

16. A. Bacchi, D. Belli Dell' Amico, F. Calderazzo, U. Giurlani, G. Pelizzi, L. Rocchi, *Gazz. Chim. Ital.*, **122**, 429 (1992).

17. K. L. Taft, G. C. Papaefthymiou, S. J. Lippard, *Inorg. Chem.*, **33**, 1510 (1994).

18. F. Calderazzo, U. Englert, A. Guarini, F. Marchetti, G. Pampaloni, A. Segre, *Angew. Chem., Int. Ed. Engl.*, **33**, 1188 (1994).

19. A. Belforte, F. Calderazzo, U. Englert, J. Strähle, *Inorg. Chem.*, **30**, 3778 (1991).

20. D. Belli Dell' Amico, F. Calderazzo, M. Dell' Innocenti, Eur. Pat. Appl. EP 394669, Oct. 31 1990; *Chem. Abs.*, **115**, 40789p (1991); D. Belli Dell'Amico, F. Calderazzo, M. Dell' Innocenti, P. Robino, US Pat. 5 198 565, Mar. 30 1993; *Chem. Abs.*, **120**, 67989u (1994).

21. Kirk-Othmer Encyclopedia of Chemical Technology, 3rd ed., vol. 20, J. Wiley, New York, p. 766.

22. F. Calderazzo, D. Belli Dell' Amico, G. Pampaloni, P. Robino, It. Pat. MI 92 A 000416, Feb. 26 1992; D. Belli Dell'Amico, F. Calderazzo, R. Caminiti, G. Pelizzi, P. Robino, A. Tomei, Abstr. Papers presented at the Meeting on Perspectives of Inorganic Chemistry, Bressanone, Italy, October, 2-7, 1994, p. 260.

COMPOSITE HYDROGELS OF THE POLYSACCHARIDE / POLYVALENT METAL TYPE

T. SPYCHAJ, A.BARTKOWIAK
Polymer Institute, Technical University of Szczecin,
70-322 Szczecin, Poland

1. Introduction

Polysaccharides bearing hydroxyl groups such as dextran are able to create gels if crosslinked with covalent bonds by bifunctional reagents e.g. epichlorohydrin or after complexation with polyvalent metal cations.

Dextran crosslinked with epichlorohydrin is widely applied as hydrogel in various kinds of chromatography under trade name Sephadex. These gels are especially useful for biological and biochemical separations [1].

Products of dextran complexation with polyvalent metal cations have found important role in medicine field as cure agents and magnetic cariers in diagnosis.

Nevertheless of wide applications of the both groups of gels our knowledge about composite gels created by parallel covalent and complex networks is rather insufficient. New possibilities of usage of the modified dextran gels are still under investigation [2].

This contribution presents some results of investigation of hydrogels prepared from dextran, epichlorohydrin and polyvalent metal salts.

2. Experimental

Composite gels with spherical shape of particles were prepared with standard ratio of epichlorohydrin to dextran: 1.5/1 mol/mol. Concentration of sulfate salts of the particular metals (Mg, Zn, Mn, Co, Ni, Fe, Al, Ti, Cu and Ag) was kept constant 5 wt.% in relation to dextran.

I. T. Horváth and F. Joó (eds.), Aqueous Organometallic Chemistry and Catalysis, 211–213.

Prepared hydrogels after proper washing procedure were vacuum dried and characterized. Swelling tests in water, inverse size exlusion chromatography (SEC), scanning electron microscopy (SEM) and differential scanning calorimetry (DSC) were applied. Some applied research was performed including investigation of rate of decomposition of hydrogen peroxide and bactericidal activity of the gel materials.

3. Results

Strong dependence between kind of the metal cation applied for the gel preparation and pore volume characteristics in swollen state was found. It may be connected with various coordination ability to link hydroxyl groups by the partical metal cations. These differences are visualized in Fig.1 presenting swellabilities of the hydrogels and Fig.2 comparing SEC calibration curves measured for poly(ethylene glycol) standards.

DSC patterns of the composite materials were compared. High temperature endothermic peaks have changed from ca. 163°C (Ni-gel) to ca. 210°C (Cu-gel). Dextran composite hydrogels containing Fe, Cu, Co and Ti exibited catalytic activity as evaluated by rate of decomposition of hydrogen peroxide (Fig.3) and by ability to catalyze acrylamide polymerization.

Bactericidal activity of the gel materials with Cu, Zn and Ag was evaluated.

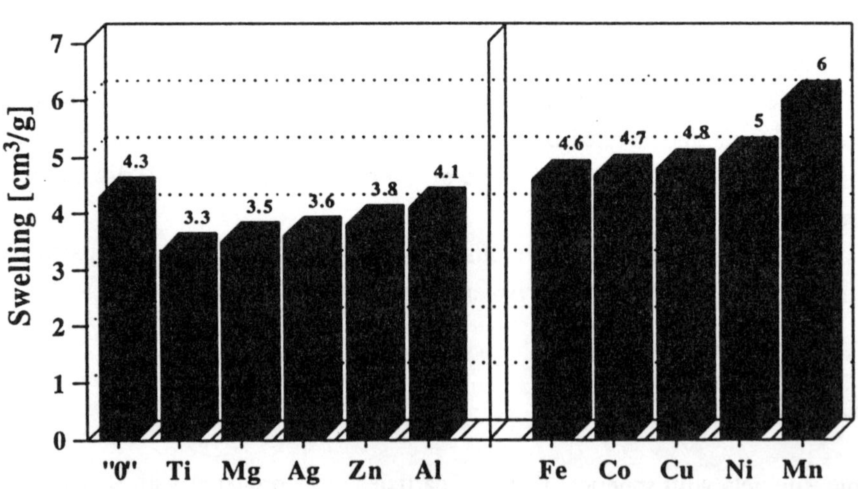

Fig.1. Comparison of the hydrogel swellabilities (water, 20 °C, dextran/epichlorohydrin 1/1.5 mol/mol ratio) for gels synthesized in presence of 5 % of the different cation sulfates (Ti - gel was prepared from tetraethyl ortotitanate).

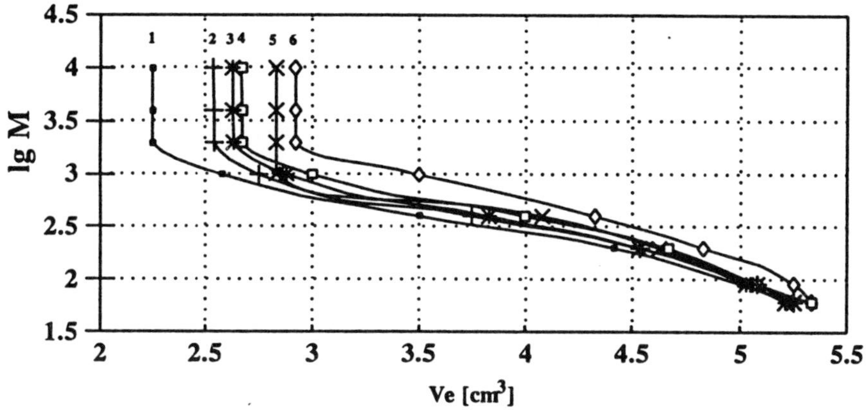

Fig.2. SEC calibration curves for dextran crosslinked with epichlorohydrin: blank sample (1) and in presence of Al^{+3} (2), Ti^{+4} (3), Mg^{+2} (4), Ag^{+1} (5) and Zn^{+2} (6). Solute: poly(ethylene glycol).

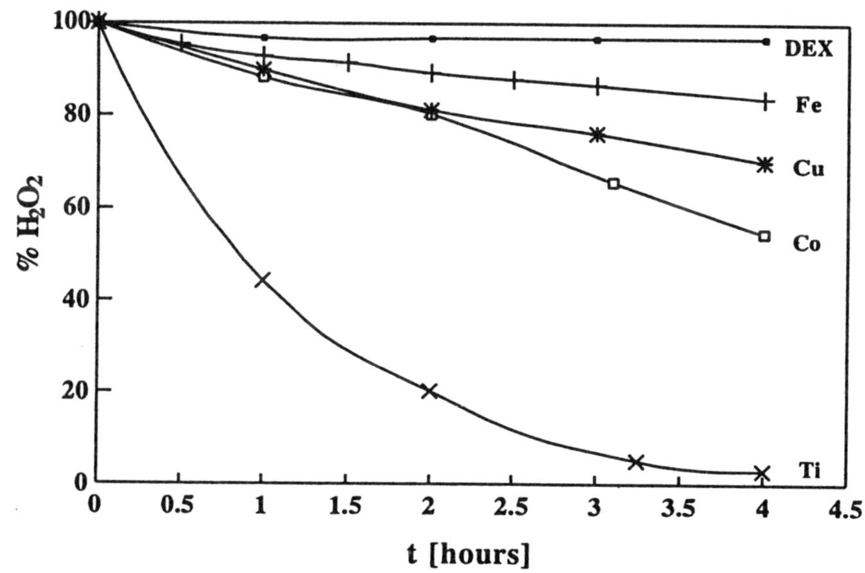

Fig.3. Rate of H_2O_2 decomposition in presence of composite gels with metal content (20°C). DEX - blank sample.

4. References

1. Robyt J.F. (1986) Dextran, Encycl. Polym. Sci. & Engn., J.Wiley Intersci. Publ., New York, vol.4, p.752-767
2. Spychaj.T, Bartkowiak.A, to be published

ORGANOMETALLIC CATALYSIS IN WATER AND IN A TWO-PHASE SYSTEM

D. SINOU

Laboratoire de Synthèse Asymétrique, associé au CNRS, Université Claude Bernard Lyon I, E.S.C.I.L., 43, boulevard du 11 Novembre 1918, 69622 Villeurbanne Cédex, France.

Abstract:

Amino acids precursors are reduced in a two-phase system ethyl acetate-water with e.e. up to 90 % in the presence of a rhodium complex associated with chiral sulfonated diphosphines. Performing the reaction in deuterium oxide allows the regiospecific introduction of a deuterium atom at the carbon α to the acetamido and the ester group.

Palladium(0)-catalyzed allylic substitution of acetates or carbonates occur in a two-phase system nitrile-water using tppts as the ligand; recycling of the catalyst is possible whatever the nucleophile used (carbo or heteronucleophile). The application of this methodology to uracil and 2-thiouracil derivatives afford the regiospecific allylated product at N-1 for the uracil and at sulfur for the thiouracil series.

1. Introduction

One of the greatest drawbacks of homogeneous metal catalysis in organic synthesis is the separation of the organic product(s) from the active catalyst, which is sometimes toxic and costly. A very elegant way to solve this problem consists in using water soluble ligands which are poorly soluble in organic media; the catalysis could be carried

I. T. Horváth and F. Joó (eds.), Aqueous Organometallic Chemistry and Catalysis, 215–230.

out in a two-phase system and decantation of the two-phases would allow an easy separation and recovery of the catalyst [1-7]. Another advantage of such a technology is the economical and safety use of water as a solvent. Consequently these systems could be used in large-scale industrial processes, such as hydroformylation, hydrogenation, carbonylation and metathesis polymerisation, and in the preparation of fine chemicals. On the other hand, carrying out an organometallic catalysis in water allowed the easy transformation of hydrosoluble substrates and could also modify the kinetics and the regio- and stereoselectivity of a given reaction. In the present paper we report on these two aspects: the asymmetric reduction of unsaturated amino acids precursors using rhodium complexes in association with chiral sulfonated phosphines and the allylic nucleophilic substitution reaction catalyzed by palladium in association with tppts or $P(C_6H_4\text{-}m\text{-}SO_3Na)_3$ for the first aspect, and the regiospecific alkylation of uracil and thiouracil derivatives for the second one.

2. Asymmetric reduction in a two-phase system.

Asymmetric reduction of prochiral substrates like amino acids precursors in the presence of rhodium or ruthenium complexes is now a well known reaction, giving enantioselectivities higher than 95 % [8-10]. However one of the problem is the cost of the chiral catalyst, the chiral ligand being more expensive than the metal. So, for an industrial point of view, recycling of the catalyst (metal + ligand) would be very important.

Chiral tetrasulfonated phosphines (*S,S*)-cyclobutanediop$_{TS}$ **1**, (*S,S*)-BDPP$_{TS}$ **2** and (*S,S*)-Chiraphos$_{TS}$ **3** (Scheme 1) were obtained by direct sulfonation of the corresponding chiral non sulfonated phosphines in concentrated sulfuric acid containing 20-30 % SO_3 for 1 day (cyclobutanediop) to 1 week (Chiraphos).

Asymmetric reduction of some amino acids precursors was studied in a two-phase system AcOEt/H_2O with the catalyst obtained from $[Rh(COD)Cl]_2$ and the tetrasulfonated chiral diphosphines **1-3** (Table 1) [11]. The results summarized in Table

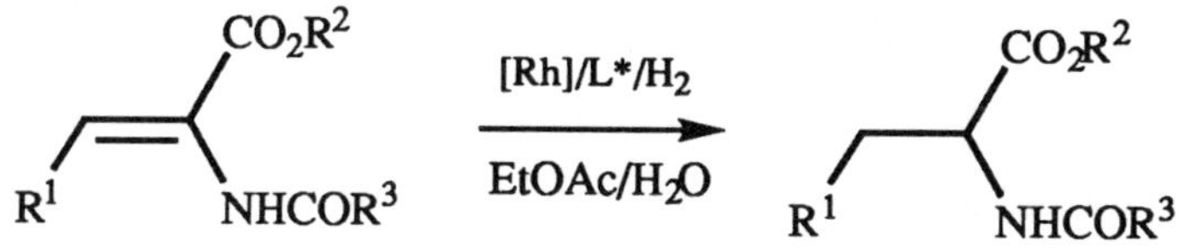

4 **5**

a R^1 = Ph ; R^2 = H ; R^3 = CH_3 ; **b** R^1 = Ph ; R^2 = CH_3 ; R^3 = CH_3
c R^1 = Ph ; R^2 = H ; R^3 = C_6H_5 ; **d** R^1 = H ; R^2 = H ; R^3 = CH_3

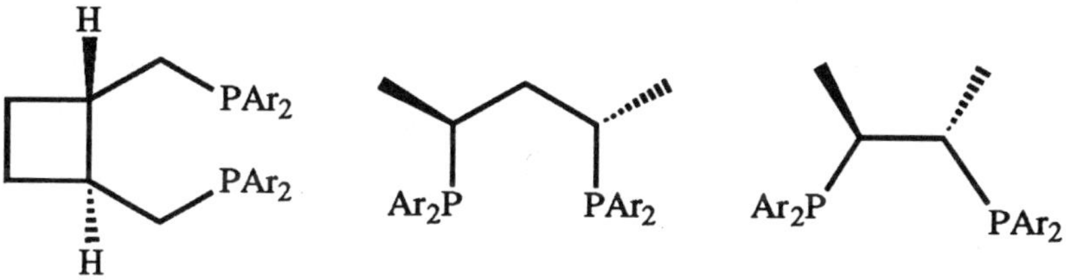

1 **2** **3**

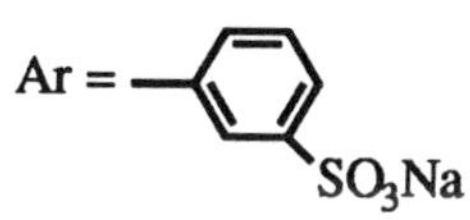

Scheme 1

TABLE 1. Reduction of various *(Z)* enamides **4** using $[Rh(CODCl]_2$ + diphosphine **1-3** in a two-phase system AcOEt/H_2O.[a]

Substrate	Ligand	P_{H_2} atm	e.e. % (config)
4 a	**1**	1	34 (S)
			37 (S)[b]
	2	15	65 (R)
	3	10	87 (R)
4 b	**1**	1	20 (S)
			23 (S)[b]
	2	15	45 (R)
	3	10	81 (R)
			88 (R)[b]
			87 (R)[c]
4 c	**1**	1	13 (S)
	2	15	44 (R)
	3	10	86 (R)

[a] [substrate]:[ligand]:[Rh] = 100:1.1:1; [substrate] = 1 M; 10 mL H_2O; 10 mL AcOEt; 25 °C; quantitative chemical yield. [b] Catalyst reused. [c] Catalyst twice reused.

1 show clearly that tetrasulfonated (*S,S*)-Chiraphos **3** lead to a very efficient catalyst, giving enantioselectivity up to 87 % in the reduction of the acid **4a**, 81 % for the methyl ester **4b** and 86 % for the acid **4c**. Reuse of the catalyst solution is possible with no loss of enantioselectivity. Tetrasulfonated (*S,S*)-BDPP **2** based catalyst gives lower enantioselectivity (e.e. ≈ 44-65 %). Tetrasulfonated (*S,S*)-cyclobutanediop **1** based catalyst, although being the most active, gives the lowest enantioselectivities with values in the range of 20-35 %.

In order to throw light on the reasons of the decrease in enantioselectivity upon going from the organic phase (usually ethanol) to the aqueous phase, we undertook some reductions of (Z)-α-acetamidocinnamic acid **4a** using solvents or mixtures of solvents with different polarity and solvophobicity parameters [12]. The plots of values of log (%*S*/%*R*) against the solvophobicity parameter S_p or the solvent polarity $E_T(30)$ obtained in the reduction of this substrate **4a** using $[Rh(COD)_2]ClO_4$ + cyclobutanediop as the catalyst are shown in Fig. 1 [13]. A linear correlation was obtained for the plot of log (%*S*/%*R*) *vs*. S_p values, and no correlation with the $E_T(30)$ was observed. A similar linear correlation between log (%*S*/%*R*) and S_p values was found in the reduction of substrate **4a** in various alcohol-water mixtures using $[Rh(COD)Cl]_2$ and **1a** or **2a** as the catalyst (Fig. 2). These results showed that one of the reasons of the lowering of enantioselectivity in the reduction of amino acids precursors in a two-phase system is the very hight solvophobicity parameter S_p of water.

The reduction of (Z)-α-acetamido cinnamic acid methyl ester **4b** was also performed under hydrogen in a two-phase system D_2O-AcOEt in the presence of the catalyst $[Rh(COD)Cl]_2$ + tppts [14]. We observed a regioselective deuterium incorporation up to 75 % at the carbon α to the acetamido and the ester group (Scheme 2) as shown by ^1H and ^{13}C NMR. Performing the same reduction under a deuterium atmosphere in a two-phase system H_2O-AcOEt indicates a regio- and a stereospecific hydrogen incorporation up to 94 % at the carbon α to the acetamido and ester group, the overall addition of HD to **4b** being a *cis*-addition. The same results were found in the reduction of α-acetamido acrylic acid methyl ester **4d**. An extensive study showed that this incorporation was only due to the presence of water, although the nature of the co-solvent and the phosphine had a great influence on the extent of incorporation.

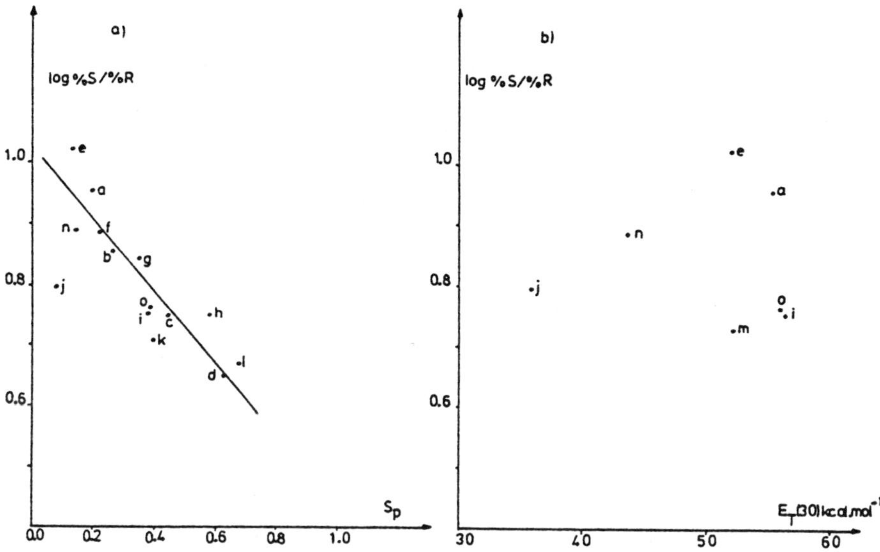

Figure 1. (a) Log (*%S*/*%R*) *vs.* solvophobicity parameter S_p for the asymmetric hydrogenation of α-acetamido cinnamic acid **4a** catalyzed by $[Rh(COD)_2]ClO_4$ + cyclobutanediop. a-d: 100, 90, 70, 50 % CH_3OH-H_2O (v/v); e-h: 100, 80, 60, 40 % C_2H_5OH-H_2O (v/v); i: CH_2OHCH_2OH; j-l: 100, 60, 40 % dioxane-H_2O (v/v); m: N-CH_3 acetamide; n: dimethylformamide; o: formamide. (b) Enantioselectivity as a function of solvent polarity $E_T(30)$.

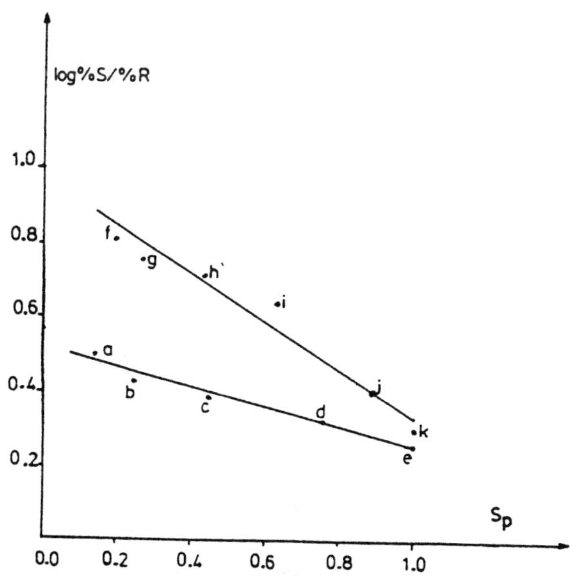

Figure 2. Log (*%S*/*%R*) *vs.* solvophobicity parameter S_p for the asymmetric hydrogenation of α-acetamido cinnamic acid **4a** catalyzed by $[Rh(COD)Cl]_2$ + **1** in 100, 75, 50, 25 and 0 % C_2H_5OH-H_2O (v/v) (a-e) or **2** in 100, 85, 75, 50, 20 and 0 % CH_3OH-H_2O (v/v) (f-k).

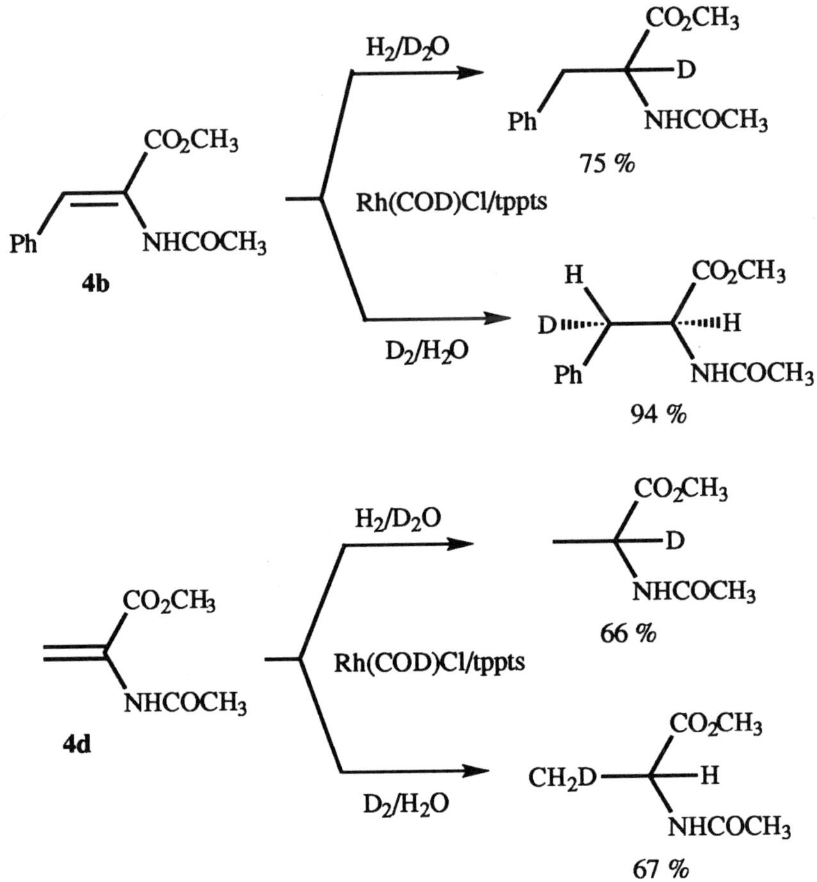

Scheme 2

The reduction of these substrates was also performed in water with the catalyst derived from $[Rh(COD)Cl]_2$ and ligand **1** or **2** and in the presence of ammonium formate as the hydrogen source (Eq. 1) [15]. The results summarized in Table 2 show that enantioselectivities up to 46 % have been achieved using $BDPP_{TS}$ **2** as the ligand.

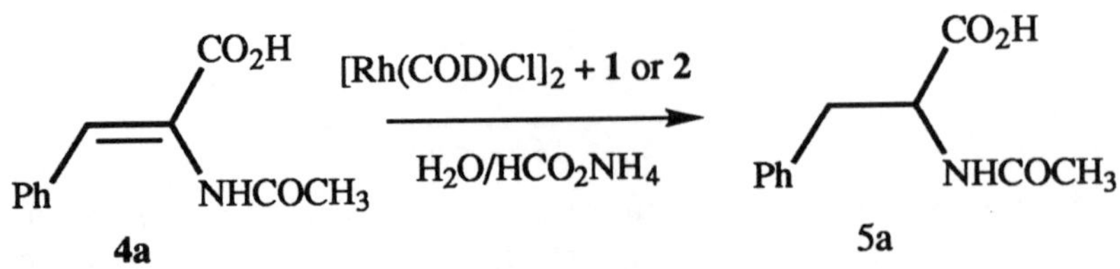

Eq. 1

TABLE 2. Reduction of enamides **4a** and **4d** using $[Rh(CODCl]_2$ + diphosphine **1-2** in water in the presence of HCO_2NH_4.[a]

Substrate	Ligand	T °C	e.e. % (yield)
4a	**1**	75	40 (100)
	1	100	32 (100)
	2	50	46 (50)
	2	100	12 (82)
4d	**1**	50	25 (100)
	1	100	23 (100)
	2	75	32 (100)
	2	100	30 (100)

[a] $[substrate]:[HCO_2NH_4]:[ligand]:[Rh] = 25:125:1.1:1$; [substrate] = 0.25 M; 10 mL H_2O; 24 h.

3. Palladium(0)-catalyzed allylic substitution in an aqueous-organic medium.

Although palladium-catalyzed nucleophilic substitution of allylic substrates is an important methodology in organic synthesis, providing mild and selective routes to a variety of valuable chemicals from basic organic precursors [16-20], the use of water-soluble palladium catalyst remained largely unexplored until the last years. Our continuing interest in the area of palladium chemistry was spured out by the idea that aqueous soluble palladium catalysts may be used in allylic substitution reactions under very mild conditions.

The allylic nucleophilic substitution reaction of allylic carbonates or acetates was carried out at 50 °C in a nitrile (acetonitrile, butyronitrile or benzonitrile)-H_2O medium in the presence of palladium complexes [$Pd_2(dba)_3$ or $Pd(OAc)_2$] associated with sulfonated phosphines such as tppts (Eq. 2) [21, 22]. Carbonucleophiles or heteronucleophiles could be used in this reaction leading to the allylated compounds with chemical yields up to 90 %. Carrying the reaction in a butyronitrile or benzonitrile-H_2O medium allowed an easy separation and recycling of the catalyst.

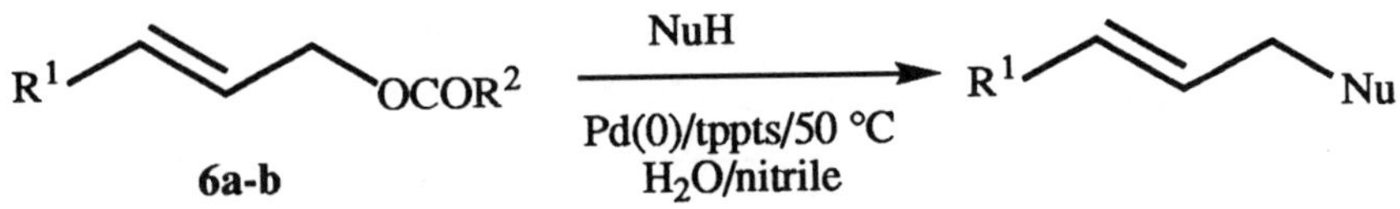

Eq. 2

$R^2 = OCH_3, CH_3$; nitrile = CH_3CN, C_3H_7CN or C_6H_5CN ;
NuH = carbonucleophile, heteronucleophile

Allylic carbonates such as cinnamyl ethyl carbonate **6a** reacted with various active methylene compounds such as ethyl acetoacetate, acetylacetone or Meldrum acid without the presence of a base to give a single regio and stereoisomer **7** having the *E* configuration (Scheme 3); malonate didn't react under these conditions, the presence of a base being necessary. So, it's possible to allylated selectively ethyl acetoacetate in the presence of malonate, even using of a large excess of the allylating reagent, which is impossible in an organic medium. Alkylation of acetates such as **8** needed the presence of a base like triethylamine or better DBU (Eq. 3).

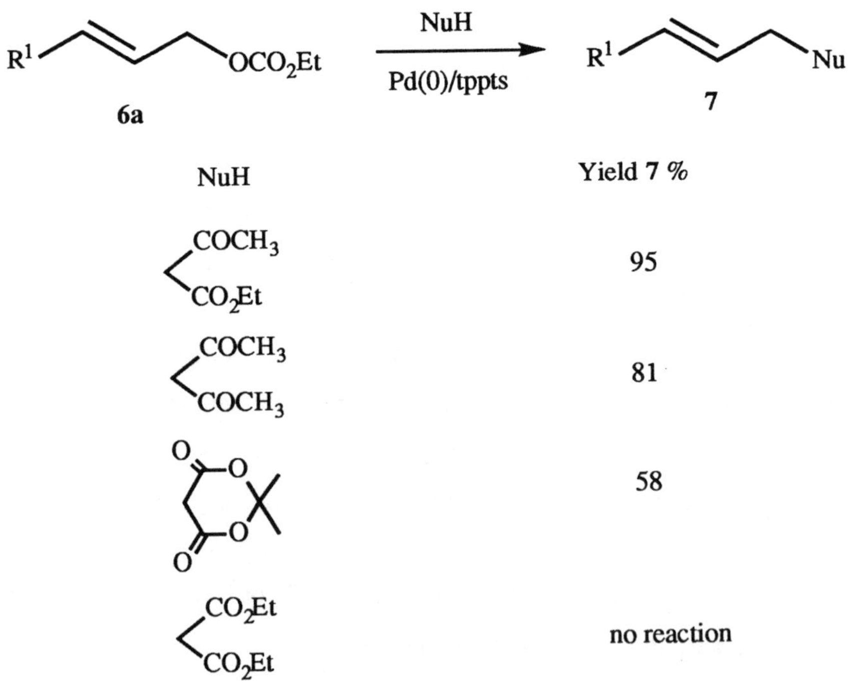

Scheme 3

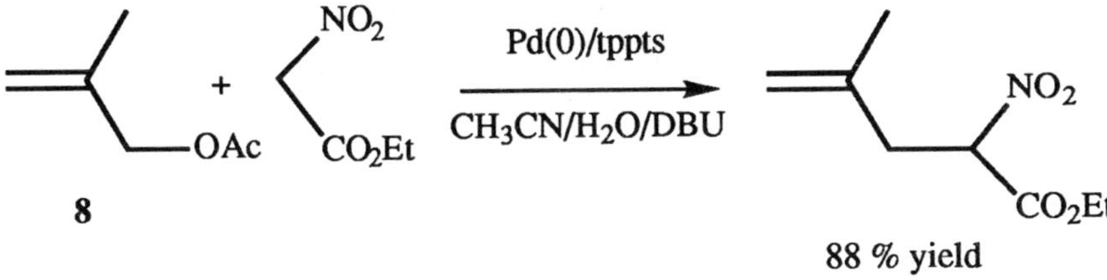

Eq. 3

Reaction of the unsaturated hydroxy acetate **9** with diethyl malonate led regiospecifically to the formation of a single product **10** having the *E* configuration (Eq. 4); the same regiospecificity was found in the reaction of ethyl acetoacetae on the epoxybutene **11** (Eq. 5).

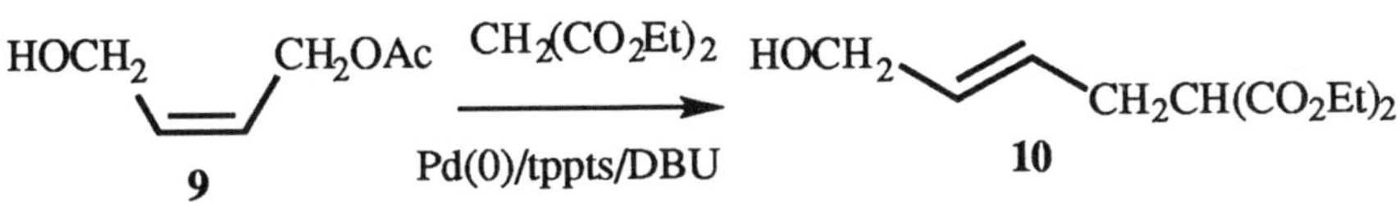

Eq. 4

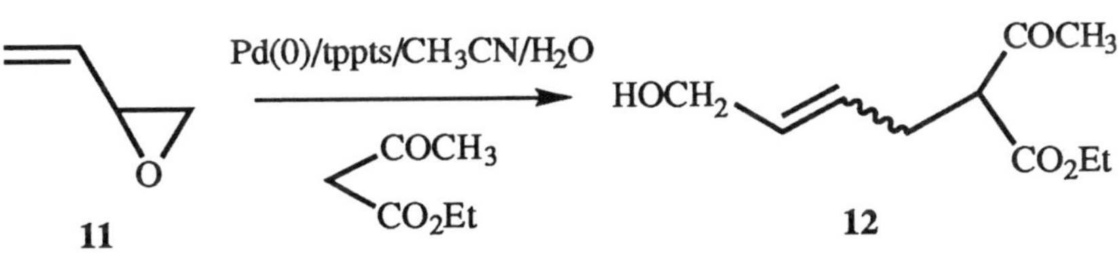

Eq. 5

Secondary amines (morpholine, benzyl methylamine) and primary amines (*n*-butylamine, 2,2-diethylpropargylamine, cycloheptylamine) reacted with cinnamyl acetate **6b** affording cleanly the product of monoalkylation **13** in quite good yields (Scheme 4).

Under the same conditions, hydroxylamine hydrochloride afforded the hydroxylamine **14** resulting from a bisallylation reaction (Eq. 6).

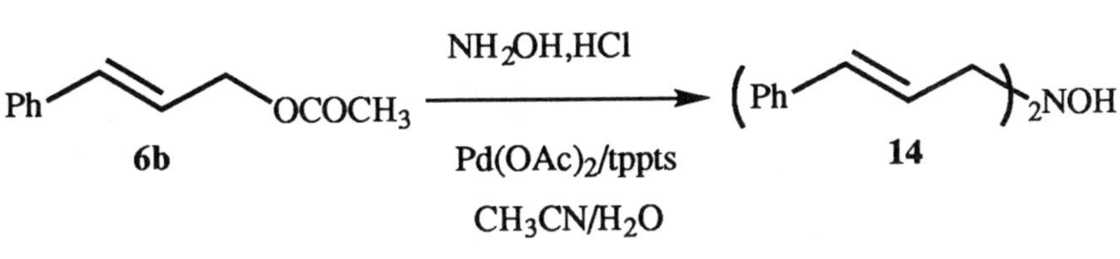

Eq. 6

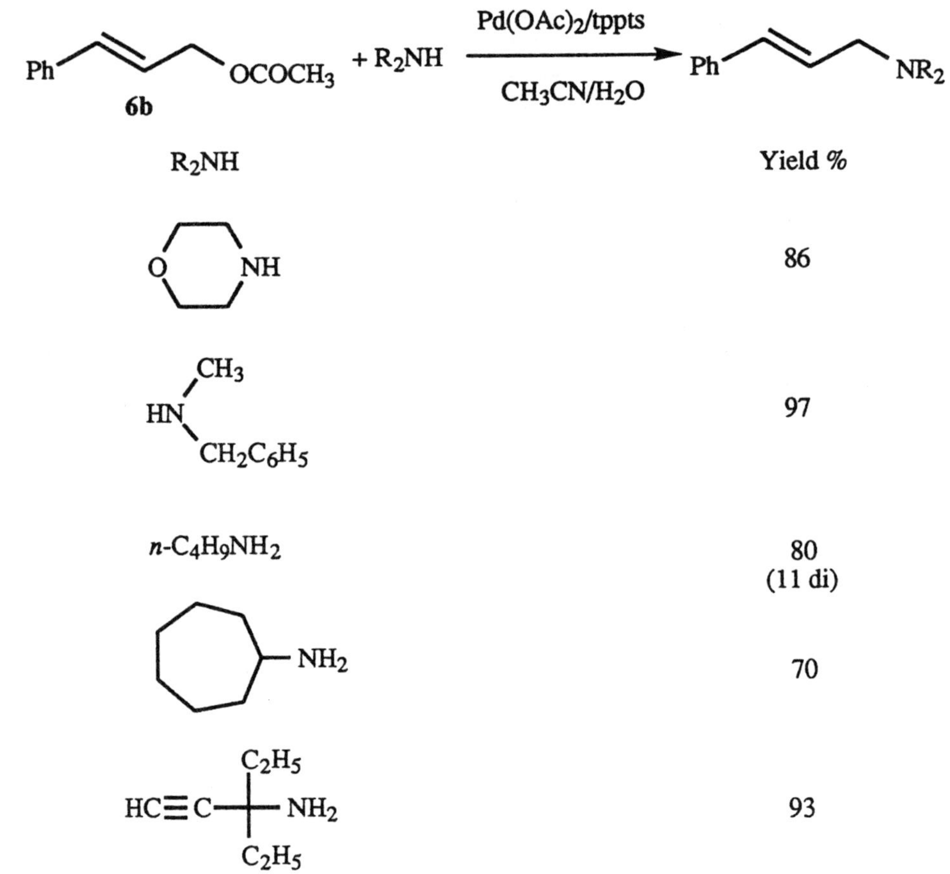

Scheme 4

Finally reaction of cinnamyl acetate **6b** with sodium azide or *p*-toluene sulfinate, which are soluble in water, gave the corresponding allyl azide and sulfone with excellent yields using the biphasic system (up to 90 % yield even after recycling of the catalyst).

It is to be noticed that such an aqueous medium was successfully used recently for the efficient removal of the allyloxycarbonyl moiety of a wide range of alcohols and secondary amines under quite mild conditions [23].

4. Allylation of uracil and 2-thiouracil derivatives.

Palladium(0)-catalyzed coupling between suitable heterocycles bases and allylic acetates has been used for the synthesis of carbocyclic nucleosides analogs such as carbovir. However some regioselectivity problems arise when two amide groups are present in the molecule, i.e. in the uracil and 2-thiouracil families: mixtures of mono and diallylated compounds at N-1 and N-3 were obtained for uracil derivatives, and allylation at N-1, N-3 and sulfur occured in the case of 2-thiouracil derivatives[24]. Moreover, uracils and 2-thiouracils are scarcely soluble in organic media and highly polar solvents at high temperature are required.

We studied the allylation of uracil **15a** and thymine **15b** in an acetonitrile-aqueous medium (1/9) by *(E)*-cinnamyl acetate **6b** in the presence of $Pd(OAc)_2$ + tppts and DBU (Scheme 5). Only 1-cinnamyl uracil **16a** and 1-cinnamyl thymine **16b** were obtained with respectively chemical yields up to 80 % and 53 % [25]. It is remarkable that the use of a large excess of cinnamyl acetate results only in an improvement of the yield, but not in diallylation. This regiospecificity might be due to a) the faster alkylation reaction at N-1 than at N-3 and b) the precipitation of compounds **16** in the reaction medium.

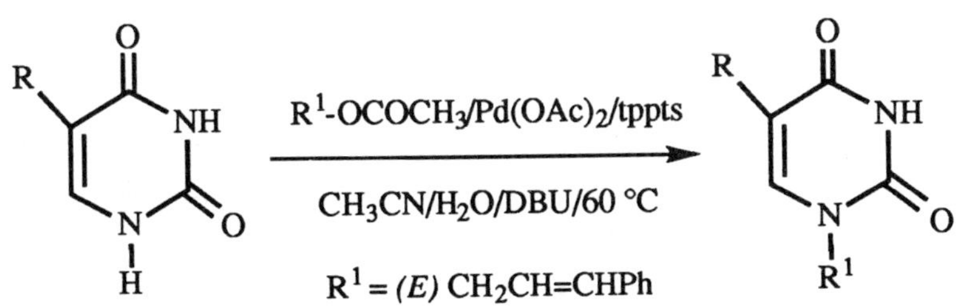

15 **16**

a R = H 80 %
b R = CH_3 53 %

Scheme 5

Usual alkylation reaction of 2-thiouracil compounds **17** catalyzed by palladium(0) afforded the four possible N-allylated and S-allylated products with low chemical yields. Again the combination water-acetonitrile/cinnamyl acetate/DBU/Pd(OAc)$_2$/tppts permitted the isolation of a single compound **18** resulting from a S-allylation reaction with quite good yields. In this case, the regiospecificity of the allylation is probably due to the fact that a) the reaction is under kinetic control and b) the sulfide formed precipitates in the reaction medium.

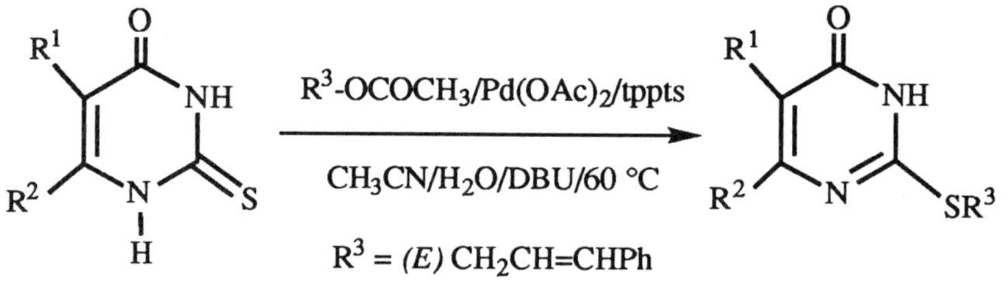

17 **18**

a $R^1 = R^2 = H$	53 %
b $R^1 = CH_3$, $R^2 = H$	43 %
c $R^1 = H$, $R^2 = CH_3$	92 %

Scheme 6

To confirm these hypotheses, we performed some complementary experiments. Thus sulfide **18a**, independently prepared, was isomerized to a mixture of N-allylated products (ratio determined by ^{1}H-NMR) in dioxane at 100 °C in the presence of Pd(PPh$_3$)$_4$ (Eq. 7). Performing the reaction of allylation in an homogeneous medium H_2O-THF (1/1) at 50 °C instead of H_2O-CH_3CN gave only the N-1 allylated compound **19** with 41 % chemical yield (Eq. 8).

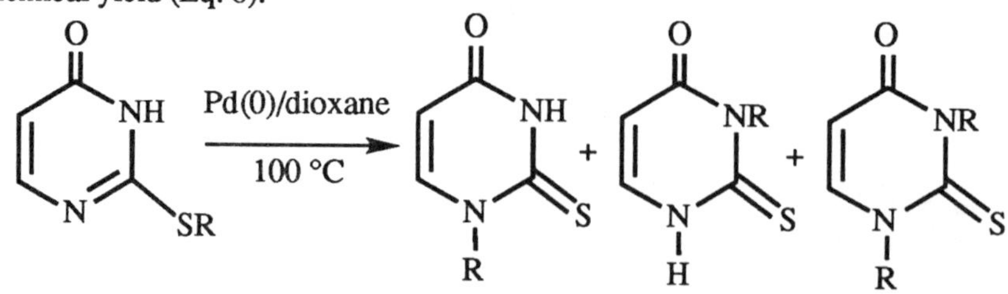

Eq. 7

18a — 57 % — 27 % — 16 %

R = *(E)* CH_2CH=CHPh

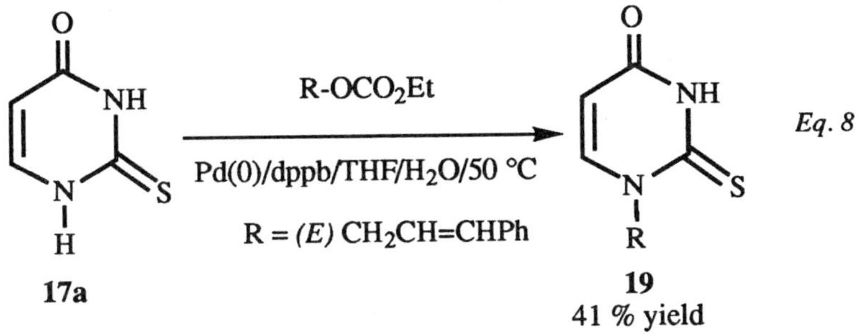

Finally this drastic effect of water on the regioselectivity of the allylation in the case of ambident nucleophiles was also observed in the case of other substrates. For exemple mercaptobenzoxazole **20** afforded the S-allylated substrate **21** in THF in the presence of palladium(0), although the N-allylated compound **22** was obtained using a H_2O-THF medium (Scheme 7) [26].

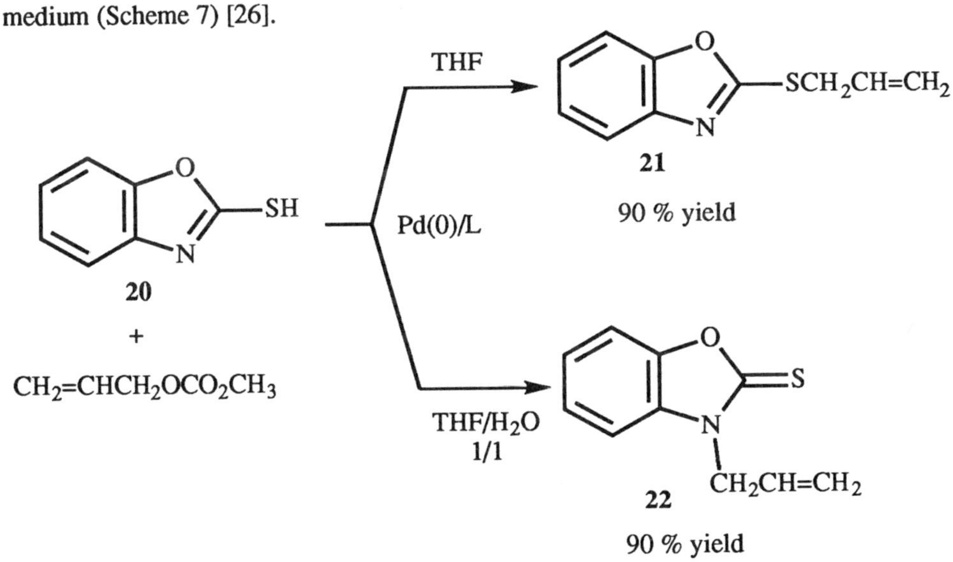

Scheme 7

5. Conclusion.

In conclusion, the use of chiral sulfonated phosphines in an aqueous-organic two-phase system allows the asymmetric reduction of prochiral compounds, the catalyst being reused with no loss of enantioselectivity; however some difference in enantioselectivity occurred, which is due to the use of water as a co-solvent. Aqueous palladium catalysts could also be used in the substitution reaction of allylic substrates. Applying this methodology to uracils and thiouracils derivatives allowed the regiospecific allylation at N-1 for uracil and at sulfur for thiouracil, when mixtures were obtained in an organic solvent. This quite young technology has to be more and more used in the near future at research at universities and in industry.

Acknowledgement.

The author express his deepest gratitude to all past and current members of his research group working in the field of organometallic catalysis in water. Some part of this work was done in collaboration with Pr. B. Heil and Dr. J. Bakos (University of Veszprem, Hungary) and Pr. M. Moreno-Mañas (University of Barcelona, Spain). Supports from C.N.R.S., M.E.S.R, Rhône-Poulenc Recherches and Région Rhône-Alpes are gratefully acknowledged.

References

1. Joò, F. and Tòth, Z. (1980) Catalysis by water-soluble phosphine complexes of transitio metal ions in aqueous and two-phase media, *J. Mol. Catal.* **8**, 369-383.
2. Sinou, D. (1987) Phosphines hydrosolubles. Synthèses et applications en catalyse, *Bull Soc. Chim. Fr.* 480-486.
3. Kuntz, E. G. (1987) Homogeneous catalysis in water, *Chemtech* **17**, 570-575.
4. Southern, T. G. (1989) Biphasic and phase transfer catalysis, *Polyhedron* **8**, 407-413.
5. Barton, M. and Atwood, J. D. (1991) Aqueous soluble organometallic complexes, *J. Coord Chem.* **24**, 43-67.
6. Kalck, P. and Monteil, F. (1992) Use of water-soluble ligands in homogeneous catalysis, *Adv. Organomet. Chem.* **34**, 219-284.
7. Herrmann, W. A. and Kohlpaintner, C. W. (1993) Water-soluble ligands, metal complexes, and catalysts: synergism of homogeneous and heterogeneous catalysis, *Angew. Chem. Int. Ed. Engl.* **32**, 1524-1544.

8. Halpern, J. (1985) Asymmetric catalytic hydrogenation: mechanism and origin of enantioselection, in J. D. Morrison (ed), *Asymmetric Synthesis*, Academic Press, Inc., London, **5**, 41-69.

9. Koenig, K. E. (1985) The applicability of asymmetric homogeneous catalytic hydrogenation, in J. D. Morrison (ed), *Asymmetric Synthesis*, Academic Press, Inc., London, **5**, 71-101.

10. Brunner, H. and Zettlmeier, W. (1993) *Handbook of Enantioselective Catalysis*, VCH, Weinheim.

11. Amrani, Y., Lecomte, L., Sinou, D., Bakos, J., Toth, I. and Heil, B. (1989) Chiral sulfonated phosphines. Syntheses and use as ligands in asymmetric hydrogenation using an aqueous-organic two-phase system, *Organometallics* **8**, 542-547.

12. Reichart, C. (1988) *Solvents and Solvents Effects in Organic Chemistry*, VCH, Weinheim.

13. Lecomte, L., Sinou, D., Bakos, J., Toth, I. and Heil, B. (1989) Chiral sulfonated phosphines. II. Influence of water on the enantioselectivity in the reduction of dehydro-aminoacids, *J. Organomet. Chem.* **370**, 277-284.

14. Bakos, J., Karaivanov, R., Laghmari, M. and Sinou, D. (1994) Chiral sulfonated phosphines. IX. Role of the water in the hydrogenation of dehydroaminoacids, *Organometallics*, under press.

15. Sinou, D., Safi, M., Claver, C. and Masdeu, A. (1991) Chiral sulfonated phosphines. Part VII. Catalytic transfer-hydrogenation of unsaturated substrates with formates in the presence of water soluble complexes of rhodium, *J. Mol. Cat.* **68**, L-9-L12.

16. Trost, B. M. and Verhoeven, T. R. (1982) Organopalladium compounds in organic synthesis and in catalysis, in G. Wilkinson, A. F. G. Stone and E. W. Abel (eds), *Comprehensive Organometallic Chemistry*, Pergamon Press, New York, **8**, 799-938.

17. Tsuji, J. (1986) New general synthetic methods involving π-allylpalladium complexes as intermediates and neutral reactions conditions, *Tetrahedron* **42**, 4361-4401.

18. Heck, R. F. (1985) *Palladium Reagents in Organic Synthesis*, Academic Press, London.

19. Goldeski, S. A. (1991) Nucleophiles with allyl-metal complexes, in B. M. Trost (ed), *Comprehensive Organic Synthesis*, Pergamon Press, Oxford, **4**, 585-661.

20. Frost, C. G., Howarth, J. and Williams, J. M. (1992) Selectivity in palladium catalyzed allylic substitution, *Tetrahedron: Asymmetry* **3**, 1089-1122;

21. Safi, M. and Sinou, D. (1991) Palladium(0)-catalyzed substitution of allylic substrates in a two-phase aqueous-organic medium, *Tetrahedron Lett.* **32**, 2025-2028.

22. Blart, E., Genêt, J. P., Safi, M., Savignac, M. and Sinou, D. (1994) Palladium(0)-catalyzed substitution of allylic substrates in an aqueous-organic medium, *Tetrahedron*, **50**, 505-514.

23. Genêt, J. P., Blart, E., Savignac, M., Lemeune, S., Lemaire-Audoire, S., Paris, J. M. and Bernard, J. M. (1993) Practical palladium-mediated deprotective method of allyloxycarbonyl in aqueous media, *Tetrahedron* **50**, 497-505.

24. Moreno-Mañas, M., Pleixats, R. and Villarroya, M. (1993) Palladium-catalyzed allylation of pyrimidine-2,4-diones (uracils) and of 6 membered heterocyclic ambident sulfur nucleophiles, *Tetrahedron* **49**, 1457-1464.

25. Sigismondi, S., Sinou, D., Pérez, M., Moreno-Mañas, M., Pleixats, R. and Villarroya (1994) Palladium(0)-catalyzed allylation of uracils and 2-thiouracils. Drastic effect of an aqueous reaction medium on the regioselectivity, *Tetrahedron Lett.* in press.

CHIRAL SULFONATED PHOSPHINES IN ENANTIOSELECTIVE CATALYSIS

J. BAKOS
University of Veszprém,
Department of Organic Chemistry, H-8200 Veszprém, P.O.B. 158, Hungary

The imines ArC(Me)=NCHPh (Ar=Ph, 2-MeO-C_6H_4, 3-MeO-C_6H_4, 4-MeO-C_6H_4) are hydrogenated to the corresponding amines with extremely high enantioselectivities up to 96 % under very mild conditions, using rhodium complexes associated with sulfonated (-)-(2S,4S)-2,4-bis(diphenylphosphino)pentane (S,S)-(BDPP). These complexes are also effective as catalyst for asymmetric hydrogenolysis of sodium cis-epoxysuccinate to sodium hydroxysuccinate in aqueous-organic two-phase solvent system or in aqueous solution. A catalytic cycle involving rhodaoxetane intermediate and a heterolytic hydrogen splitting, is suggested. Catalytic hydrogenolysis of sodium salt of trans-phenylglycidic acid proceeds via a regioselective kinetic resolution to give sodium salt of α-hydroxycarboxylic acid.

1. Introduction

Water soluble organometallic compounds have attracted considerable interest as catalysts [1-3]. Recently a great deal of work has appeared in which highly polar functional groups as sulfonate [4-6], carboxylic [7-9], hydroxide [10-11], quaternized and protonated amino group [13-14] were incorporated into monotertiary or ditertiary phosphines to obtain water-soluble ligands. Transition-metal complexes of these phosphines have been applied to hydrogenation [5-6], hydroformylation [1], carbonylation [15], hydrocyanation [16], alkylation [17] and nucleophilic substitution [18]. Reported examples of highly enantioselective catalysis in aqueous system using water-soluble sulfonated phosphines as ligands include hydrogenation of dehydroaminoacids [6, 19-23] and imines [24]. The development of water-soluble catalyst is mainly due to the simplification of the catalyst-product separation, the economy and safety of using water as a solvent. There is also an expectation of other advantages with the use of this unusual solvent.

2. Chiral sulfonated phosphines

Several chelating chiral water-soluble phosphines have been reported in the literature. Sulfonated phosphines at present constitute the most widely used class of chiral ligands in water-soluble chiral metal complexes, especially in asymmetric catalysis. These in-

I. T. Horváth and F. Joó (eds.), Aqueous Organometallic Chemistry and Catalysis, 231–244.

clude (Figure 1) the sulfonated BDPP, -chiraphos, -prophos, -cyclobutaneDIOP [6], NMDPP [25] and -Binap [22].

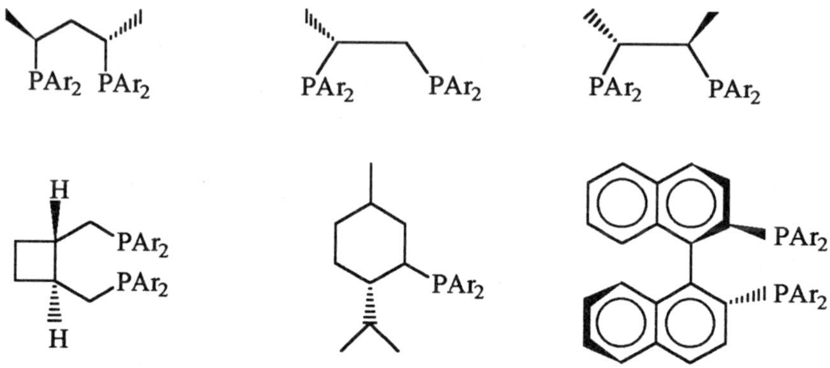

Figure 1. Water-soluble chiral sulfonated phosphines obtained by direct sulfonation with oleum.

All of these compounds (sodium salt) can be obtained by direct sulfonation of the parent ligand with oleum and subsequent neutralisation with NaOH. Sulfonation of (S,S)-BDPP was carried out in concentrated sulfuric acid containing 20% SO_3 in the manner described for triphenylphosphine [26]. To know the time and conditions necessary for tetrasulfonation of the diphosphines, we followed the reaction by ^{31}P NMR and HPLC by using various concentrations of the phosphine in oleum. The sequence of sulfonation of the diphosphine starts with the protonation of the phosphorus followed by the successive sulfonation of the phenyl ring giving rise to the formation of monosulfonated, disulfonated, trisulfonated, and tetrasulfonated diphosphines. Due to the protonation of the phosphorus the sulfonate group mostly occupies the position meta to the phosphorus atom. The tetrasulfonated derivatives are most easily obtained with 1,4-diphosphines followed by the 1,3- and 1,2-diphosphines.

Partial sulfonation of (S,S)-BDPP leading to new chiral center at the phosphorus changes the C_2 symmetry of the ligand to C_1 in the case of $BDPP_{MS}$ and $BDPP_{TRS}$

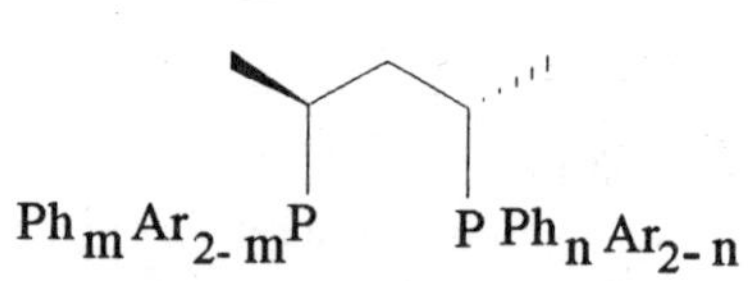

Nonsulfonated BDPP:	(2S,4S)	BDPP
Monosulfonated BDPP:	(PS,2S,4S)	$BDPP_{MS}$
	(PR,2S,4S)	
Disulfonated BDPP:	(PS,2S,4S,P'S)	$BDPP_{DS}$
	(PR,2S,4S,P'S) ≡ (PS,2S,4S,P'R)	
	(PR,2S,4S,P'R)	
Trisulfonated BDPP:	(PS,2S,4S)	$BDPP_{TRS}$
	(PR,2S,4S)	
Tetrasulfonated BDPP:	(2S,4S)	$BDPP_{TS}$

Figure 2. Formation of possible epimeric diphosphines in the sulfonation of (S,S)-BDPP.

(Figure 2). The $BDPP_{DS}$ consists of three epimers in 1 : 2 : 1 ratio due to the chiral phosphorus atoms. The sulfonation of (S,S)-BDPP was monitored using RP-HPLC analysis of the sodium salt on a C8 reverse phase column (25 cm x 4.6 mm), flow rate 1 ml/min, eluent gradient 0.5 % $(NH_4)_2CO_3$ buffer/CH_3CN (5-100 % in 8 min) [6, 25].

3. Rhodium(I)-Sulfonated BDPP Catalysed Asymmetric Hydrogenation of Imines in Aqueous-Organic Two-phase Solvent Systems

Catalytic asymmetric hydrogenation of prochiral alkenes and ketones have been extensively investigated and various homogeneous catalysts have been found to be efficient in the asymmetric hydrogenation of imines [27-33]. Our work in this area began with the search for a catalyst for the asymmetric reduction of acetophenonebenzylimine [27]. Optical yields up to 83 % were achieved by using low temperature (0° C), H_2 pressure of 70 atmosphere, methanol as a solvent, and a particular bidentate phosphine BDPP, in conjunction with the $[Rh(COD)Cl]_2$ precursor [29]. A strong dependence of the e.e. on the solvent and temperature was observed. Increase in the ratio of C_6H_6 is accompanied by a large fall in catalytic activity, and there is a reversal of stereoselectivity. The rate of the hydrogenation was about thirty times larger in methanol than in benzene.

When the reaction is carried out by using nonsulfonated and sulfonated BDPP in various alcohols and in a polar but non-hydrogen-bonding solvent (DMF) the fastest rate and the highest enantioselectivities are achieved in methanol (Table 1). An explanation for the poor rate in propan-2-ol, DMF and benzene is provided by the mechanism proposed by Wilkinson et al. [30]. It is implied that the function of the alcohol is to assist change from η^1 to η^2 of the imine binding via an intramolecular hydrogen bonding. This mechanism is adapted by James et al. [31] with the modification that hydrogenation probably occurs via an unsaturate route in the case of bidentate ligands. It is more difficult to rationalise the high enantioselectivity in methanol and ethanol. The alcohol perhaps occupies specific coordination sites on the metal and facilitates the binding of one of the two enantiotopic faces of the substrate, thus leading to high enantioselectivity. The steric crowding around the hydroxyl group probably restricts hydrogen bonding to the imine in propan-2-ol leading to poorer activity and optical induction.

We have recently described the first example of asymmetric catalytic hydrogenation of imines in a two-phase system using water-soluble catalyst to give secondary amines with practical enantioselectivity [24].

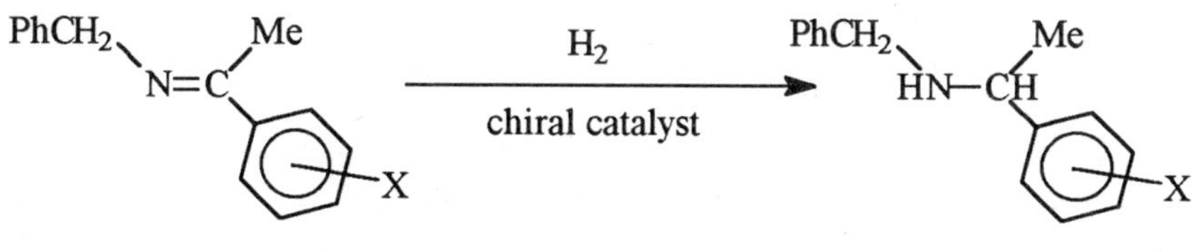

1

a, X = H, b, X = 4 - OCH3, c, X = 3 - OCH_3, d, X = 2 - OCH_3,

The catalytic hydrogenation of imines (a-d) was carried out as follows. The cyclooctadiene dimer $[Rh(COD)Cl]_2$ and an appropriate amount of sulfonated BDPP were dissolved in degassed H_2O/AcOEt (1:1) under an argon atmosphere. The substrate was then added. The two-phase liquid mixture was transferred to a stainless steel autoclave under argon, degassed three times with H_2 (70 bar) then pressurised to 70 bar, and the contents stirred for 6h at 20 °C. The amines were recovered in almost quantitative yield from reaction mixture by separation of the organic phase followed by pumping off the solvent and distillation of the residue under vacuum.

The hydrogenation is quite chemoselective, although a small amount of primary amine and ketone generated by hydrolysis of the imine are observed.

Table 1. Influence of solvent on the asymmetric hydrogenation of acetophenonebenzylimine[a]

Sulfonation degree of the ligand[b]	Solvent	Yield of amine (%)	E.e. (%)[g]
0	MeOH	96	74
0	EtOH	86	71
0	n-PrOH	65	44
0	i-PrOH	15	18
0	DMF	11	0
2.21[d]	MeOH	96	70
2.21	EtOH	86	70
2.21	n-PrOH	80	67
2.21	i-PrOH	65	9
4.00	MeOH	55	34
3.75[c]	MeOH	96	31
2.21	MeOH	96	70
1.65[e]	MeOH	98	77
1.41[f]	H_2O-CH_2Cl_2	84	7
1.41[f]	H_2O-C_6H_5Me	34	55
1.41[f]	H_2O-C_6H_6	67	75
1.41[f]	H_2O-AcOEt	96	96 (88)

[a]Conditions: Rh(I) : sulfonated BDPP : substrate = 1 : 1.1 : 100, at 20 °C, 70 atm of H_2, reaction time is 6 h. [b]Composition of the ligand used was determined by HPLC: [c]25 % $BDPP_{TRS}$ and 75 % $BDPP_{TS}$; [d]79 % $BDPP_{DS}$ and 21 % $BDPP_{TRS}$; [e]35 % $BDPP_{MS}$ and 65 % $BDPP_{DS}$; [f]59 % $BDPP_{MS}$ and 41 % $BDPP_{DS}$. [g]The optical yield of amines was determined by ^1H NMR. (300 MHz) spectroscopy. Optically pure mandelic acid as a chiral solvating agent induced non-equivalence = 0.25 ppm) in the signals of methyl and methin protons of amines and by HPLC analysis (values in parenthesis) of the Mosher derivative using Spherisorb S5OD2 as the column (150 x 4.6 mm), and CH_3CN:H_2O (60 : 40) as the eluent. Product configuration is (R) in all cases.

The data of Table 2 show that the in situ catalyst is most effective for hydrogenation of imines (a-d) when the sulfonation degree of the phosphine is below 2, and when the reaction is performed in H_2O-AcOEt two-phase solvent system. This unprecedent high enantioselectivity (96 %) and mild reaction condition appears to be superior to that known for hydrogenation of C=N double bond of imines. Kang et al. [32] have reported 91 % e.e. for the hydrogenation of acetophenonebenzylimine (a) using an in situ Rh(I)-cycphos system in the presence of iodide cocatalyst, but the reduction was carried out at -25 °C and took 144 h.

Table 2. The asymmetric hydrogenation of imines in two-phase solvent system[a]

Substrate	Sulfonation degree of the ligand[b]	Solvent	Yield of amine (%)	E.e. (%)[g]
1a	3.75[c]	H_2O-AcOEt	55	19
1a	2.21[d]	H_2O-AcOEt	41	32
1a	1.65[e]	H_2O-AcOEt	94	96 (95)
1b	1.65[e]	H_2O-AcOEt	96	95 (86)
1c	1.65[e]	H_2O-AcOEt	93	89 (86)
1d	1.65[e]	H_2O-AcOEt	94	91 (92)
1a	1.41[f]	H_2O-AcOEt	96	96 (88)

[a]Conditions: see in Table 1.

Although the number of water-soluble organometallic catalysts [34] is steadily growing, to our knowledge, this is the first example for a systematic comparison of the performance of the chiral ligand containing various numbers of sulfonate groups. The epimers of the ligands have not been separated, however, determination of the ligand composition has been made by HPLC. Thus, degree of sulfonation could be calculated. The extent of asymmetric induction as a function of sulfonation degree or composition of the ligand shows that the presence of monosulfonated derivative is important for high chiral discrimination, in spite of that, this is a mixture of two diastereomers.

Mixture of $BDPP_{MS}$ and $BDPP_{DS}$ forms complexes when reacted in a 2.2 : 1 ratio with $[Rh(COD)Cl]_2$ in D_2O/EtOAc. ^{31}P NMR spectroscopic studies of the two phases show a selective separation of the rhodium complexes containing $BDPP_{MS}$ from that containing $BDPP_{DS}$.(Figure 3). The complex associated with $BDPP_{MS}$ yields two different complexes in equal amount. Each complex gives rise to an eight line pattern due to non-equivalence of the phosphorus atoms (in CD_2Cl_2: δ 28.5 ppm, dd, $^1J_{Rh\text{-}P}$ = 143 Hz, δ 26.5 ppm, dd, $^1J_{Rh\text{-}P}$ = 141 Hz, $^2J_{P\text{-}P}$ = 46 Hz; δ 27.0 ppm, dd, $^1J_{Rh\text{-}P}$ = 143 Hz, δ 26.4 ppm, dd, $^1J_{Rh\text{-}P}$ = 143 Hz, $^2J_{P\text{-}P}$ = 46 Hz). Due to the excess of the ligand used a small amount of $Rh(BDPP_{MS})_2Cl$ is formed (δ 25.0 ppm, d, $^1J_{Rh\text{-}P}$ = 134 Hz).

The $[Rh(COD)Cl]_2$ dimer associated with $BDPP_{DS}$ gives three different complexes in 1:2:1 ratio (in D_2O: δ 27.1 ppm, d, $^1J_{Rh\text{-}P}$ = 142 Hz, δ 27.6 ppm, d, $^1J_{Rh\text{-}P}$ = 142 Hz, $^2J_{P\text{-}P} \approx 0$ Hz; δ 28.1 ppm, d, $^1J_{Rh\text{-}P}$ = 144 Hz) due to the configuration of the phosphorus atoms in the isomers (R,R; R,S≡S,R; S,S).

Experimental catalytic results confirm the spectroscopic evidences: Rh complexes of $BDPP_{MS}$ in the EtOAc phase of the two-phase system give not only a very active catalyst but also leads to very high enantioselection in the hydrogenation of imines, whereas Rh complexes of $BDPP_{DS}$ are only soluble in the aqueous phase leading to low hydrogenation activity and enantioselectivity. The same phenomena were observed by de Vries et al. when $BDPP_{MS}$ and $BDPP_{DS}$ were separated by column chromatography and used for the same catalytic reaction [35].

Thus, the present reaction not only introduces a novel class of enantioselective reduction of C=N double bond using a water-soluble transition metal complex as catalyst, but suggests that this approach to the synthesis of new catalysts has great promise.

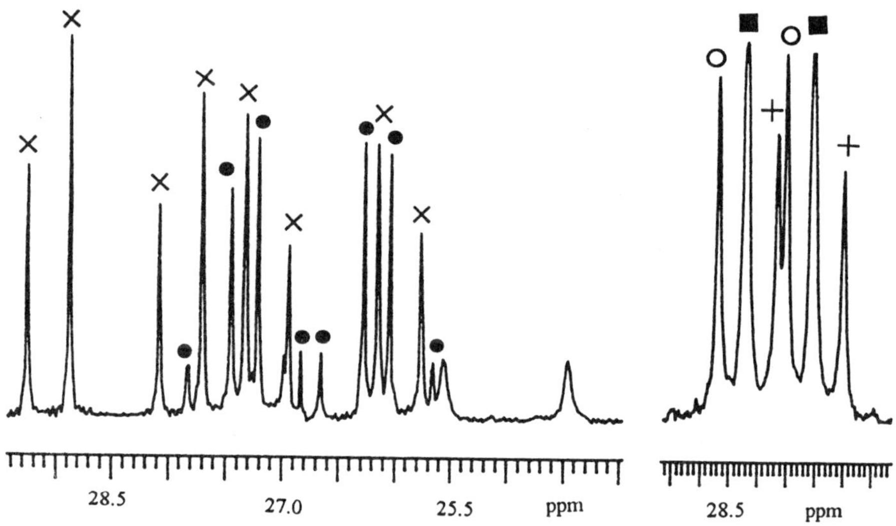

Figure 3. ^{31}P NMR spectra of the rhodium-cyclooctadiene comlexes of $BDPP_{MS}$ and $BDPP_{DS}$.

4. Rhodium(I)-Catalysed Asymmetric Hydrogenolysis

4.1. HYDROGENOLYSIS OF SODIUM-EPOXYSUCCINATE

Rhodium(I) catalysts formed with sulfonated diphosphines are efficient for the asymmetric hydrogenation of enamides (88% e.e.) [6] and Schiff-bases (96 % e.e.) [24] in aqueous-organic two-phase solvent system. In the latter case sulfonated (S,S)-BDPP catalyst gave higher enantioselectivity than the analogous lipophilic catalyst containing the nonsulfonated (S,S)-BDPP.

We extended our study for the asymmetric hydrogenolysis of a symmetrical epoxide using rhodium(I) complexes of sulfonated (S,S)-BDPP. Thus obtained malic acid is a versatile building block for natural product syntheses.

The first example of homogeneous catalytic asymmetric hydrogenolysis of an epoxide was reported only recently [36]. A number of most commonly used lipophilic (bisphosphine)rhodium(I) catalysts were tested. It was pointed out that the substrate sodium salt of epoxysuccinic acid is not soluble in most organic solvents, therefore the reaction was accomplished in a solvent mixture of H_2O-MeOH (1:2~1:3), but even in this polar solvent mixture the solubility of such a hydrophilic substrate is rather limited (~ 40 g/l). In order to expand the methodology for this reaction, we present herein our results obtained by the use of water-soluble catalyst [30].

The cyclooctadiene dimer $[Rh(COD)Cl]_2$ (0.0250 mmol) and an appropriate amount of sulfonated BDPP were dissolved in 10 ml of degassed H_2O/AcOEt (1 : 1) in a Schlenk tube, under an argon atmosphere. The two-phase system was stirred for 20 minutes and then 0.88 g (5 mmol) of disodium cis-epoxysuccinate was added. The mixture was injected to a 20 ml stainless steel autoclave under argon. The reaction vessel was then flushed with hydrogen and pressurised with the same gas to, usually, 70 atmosphere, and the contents were stirred for 6 h.

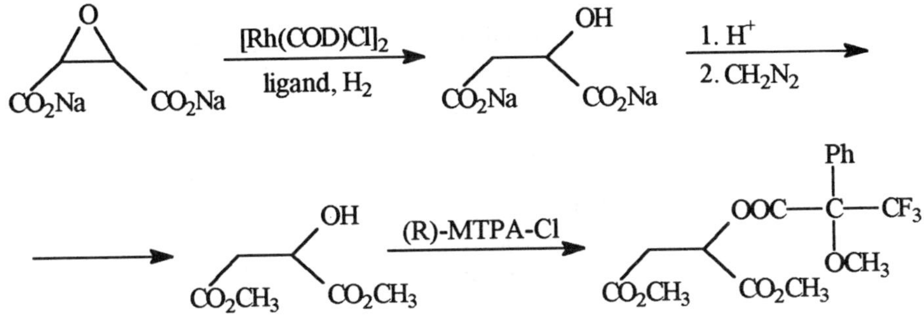

Scheme 1. An outline of the hydrogenolysis and the workup of the product for the determination of enantiomeric excess.

Table 3. Hydrogenolysis of sodium cis-epoxysuccinate[a]

Entry	Ligand	Solvent	$P(H)_2$ (bar)	E.e.* (%)
1	BDPP	H_2O/MeOH	70	39 (41)
2	$BDPP_{MS}$	H_2O/EtOAc	100	36 (34)
3	$BDPP_{MS}$	H_2O/EtOAc	70	34
4	$BDPP_{MS}$	H_2O/EtOAc	35	34
5	$BDPP_{DS}$	H_2O	100	30
6	$BDPP_{DS}$	H_2O	70	32
7	$BDPP_{DS}$	H_2O/EtOAc	70	31
8	$BDPP_{TS}$	H_2O	70	26
9	$BDPP_{TS}$	H_2O	35	27

[a]Rh : P : substrate (1/2.2/100); 20 °C, 6 h; Chemical yield > 90 %

*Determined by chiral GLC and ^1H NMR spectrum (400 MHz) of diastereomeric esters of (R)-(+)-α-methoxy-α-(trifluoromethyl)phenylacetic acid. For the determination of enantiomeric excess (e.e.), the aqueous phase of the reaction mixture was separated and then passed down a column packed with ion-exchange resin (H^+ form). The column was washed with water. The water was then evaporated under reduced pressure and the solid residue was dissolved in methanol and treated with diazomethane. The conversion of hydrogenolysis was determined by capillary glc analysis of the corresponding methyl esters. Then the hydroxyl group was esterified with (S)-MTPA chloride to give the Mosher ester derivatives [38]. The e.e. value of this ester was obtained from ^1H NMR examination of its methyl protons. The absolute configuration of the product acid was assigned on the basis of the spectra of the Mosher derivatives obtained from the commercially available enantiomerically pure malic acid. Both the percentage conversion and the e.e. of the malic acid in the form of methyl ester were determined also by chiral GLC on CP-cyclodextrin-β -2,3,6-M-19 column (50 m, 0.25 mm i.d., Chrompack). The enantiomers of 2-hydroxy-succinic acid dimethyl ester were base-line resolved. The configuration of the product was R giving smaller retention time.

As shown in Table 3 among ligands examined, monosulfonated BDPP was the best. The sulfonation degree of the ligand was found to have small influence on enantioselectivity.

To gain insight into the features of the reaction by water-soluble catalysts deuterium labelling study was carried out. The reaction of sodium cis-epoxysuccinate was carried out in deuterium oxide at 25 °C under 20 atm of deuterium with $[Rh(COD)Cl]_2$ in association with $BDPP_{TS}$ as catalyst (Scheme 2). The dimethyl ester of the product was analysed by NMR.

The ^1H NMR of the sample shows incorporation of deuterium on the β-carbon, and no deuterium incorporation into the CH-OH group: the doublet of triplet at δ 2.88 ppm ($^2J_{H\text{-}H}$ = 4.1 Hz, $^2J_{H\text{-}D}$ = 2.1 Hz) corresponds to a β-H, whereas a doublet corresponding to the-CH< group of the labelled ester appears at δ 4.51 ppm ($^2J_{H\text{-}H}$ = 4.1 Hz). The other diastereotopic proton of the -CH_2- is missing owing to the replacement by deuterium. The unlabelled 2-hydroxy-succinic acid dimethyl ester gives two doublets of doublet at δ 2.80 ppm (^2J = 16.5 Hz, ^3J = 6.1 Hz) and δ 2.88 ppm (^2J = 16.5 Hz, ^3J = 4.1 Hz) corresponding to the diastereotopic protons of -CH_2- and a doublet of doublet at δ 4.51 ppm ^3J=6.1 Hz, ^3J = 4.1 Hz) corresponding to the -CH< group. The $^{13}C\{^1H\}$ NMR spectrum of the sodium salt in D_2O shows a singlet at δ 73.0 ppm for the -CH< group and a triplet at δ 45.0 ppm ($^3J_{C\text{-}D}$ = 20 Hz) for the -CHD< group.

These results indicate that the incorporation of deuterium occurs regiospecifically at the β position and is only due to the dominant hydrogenolysis route. These spectral evidences unambiguously show that hydrogenolysis occurs and the reaction proceeds via a direct C-O bond cleavage, instead of an epoxide to ketone isomerization followed by ketone or enol hydrogenation (Scheme 2). The same phenomena were observed by Chan et al. using lipophilic (bisphosphine)rhodium(I) catalyst systems [36].

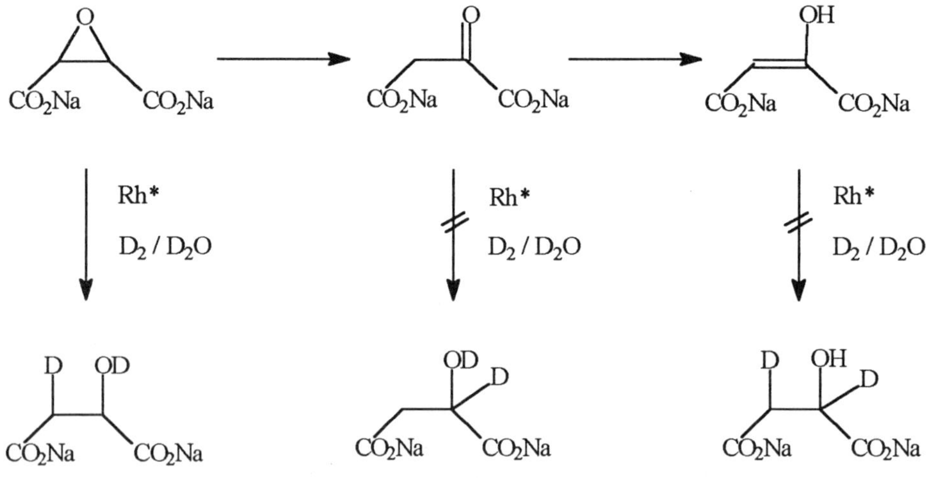

Scheme 2. An outline of the possible pathways in the hydrogenolysis of sodium cis-epoxysuccinate.

When $[Rh(COD)Cl]_2$ is treated with (S,S)-$BDPP_{TS}$ under argon atmosphere in D_2O/EtOAc two-phase solvent system at room temperature for 10 min, ^{31}P NMR spectrum of the aqueous phase reveals a doublet at δ 29.3 ppm ($^1J_{Rh\text{-}P}$ = 144 Hz) due to the formation of $[Rh(COD)(S,S)\text{-}BDPP_{TS}]^+$. The active aqua catalyst $[Rh(S,S)\text{-}BDPP_{TS}(D_2O)_2]^+$ is formed (δ 53.2 ppm, d, $^1J_{Rh\text{-}P}$ = 185 Hz) within 40 min when the aqueous solution contacted with 30 atmosphere of hydrogen, no dihydride being detected. Further addition of two equiv. of (S,S)-$BDPP_{TS}$ or initial mixing of 4 equiv. with $[Rh(COD)Cl]_2$ gives the inactive $[Rh\{(S,S)\text{-}BDPP_{TS}\}_2]^+$ (δ 24.9 ppm, d, $^1J_{Rh\text{-}P}$ = 131 Hz).

High pressure NMR studies showed that the catalytic hydrogenolysis of cis epoxysuccinate to sodium salt of malic acid in aqueous solution, using Rh dimer associated with $BDPP_{TS}$ involves formation of a rhodaoxetane complex. Characterisation of the intermediates in these reactions is important for an understanding of the chemistry in aqueous system.

^{31}P NMR spectra of the catalytic and stoichiometric reaction of the active aqua catalyst in D_2O with the substrate sodium cis-epoxysuccinate show the presence of four isomers (P_1 δ 35-40 ppm, P_2 40-44 ppm, $^1J_{Rh\text{-}P}$ = 126, 136 Hz; 129, 131 Hz; 126, 130 Hz; 120, 139 Hz; $^2J_{P\text{-}P}$ = 42 Hz). Each isomer gives rise to a doublet of doublets for each phosphorus atom resulting in a thirty-two line spectrum. Although definitive assignments of the corresponding chemical shift values and coupling constants are not yet possible owning to the complexity of the ^{31}P NMR spectra, ^{31}P detected ^{103}Rh NMR data (δ 1954, 1921, 1920, 1887 ppm) are consistent with the premise that four isomers are present due to the two chiral centres in the ligand and the rhodaoxetane ring.

^{13}C NMR spectrum defines the presence of a mutually trans alkyl and phosphine ligand. This conclusion is based on the usual assumption that a large ^{13}C-P coupling (δ 76.2 ppm, d of mp., $J_{C\text{-}P}$ = 83 Hz) is associated with the trans related ligand pair [39-40].

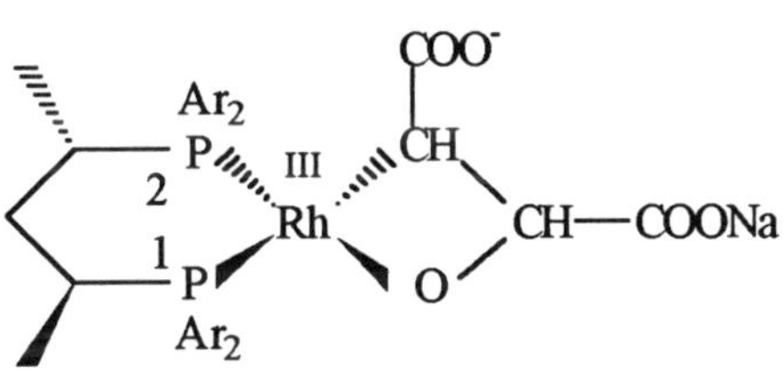

Four diastereomers: 2S, 3S, S, S
2S, 3R, S, S
2R, 3S, S, S
2R, 3R, S, S

Figure 4. Proposed intermediate of hydrogenolysis.

The proposed mechanism outlined in the Scheme 3 implies that the first step is the formation of a substrate complex through the lone pair of the oxygen of the oxirane ring. Apparently, a very narrow reactivity window exists for this reaction. Epoxides without carboxy functionality are not hydrogenolysed under similar conditions. The presence of a carboxylate anion probably helps to bring the substrate to the catalyst centre via the coordination of the oxygen of the oxirane ring. Our studies indicate that the oxidative addition of the C-O to the rhodium is a more facile process than the hydride formation. Rhodium hydrides are not readily produced from 1 : 1 rhodium(I) complexes of ligands such as diop and diphos [39-40].

Metallaoxetanes, analogous to our intermediate, are often postulated as intermediates in various metal-mediated reactions of epoxides, in olefin methathesis, and in conversion of carbonyls to alkenes by metal alkylidenes [41]. Rate determinig reaction between the rhodaoxetane and hydrogen involves the heterolytic splitting of H_2 with the formation of a hydrido rhodium(III) complex.

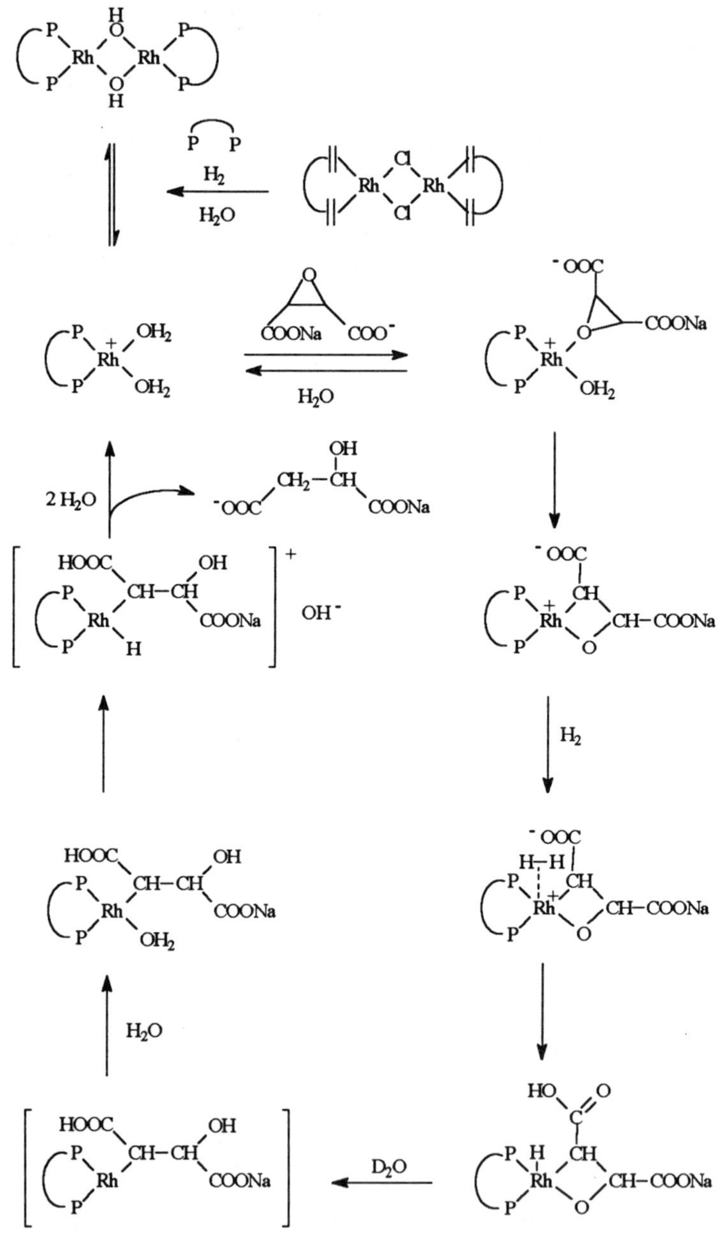

Scheme 3. Proposed cycle for catalytic hydrogenolysis of sodium cis-epoxisuccinate.

The heterolytic splitting of hydrogen is assumed to be the consequence of the parallel action of the positively charged central metal ion (Rh(III)) and the carboxylate anion. The Rh(III) can be considered as a weak acid and the carboxylate anion is a relatively strong base. This may be regarded as the driving force for the heterolytic bond-breaking of hydrogen.

The hydrogen transfer could lead to rhodium alkyl intermediate (assuming this takes place in the direction shown in Scheme 3). The oxidative addition of H_2O (or D_2O) could give hydridoalkylrhodium complex. The second hydride transfer could then give the product and the aqua catalyst. Using Rh dimer associated with $BDPP_{TS}$ involves the formation of a binuclear μ-hydroxo rhodium complex under hydrogen pressure. The formation of a binuclear μ-hydroxo rhodium complex $[Rh(\mu\text{-}OH)(TPPTS)_2]_2$ has been suggested by Herrmann et al. in the reaction of $RhCl(TPPTS)_3$ with hydrazine [42].

Performing the reaction under hydrogen in D_2O or under deuterium in H_2O ~100 % incorporation of deuterium or hydrogen to the β position from the solvent was observed, respectively. However a possible explanation for the deuterium incorporation could be a H/D exchange reaction between deuterium oxide and hydrogen (or water and deuterium) in the presence of rhodium complex leading to the formation of HD. The fact that the amount of deuterium incorporation (using $BDPP_{TS}$) in the β position of the product is significantly higher than can be expected from the isotopic compositions of the aqueous phase at the end of the reaction indicates the participation of the water in the reaction.

4.2. KINETIC RESOLUTION OF RACEMIC TRANS-PHENYLGLYCIDATE

The results obtained in the hydrogenolysis of sodium cis epoxysuccinate prompted us to examine the bond cleavage in the case of a racemic substrate. One would expect such an asymmetric catalytic system to be sensitive to pre-existing chirality in a substrate. Efficient chemical kinetic resolution of functionalised epoxide has been achieved by the use of water-soluble catalyst.

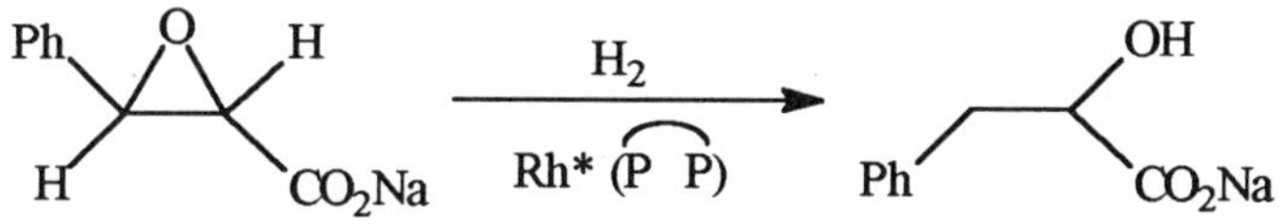

2

Several notable observations in this study are: (i) the bond cleavage takes place regioselectively according to the Equation 2 in the β position to the ester function, (ii) the enantiomeric excess of the product depends on the conversion, (iii) the (S,S)-$BDPP_{TS}$ modified catalyst favours R-product.

There is a mathematical treatment of three variables: the % e.e. of the product α-hydroxycarboxylic acid, the extent of conversion (c) and the relative rate (k_{fast}/k_{slow}). Knowledge of any two of the variables allows the definition of the third. Determination of the extent of conversion and enantiomeric excess of the product allowed us to calculate the relative rate (k_R/k_S).

The first thing to note are the entries in columns headed (k_{fast}/k_{slow}). These are the relative rates for the enantiomeric pairs. Under same experimental conditions these values do not depend on the conversion indicating that the hydrogenolysis proceeds via kinetic resolution (Scheme 4) of the sodium salt of trans-phenylglycidic acid (Table 4).

Table 4. Kinetic resolution of racemic sodium trans-phenylglycidate

Conversion (%)	E.e. (%)	k_R/k_S
15.7	23.0	1.66
38.4	18.4	1.61
48.4	16.2	1.58
63.1	15.0	1.68
83.8	10.0	1.81
96.1	2.1	1.40

Reaction conditions: 20 °C, 70 bar H_2, H_2O/EtOAc
Substrate/Rh/P=100/1/2.2. Ligand: $BDPP_{TS}$
Workup of the product: see footnote in Table 3

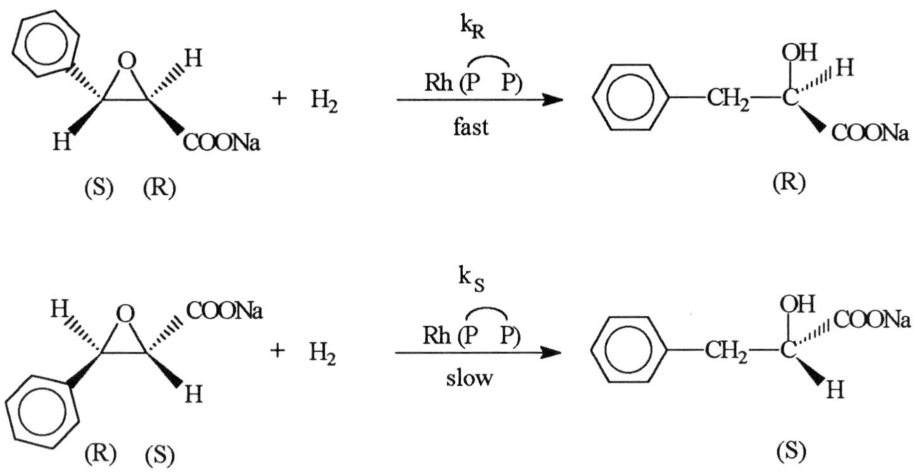

Scheme 4. Differential R and S selectivity for enantiomers of sodium trans-phenylglycidate.

The present method would provide a useful method for the preparation of chiral trans-disubstituted epoxide as well as chiral α-hydroxycarboxylic acids.

Acknowledgement

Financial support from the Hungarian National Science Foundation (OTKA-2320 and T-4293) and OMFB (H 9112-0276) in the framework of ACCORD Programme is gratefully acknowledged. Part of this work results from a collaboration with Professor D. Sinou (Lyon). I am also grateful to Professor L. Markó (Veszprém), Professor C. J. Elsevier and Dr. I. Tóth (Amsterdam) for helpful discussions.

References

1. Kuntz, E.G. (1987) Homogeneous catalysis in water, Chemtech. 570-575.
2. Joó, F., Tóth, Z. (1980) Catalysis by water-soluble phosphine complexes of transition metal ions in aqueous and two-phase media, J. Mol. Catal. **8**, 369-383.
3. Sinou, D. (1987) Water-soluble phosphines. Synthesis and use in catalysis. Bull. Soc. Chim. Fr. 480-486.
4. Ahrland, S., Chatt, J., Davies N.R., and Williams, A.A. (1958) The relative affinities of co-ordinating atoms for silver ion. Part II. Nitrogen, phosphorus, and arsenic, J. Chem. Soc. (London), 276-288.
5. Borowski, A.F., Cole-Hamilton D.J., and Wilkinson, G. (1978) Water-soluble transition metal phosphine complexes and their use in two-phase catalytic reactions of olefins, Nouv. J. Chim., **2**, 137-144.
6. Amrani, Y., Lecomte, L., Sinou, D., Bakos, J., Toth, I., and Heil, B. (1989) Chiral sulfonated phosphines. Syntheses and use as a ligand in asymmetric hydrogenation using an aqueous-organic two-phase solvent system, Organometallics **8**, 542-547.
7. Petit, L. D. and Irving, H.M.N.H. (1964) Ligand containing elements of group V. Part I. J. Chem. Soc. (London), 5336-5343.
8. Russel M.J.H. (1988) Water soluble rhodium catalysts Platinum Metal Rev., **32**, 179-186.
9. Murrer, B.A. (1982) D.E. 3.135.127A1 (Johnson Matthey Corp.).
10. Schwarzenbach, G. and Schellenberg, M. (1965) Die Komplexchemie des Methylguecksilber-Kations, Helv. Chim. Acta, **48**, 28-46.
11. Harrison, K.N., Hoye, P.A.T., Orpen, A.G., Pringle, P.G. and Smith M.B. (1989) Water soluble, zero-valent, platinum-, and nickel-$P(CH_2OH)_3$ complexes: catalysts for addition of PH_3 to CH_2O, 1096-1097.
12. Smith, R.T., Ungar, R.K., and Baird, M.C. (1982) Rhodium complexes of water-soluble phosphine(2-diphenylphosphinoethyl)trimethylammonium nitrate: theirchemistry in polar solvents, and their use as catalysts in aqueous/organic two-phase systems and absorbed on a cation exchange resin, Trans. Met. Chem., **7**, 288-289.
13. Toth, I., and Hanson, B.E. (1990) Novel chiral water soluble phosphines I. Preparation and characterization of amine functionalized DIOP, chiraphos, and BDPP derivatives and quaternization of their rhodium complexes, Tetrahedron: Asymmety **1**, 895-912.
14. Toth, I. and Hanson, B.E. and Davis, M. (1990) Novel chiral water soluble phosphines II. Application in catalytic asymmetric hydrogenation, Tetrahedron: Asymmetry **1**, 913-930.
15. Okano, T.,. Uchida, I., Nakagaki, T., Konishi, H., and Kiji, J. (1989) Carbonylation of benzyl chloride catalyzed by water-soluble palladium phosphine complex in two-phase system, J. Mol. Catal. **54**, 65-71.
16. Kuntz, E. (1977) Addition of hydrogen cyanide to unsaturated organic compounds with et least one ethylenic double bond, Ger. Offen 2.700.904 (Rhone-Poulenc Industries).
17. Mignani, G., Morel, D. and Colleuille, Y. (1985) A novel method for the isoprenylation of β-dicarbonyl compounds, Tetrahedron Letters, **26**, 6337-6340.
18. Safi, M., and Sinou, D. (1991) Palladium(0)-catalyzed substitution of allylic substrates in two-phase aqueous-organic medium, Tetrahedron Letters **32**, 2025-2028.
19. Alario, F., Amrani, Y., Colleuille, Y., Dang, T.P., Jenck, J., Morel, D., and Sinou, D. (1986) Asymmetric hydrogenationin aqueous-organic two-phase systems using rhodium complexes of sulfonated phosphines, J. Chem. Soc., Chem. Commun. 202-203.
20. Lecomte, L., Sinou, D., Bakos, J., Toth, I. and Heil, B. (1989) Chiral sulphonated phosphines II. Influence of water on the enantioselectivity in the hydrogenation of dehydro-aminoacids, J. Organomet. Chem. **370**, 277-284.
21. Bakos, J., Karaivanov, R., Laghmari, M., and Sinou, D. (in press) Chiral sulfonated phosphine. IX. Role of the water in the hydrogenation of dehydroaminoacids, Organometallics.
22. Wan, K., Davis, M.E. (1993) Asymmetric hydrogenation in water by a rhodium complex of sulfonated 2,2'-bis(diphenylphosphino)-1,1'-binaphthyl (binap) J. Chem. Soc., Chem. Commun. 1262-1264.
23. Wan, K., Davis, M.E. (1993) Rhutenium (II)-sulfonated BINAP: a novel water-soluble asymmetric hydrogenation catalyst, Tetrahedron: Asymmetry **12**, 2461-2468.
24. Bakos, J. Heil, B., Orosz, Á., Laghmari, M., Lhoste, P. and Sinou, D. (1991) Rhodium(I)-Sulfonated-bdpp Catalysed Asymmetric Hydrogenation of Imines in Aqueous-Organic Two-Phase Solvent Systems, J. Chem. Soc., Chem. Commun., 1684-1685.

25. Lecomte, L., Triolet, J. and Sinou, D. (1987) High performance liquid chromatographic separation of sulphonated phosphines, J. Chromatogr. **408**, 416-419.
26. Kuntz, E.G. (1976) Aldehydes by hydroformylation of olefins Br. Pat. 1540242 (Rhone-Poulenc Industries).
27. Vastag, S., Bakos, J., Tõrös, S., Takach, N.E., King, R.B., Heil, B. and Markó, L. (1984) Rhodium complexes as homogeneous catalysts. 14. Asymmetric hydrogenation of Schiff base of acetophenone - effect of phosphine and catalyst structure on enantioselectivity, J. Mol. Catal. **22**, 283-287.
28. Bakos, J., Toth, I., Heil, B. and Markó, L. (1985) A facile method for the preparation of 2,4-bis(diphenylphosphino)pentane (BDPP) enantiomers and their application in asymmetric hydrogenation, J. Organomet. Chem. **279**, 23-29.
29. Bakos, J., Toth, I., Heil, B., Szalontai, G., Párkányi, l., Fülöp, V. (1989) Catalytic and structural studies of Rh(I) complexes of (2S,4S)-2,4- bis(diphenylphosphino)pentane. Asymmetric hydrogenation of acetophenonebenzylimine and acetophenone, J. Organomet. Chem. **370**, 263-276.
30. Longley, C.J., Goodwin, T.J. and Wilkinson, G. (1986) Hydrogenation of imines by rhodium-phosphine complexes, Polyhedron **5**, 1625-1628.
31. Becalski, A.G., Cullen, W.R., Fryzuk, M.D., James, B.R., Kang, G.-J. and Rettig, S.J. (1991) Catalytic asymmetric hydrogenation of imines. Use of rhodium(I)/phosphine complexes and characterisation of rhodium(I)/imine complexes, Inorg. Chem. **30**, 5002-5008.
32. Kang, G.-J., Cullen, W.R., Fryzuk, M.D., James, B.R. and Kutney, .J.P. (1988) Catalytic asymmetric hydrogenation of imines. Use of rhodium(I)/phosphine complexes and characterisation of rhodium(I)/imine complexes, J. Chem. Soc., Chem. Commun. 5008-5009.
33. Spindler,F., Pugin, B. and Blaser, H.-U., (1990) Angew. Chem., Int. Ed. Eng. **29**, 558-559.
34. Herrmann, W.A., Kohlpaintner, C.W. (1993) Water-soluble ligands, metal complexes, and catalysts: synergism of homogeneous and heterogeneous catalysis, Angew. Chem. Int. Ed. Engl. **32**, 1524-1544.
35. Lensink, C., de Vries, J.G. (1992) Improving enantioselectivity by using a mono-sulfonated diphosphine as ligand for homogeneous imine hydrogenation, Tetrahedron: Asymmetry 3, 235-238.
36. Chan A.S.C. and Coleman J.P. (1991) Homogeneous asymmetric hydrogenolysis of sodium epoxysuccinate: the first example of asymmetric hydrogenolysis of an epoxide, J. Chem. Soc., Chem. Commun. 535-536.
37. This work has recently been reported. Bakos, J. Orosz, A., Sinou D. (1990) Rhodium(I) sulphonated phosphine complexes as catalysts for asymmetric hydrogenolysis of epoxides, Abstract of Papers, ISHC-7, Lyon, , P-43.
38. Dale, J.A.,Dull, D.L. and Mosher, H.S. (1969) J. Org. Chem. **34**, 3543.
39. Chan, A.S.C., Halpern, J. (1980) Interception and characterization of a hydridoalkylrhodium intermediate in a homogeneous catalytic hydrogenation reaction, J. Am. Chem. Soc. **102**, 838-840.
40. Brown, J.M. and Chaloner, P.A. (1980) Structural characterisation of a transient intermediate in rhodium-catalysed asymmetric homogeneous hydrogenation, J. Chem. Soc., Chem. Commun. 344-346.
41. Calhorda, M.J., Galvao, A.M., Ünaleroglu, c., Zlota, A.A., Frolow, F., and Milstein, D. (1993) Rhodaoxetane: synthesis, structure, and theoretical evaluation, Organometallics **12**, 3316-3325.
42. Herrmann, W.A., Kulpe, J.A., Konkol, W., Bahrmann, H. (1990) Wasserlösliche Metallkomplexe und Katalysatoren II. Tris(natrium-m-sulfonatophenyl)phosphan (TPPTS) und katalyserelevanter Rhodium-Komplexe, J. Organomet. Chem. **389**, 85-101.
43. Martin, V.S., Woodard, S.S., Katsuki, T., Yamada, Y., Ikeda, M., and Sharpless, B. (1981) Kinetic resolution of racemic alcohols by enantioselective epoxidation. A route to substances of absolute enantiomeric purity, J. Am. Chem. Soc. **103**, 6237-6240.

ASYMMETRIC COMPLEX CATALYSIS IN MICELLAR SYSTEMS

G. OEHME, I. GRASSERT, N. FLACH
Institut für Organische Katalyseforschung an der Universität Rostock e.V.
Buchbinderstraße 5-6, 18055 Rostock, Germany

Introduction

The development of organometallic chemistry is closely connected to the application of organic solvents especially solvents with coordinating properties.
Water as solvent became interesting with the idea of phase separation. It seems to be attractive to transfer typical advantages of homogeneously catalytic systems, like high activity, selectivity, and reproducibility, into a two- or multiphase-system with the same achievements and the advantage of a convenient separation of products and catalysts.
Usually, four types of multiphase systems are to discuss
1. supported catalysts
2. clusters as catalysts
3. biphasic and phase-transfer catalysis (liquid-liquid)
4. catalysis in micellar and vesicular systems.

A lot of work has been done in the immobilization of homogeneous catalysts on inorganic and organic supports [1] but the expense of preparation is often high and the lifetime too short. Nevertheless, immobilization is the standard method to combine the advantages of homogeneous and heterogeneous catalysis.
Catalysis by clusters can be seen as a microheterogeneous system. Actually, only a few examples are really successful [2] in catalysis.
Biphasic catalysis and phase transfer catalysis have developed into a very important field within coordination catalysis with the premise, that the solubilities of catalyst and educts or products, respectively, are quite different [3]. Two processes of industrial importance are to be mentioned as biphasic catalytic systems: the Shell Higher Olefin Process [4] and hydroformylation of olefins by Rhone Poulenc/Ruhrchemie [5]. The latter process stimulated the development of water soluble phosphins [6]. We also were interested in the synthesis of water soluble phosphines [7] and their use in coordination catalysis [8].
Transfer of coordination catalysis with transition metal complexes into colloidal systems with micelles or vesicles in an aqueous medium is not very common [9] although there is a strong resemblance to enzymes as very effective natural systems [10].
The comparison of homogeneous and micellar systems shows us a new dimension: the voluntary organization of amphiphilic molecules to special assemblies.
Some principles of aggregation are shown in Fig. 1.
The pictures are taken from a review by J. Fendler, one of the pioneers in membrane mimetic chemistry [11]. An amphiphilic molecule is built of a highly hydrophilic

I. T. Horváth and F. Joó (eds.), Aqueous Organometallic Chemistry and Catalysis, 245–257.

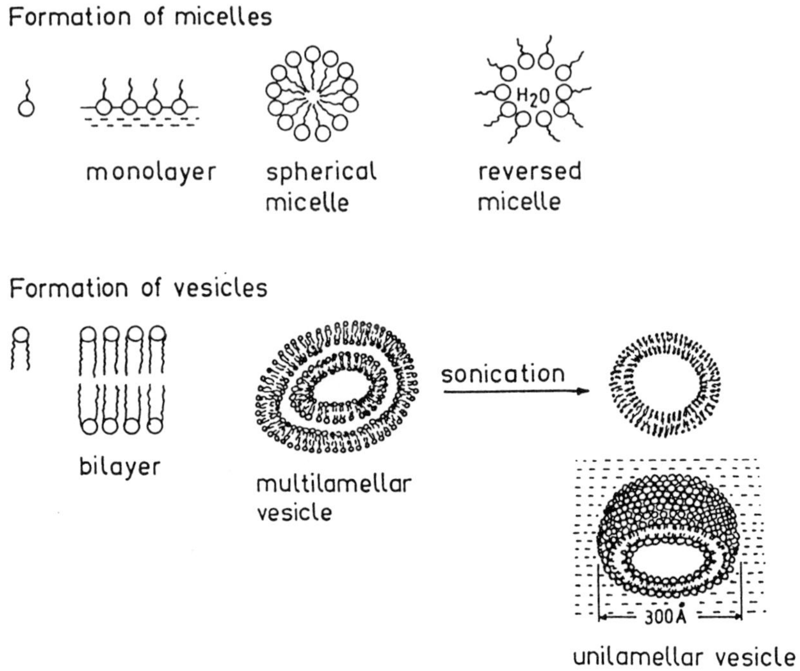

Figure 1. Formation of micelles and vesicles.

headgroup drawn as a circle and one or two hydrophobic (lipophilic) tails. The headgroup can be anionic, cationic, zwitterionic or nonionic, the hydrophobic tail should be an alkyl chain between C12 and C18 resulting in an optimal hydrophilic-lipophilic balance (HLB). Typically, one-tail amphiphiles assemble into monolayers and spherical micelles and two-tails amphiphiles assemble to bilayers and spherical multilamellar and unilamellar vesicles. The driving force for an association in water is the hydrophobic effect based on an increase of entropy [12], the driving force of an association in nonpolar solvents around a microdrop of water seems to be the gain of enthalpy (reversed micelles).
Some organic reactions are significantly influenced in rate and selectivity if conducted in micellar systems and the field of micellar effects is well investigated [13] especially in photochemistry [14].

Hydrogenation in micelles

Our interest was focussed on the asymmetric hydrogenation of amino acid precursors by optically active Rhodium(I) complexes. The common solvent for this reaction is

methanol. As described by Sinou et al. [15] the use of water as medium decreases both, activity and enantioselectivity, significantly. The reason may be low solubility as well as lack of hydrophobicity.
In a first communication [16] we showed, that addition of a small amount of the anionic surfactant SDS (sodium dodecyl sulfate) or the nonionic surfactant Triton X-100 both lead to encouraging enhancements in activity and enantioselectivity. The systems used are given in Fig. 2.

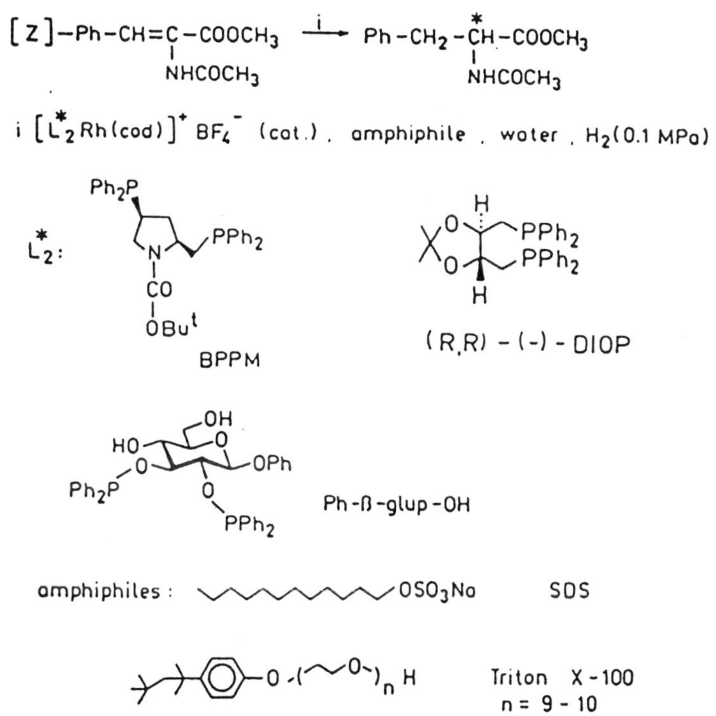

Figure 2. Asymmetric hydrogenation systems used in our investigations [16].

With one exception we used in situ-systems starting from $[Rh(cod)_2]BF_4$ and a small excess of an optically active chelating phosphine. Three examples of very common ligands are BPPM, introduced by Achiwa [17], DIOP, at first synthesized by Kagan [18], and Ph-ß-glup-OH, developed by Selke [19].
With all catalytic systems we found a significant increase of activity and enantioselectivity in water upon addition of a low amount of surfactant. An optimal ratio of Rh:amphiphile:substrate seems to be 1:20:100, that means the amount of amphiphile is much lower than the amount of substrate.
Actually, activation and enhancement of enantioselectivity takes place in a very small range of concentration (Fig. 3).
To test several kinds of amphiphiles we used preferently the system with BPPM [20].
In table 1 some results with sulfobetaines are summarized as examples of zwitterionic amphiphiles.

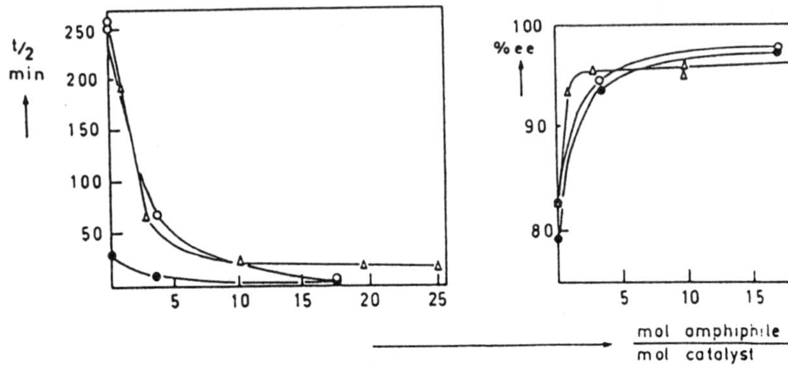

Figure 3. Influence of surfactants on the halftime and enantioelectivity of asymmetric hydrogenation of $H_2C{=}C(COOMe)NHCOCH_3$: (•) SDS, and of (Z)-PhCH=CH(COOMe)$NHCOCH_3$: (o) SDS, ( ) Triton X-100. Catalyst: [(Ph-ß-glup-OH)Rh(cod)]BF_4. Reaction conditions see table 1.

TABLE 1. Hydrogenation of (Z)-methyl-α-acetamidocinnamate in water with the catalytic system $[Rh(cod)_2]BF_4$ and BPPM. Effect of zwitterionic surfactants. Reaction conditions: 25 °C, 0.1 MPa H_2; 15 ml H_2O; 1 mmol substrate; 0,01 mmol $[Rh(cod)_2]BF_4$; 0.011 mmol BPPM. Without surfactant in water: t/2=90 min, % ee=78 % (R); without surfactant in methanol: t/2=2 min, % ee=90 (R)

Rh-Betaine	DeDAPs		DDAPs		HDAPs		ODAPs	
(mol/mol)	t/2 (min)	% ee (R)	t/2 (min)	% ee (R)	t/2 (min)	% ee (R)	t/2 (min)	% ee (R)
1 : 2	45	82	45	84	12	90	6	91
1 : 5	45	82	10	91	8	91	4	90
1 : 10	40	82	7	93	7	93	8	92
1 : 20	40	83	5	93	6	94	8	92
1 : 50	7	92	4	93	5	94	6	92
1 :100	3	94	3	95	4	95	5	94

DeDAPs N-decyl-N,N-dimethyl-3-ammonio-1-propanesulfonate
DDAPs N-dodecyl-N,N-dimethyl-3-ammonio-1-propanesulfonate
HDAPs N-hexadecyl-N,N-dimethyl-3-ammonio-1-propanesulfonate
ODAPs N-octadecyl-N,N-dimethyl-3-ammonio-1-propanesulfonate

The halftime t/2 represents the activity and the enantiomeric excess ee the enantioselectivity.
The amphiphiles are ordered by growing chain lengths. Obviously, the decylcompound shows the lowest effect because the activity and selectivity jump takes place at much higher concentration. This is clearly visible in a graphical representation (Fig. 4).

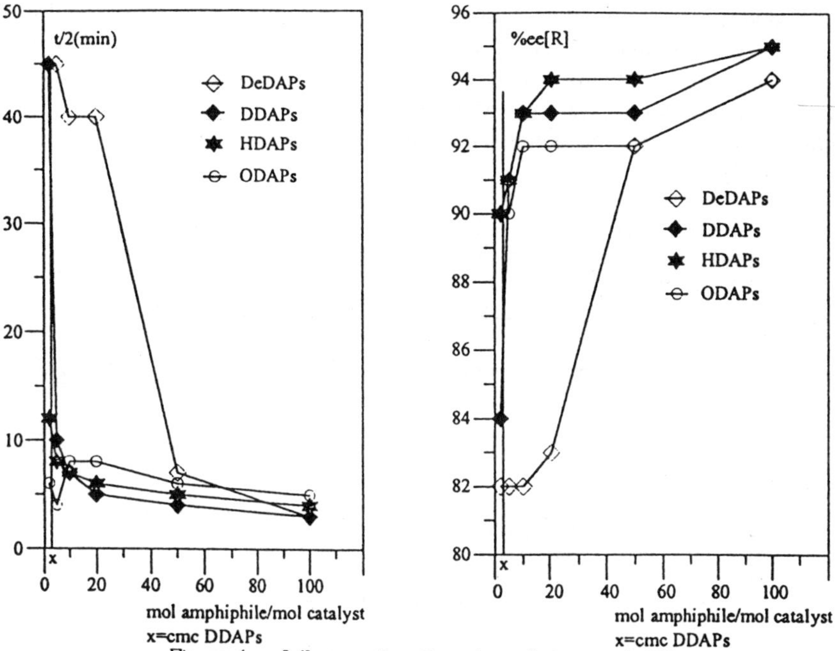

Figure 4. Influence of zwitterionic surfactants on halftime and enantioselectivity of asymmetric hydrogenation (see legend of Table 1).

Our interpretation is based on a connection between cmc (critical micelle concentration) and micellar effect.

Table 2 summarizes selected cmc's at 25 °C.

TABLE 2. Selected cm's at 25 °C [20]

		cmc (mol/l)
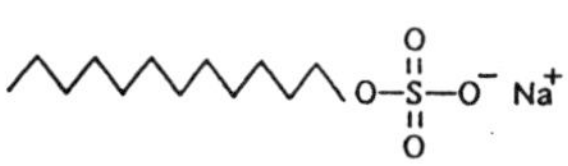	(SDS)	8.1×10^{-3}
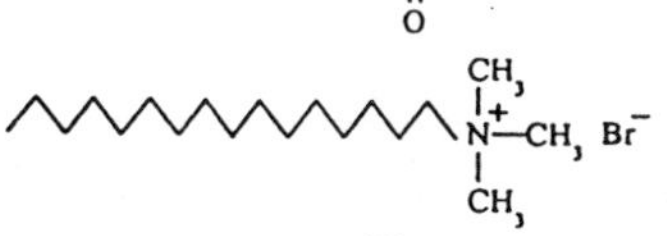	(CTAB)	9.2×10^{-4}
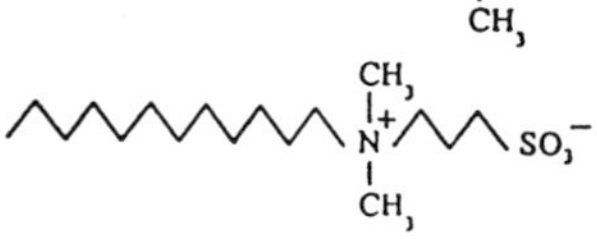	(DDAPs)	1.2×10^{-3}
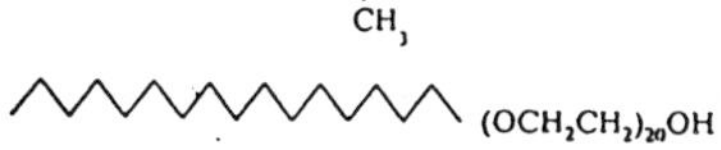	(Brij 58)	7.7×10^{-5}
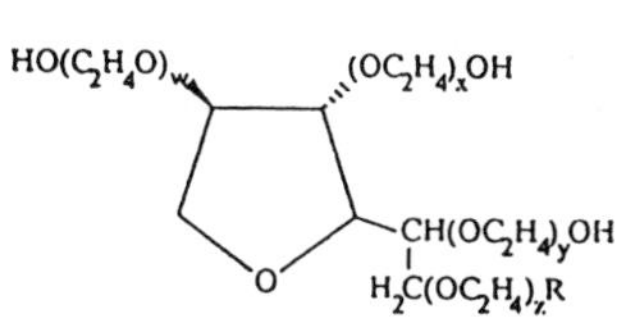	(Tween 20) R is $C_{11}H_{23}COO$	5.9×10^{-5}
	(Tween 80) R is $C_{17}H_{33}COO$	1.0×10^{-5}
	(Sum of w, x, y, z is 20)	

The cmc of N-dodecyl-N,N-dimethyl-3-ammonio-1-propansulfonate is relatively high, and in Fig. 4 has been so marked that the change of activity and enantioselectivity is maximal near the cmc. Probably, there is no good hydrophobic-lipophilic balance (HLB) in the decylcompound as an exception in this series.
Ideally suitable are nonionic surfactants like the Brij- and Tween-series because of their very low cmc [21].

TABLE 3. Hydrogenation of (Z)-methyl-α-acetamidocinnamate in water with the catalytic system $[Rh(cod)_2]BF_4$ and BPPM. Effect of neutral surfactants (Tween compounds). Reaction conditions see Table 1.

Rh:Tween	Tween 20		Tween 40		Tween 60		Tween 80	
(mol/mol)	t/2 (min)	% ee (R)	t/2 (min)	% ee (R)	t/2 (min)	% ee (R)	t/2 (min)	% ee (R)
1 : 1	35	87	60	88	40	87	40	87
1 : 2	20	89	30	87	11	90	25	93
1 : 5	9	90	9	93	8	92	11	94
1 : 10	7	92	7	94	11	93	9	94
1 : 20	6	93	5	95	4	94	7	94
1 : 50	5	93	5	96	4	93	4	94

Tween 20 polyoxyethylenesorbitan monolaurate
Tween 40 polyoxyethylenesorbitan monopalmitate
Tween 60 polyoxyethylenesorbitan monostearate
Tween 80 polyoxyethylenesorbitan monooleate

Another type of Tween-compounds (e.g. Tween 65 and Tween 85) with three hydrophobic tails was not sufficiently dispersable in water and was only effective in mixture with Tween 60.
Anionic amphiphiles like SDS or sodium dodecanesulfonate influences significantly activity and enantioselectivity of the asymmetric hydrogenation, but cationic amphiphiles on the basis of tetraalkyl-ammonium salts show a more complicate dependence.

TABLE 4. Hydrogenation of (Z)-methyl-α-acetamidocinnamate in water with the catalytic system $[Rh(cod)_2]BF_4$ and BPPM. Effect of cationic surfactants. Reaction conditions see Table 1.

Surfactant	Rh : surf. (mol:mol)	t/2 (min)	% ee (R)
CEDA BF_4	1 : 10	15 h	83
CEDA BF_4	1 : 20	85	80
CTA CF_3SO_3	1 : 20	110	79
CTA HSO_4	1 : 20	5	95
CTA HSO_4	1 : 5	5	94
CTA HSO_4	1 : 2	7	86
CEDA Br	no reaction		

CEDA BF_4 Cetyl-ethyl-dimethylammonium tetrafluoroborate
CTA CF_3SO_3 Cetyl-trimethylammonium triflate
CTA HSO_4 Cetyl-trimethylammonium hydrogensulfate
CEDA Br Cetyl-ethyldimethylammonium bromide

Cationic amphiphiles are influenced by the anion. Whereas the hydrogensulfate acts in water like SDS, the bromide is an inhibitor of the reaction. We explain this by an occupancy of free coordination sites in the rhodium complex by bromide. Replacement of bromide by tetrafluoroborate or triflate led to nonactive but not inhibiting amphiphiles. The mixtures of the latter surfactants with hydrogensulfate became active.

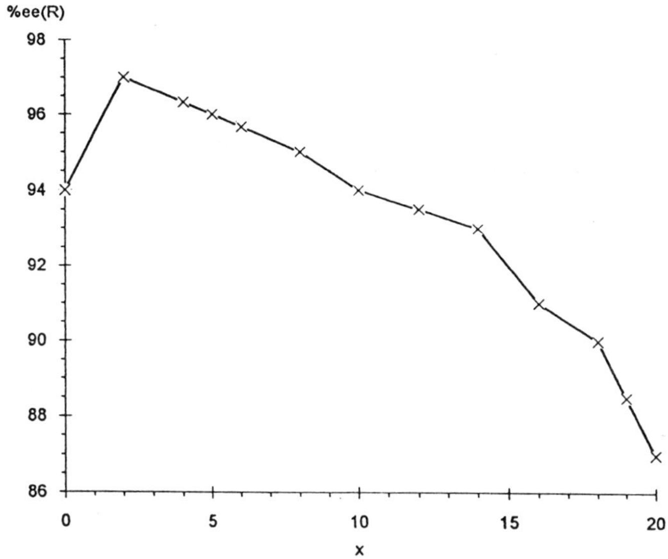

Figure 5. Effect of a mixture of $CTAHSO_4$ + SDS on the enantioselectivity of the hydrogenation of (Z)-methyl-α-acetamidocinnamate in water with the catalytic system $[Rh(cod)_2]BF_4$ and BPPM. Rh:CTA HSO_4:SDS=1:20:x

Interesting phenomena have been observed in mixtures of surfactants. Fig. 5 shows that small amounts of SDS in micelles of cetyl-trimethylammonium hydrogensulfate led to an increase of stereoselectivity but with increasing parts of SDS decreases the enantioselectivity below the starting value. Further investigations with mixed surfactants would be interesting.

Ueoka et al. [22] found in the kinetic resolution of activated amino acid esters in micelles a high dependence of the enantioselectivity on the composition of the micelle-(vesicle-)forming amphiphiles.

Separation of products and catalyst in our hydrogenation experiments is very simple: we extracted two times with chloroform and found product and a low amount of amphiphile in the organic phase.

Hydrogenation in micelle-like polymers

The excellent results with monomeric surfactants led us to the idea, to synthesize polymeric, insoluble surfactant systems to be used in hydrogenation in water which could serve as a carrier for the catalyst, be filtered off after reaction and be reused after separation from the product.

Here it should be differentiated between micelle-like polymers (polysoaps) and

polymerized micelles. Polymerized micelles have the dimension of spherical monomeric micelles, but are stabilized by the reaction of polymerizable groups [23]. These micelles act in catalysis like monomeric micelles [24] and give no advantage in practical working off. Polysoaps, however, are polymers of high molecular weight. Polysoaps are interesting for practical use and have been reviewed recently [25].
The types are differentiated due to the distance between hydrophilic head group and polymeric chain. We couldn't find any work in the literature about the binding of metal complexes in a micelle-analogous insoluble polymer.
We synthesized quite different kinds of polymers with amphiphilic brushes [26] (Fig 6).

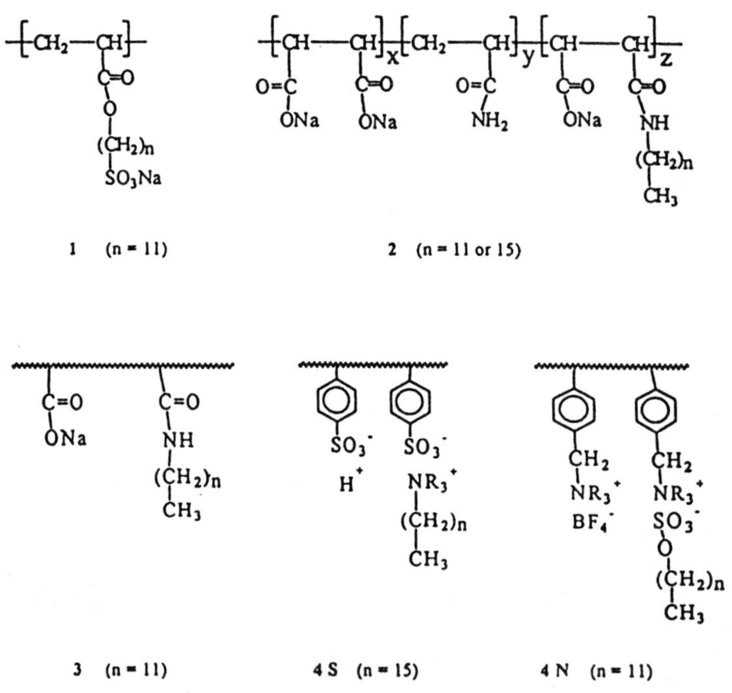

Figure 6. Polysoaps (micelle-like polymers) synthesized in our laboratory [26].

Polymer 1 was obtained by conversion of acrylic acid with sodium 11-hydroxy-undecanesulfonate followed by polymerization of the obtained surfactant monomer. Polymer 2 is derived from a 1:1 copolymer of acrylamide and maleic anhydride by partial reaction with dodecylamine or hexadecylamine, respectively, and

saponification of remaining anhydride groups. Crosslinking could be realized by the adding of tetraethyleneglycol diacrylate or alternatively N,N'-methylene bisacrylamide.
Polymer 3 was synthesized by partial amidation of the weak acid ion exchanger resin Amberlite IRC-50.
Polymers 4S and 4N are also basically ion exchange resins, loaded partially with hydrophobic counterions. The structure is inverse with respect to the acid and base group.
The table 5 shows hydrogenation results in water in presence of the polymers 1 to 4N.

TABLE 5. Hydrogenation of (Z)-methyl-α-acetamidocinnamate in water with the catalytic system $[Rh(cod)_2]BF_4$ and BPPM. Effect of polymeric amphiphiles. Reaction conditions see TABLE 1.

Amphiphile	Mole ratio amphiphile:Rh	Optical yield in % ee R	Half-time of hydrogenation t/2 in min	remarks
Monomeric				
none		78	~90	
C_{12}-SO_4 Na (SDS)	50 : 1	94	3.5	
HO-C_{11}-SO_3Na	50 : 1	83	45	
Linear polymers [a)]				
1	20 : 1	85	6	
2 C_{12} 0.7	20 : 1	89	160	
2 C_{12} 1.9	20 : 1	92	145	
2 C_{12} 2.2	20 : 1	94	960	
2 C_{16} 1.1	20 : 1	93	270	
2 C_{16} 1.4	20 : 1	92	140	
Crosslinked polymers [a)]				
2 C_{12} 1.2 [b)]	20 : 1	92	35	soluble
2 C_{12} 1.8	20 : 1	91	300	20% crosslinked
2 C_{16} 1.7	20 : 1	90	120	10% crosslinked
Ion exchangers				
3 1.3	20 : 1	49	1200	
A27-BF_4^- [c)]	60 : 1 (BF_4^- : Rh)	78	35	
4 N 0.7	20 : 1	91	15	
4 N 0.9	20 : 1	87	7	
4 S 0.5	40 : 1	93	15	
4 S 0.5	30 : 1	95	37	
4 S 0.8	30 : 1	92	14	

a) The linear and crosslinked polymers C_{12} and C_{16} are anhydride based polymers containing dodecyl- and hexadecyl chains, respectively. The numbers give the calculated amount of alkyl chains in mmol/g product.

b) The soluble fraction of a crosslinked polysoap.

c) Commercial Amberlyst A-27 is loaded with Cl^--ions. Here,Amberlyst with BF_4^--ions was used. Sample A-27 BF_4^- contained no amphiphilic chains.

The linear polymer 1 caused a high hydrogenation activity and an unexpectedly low enantioselectivity. But we should also note, that the monomeric part sodium 11-hydroxyundecanesulfonate is much less active than SDS. Probably the formation of micelles is very sensitive to a disturbance of the hydrophobic tail by polar groups.
Polymer 2 in the linear form differently loaded with hydrophobic groups retarded the hydrogenation, however, the enantioselectivity was greatly enhanced. The same results gave crosslinked polymers, except for the lowest crosslinked polymer 2, which appears to be dispersible in water.
Polymer 3 did inhibit the reaction and gave lower enantioselectivities than the reaction in pure water suggesting a lack of hydrophobic-lipophilic balance (HLB). The best results could be observed with amphiphilized ion exchange resins like 4N and 4S. Some examples are comparable in activity and selectivity with that of micelles formed by monomers. The effect seems to be extremely dependent on the degree of hydrophobic covering.
Concluding our experiences with polymers we can recognize micelle-analogous reactions. It seems to be difficult to synthesize polymers with an optimal HLB. The best results could be attained with ion exchange resins loaded with amphiphilic counterions.
The use of polysoaps means an advantage in the practice of separating catalyst and product. The resin was filtered off and contained the rhodium complex practically quantitative.
In the next table selected recycling experiments are summarized.

TABLE 6. Repeated hydrogenation with the same rhodium catalyst on an ion exchanger with amphiphilic counterions

Amphiphile	Mole ratio amphiphile : Rh	No. of cycle	Enantiomeric excess (e. e.) in %	Half-time of hydrogenation $t_{1/2}$/min	Number of observations
4N 0,9	20 : 1	1.	86,8 ± 1,5	6,5 ± 1,4	6
		2.	89,5 ± 0,5	4,9 ± 1,4	4
		3.	89,9 ± 2,4	5,3 ± 0,5	3
		4.	90,7 ± 2,8	6,1 ± 0,1	2
		5.	92	37	1
4S 0,5	30 : 1	1.	95	37	1
		2.	92	45	1
		3.	93	182	1
4S 0,8	20 : 1	1.	89	12	2
		2.	91	11	2
		3.	87	9	2
		4.	87	22	2
		5.	89	24	2
		6.	86	25	2
		7.	87	32	2
		8.	87	34	1
		9.	96[a]	43	1

[a] After this cycle, the granulate did not sedimentate anymore. The aqueous phase was extracted without prior separating of the rhodium-containing ion exchanger.

It was shown, that the hydrogenation could be repeated up to 9 times.
Let us note, that the enantioselectivities had a tendency to rise after reaction cycles but for the moment we don't have any explanation for this observation.

Future investigations

It seems to be necessary to characterize our systems with respect to the morphology of the assemblies and to find out the reaction place in the micelle. Perhaps there is a possibility to transfer chiral information from optically active micelle-formers to prochiral substrates. To date we could observe only small effects with different types of chiral amphiphiles. Experiments with reversed micelles and microemulsions remain to be done. The principle of transferring highly sensitive catalytic reactions into colloidal dimensions seems to be fruitful and one way to perform organometallic reaction in water as medium.
Recently we could show that the heterocyclization of two molecules of acetylene and one molecule of nitrile to 2-substituted pyridines catalyzed by cobalt(I)-complexes and promoted by light proceeds very successfully with high selectivity in water [27].
New experiments bring new expectations.

Acknowledgements

Authors want to thank Prof. R. Selke and Dr. E. Paetzold for cowork in an earlier phase. Thanks to the Fonds der Chemischen Industrie and Deutsche Forschungsgemeinschaft for financial support.

References

1. Hartley, F. R. (1987) Supported metal complex catalysts in F. R. Hartley (ed.), *The Chemistry of the Metal-Carbon Bond*, Wiley & Sons Inc., Chichester, chapter 14
2. Gates, B. C., Guczi, L., Knözinger, H. (1986) *Metal Clusters in Catalysis*, Elsevier, Amsterdam
3. Southern, T. G. (1989) Bi-phasic and phase transfer catalysis, *Polyhedron* **8**, 407-413
4. Keim, W. (1984) Pros and cons of homogeneous transition metal catalysis, illustraded for SHOP [Shell higher olefin process], *Chem.-Ing.-Tech.* **56**, 850-853
5. Kuntz, E. G. (1987), Homogeneous catalysis in water, *CHEMTECH* 570-575
6. Herrmann, W. A., Kohlpainter, C. W. (1993), Wasserlösliche Liganden, Metallkomplexe und Komplexkatalysatoren: Synergismen aus Homogen- und Heterogenkatalyse, *Angew. Chem.* **105**, 1588-1609
7. Paetzold, E., Kinting, A., Oehme, G. (1987), Synthesis of phosphino alkane sulfonates and their corresponding sulfonic acids by reaction of alkalimetal-phosphides with sultones, *J. Prakt. Chem.* **329**, 725-731

8. Paetzold, E., Oehme, G. (1993) Water soluble palladium(II) phosphine complexes as catalysts in the hydrodehalogenation of allyl and benzyl halogenides under biphasic and phase transfer conditions, *J. Prakt. Chem./Chem. Ztg.* **335**, 181-184
9. Brown, J. M., Baker, S. K., Colens, A., Darwent, J. R. (1980) Micellar catalysis, in P. Dunnill, A. Wiseman, N. Blakebrough (eds.), *Enzymic and Non-Enzymic Catalysis*, Horwood, Chichester, chapter 5
10. Kunitake, T., Shinkai, S. (1980) Catalysis by micelles, membranes and other aqueous aggregates as models for enzyme action, *Adv. Phys. Org. Chem.* **17**, 435-487
11. Fendler, J. H. (1983) Photochemistry in organized media, *J. Chem. Educ.* **60**, 872-876
12. Blokzijl, W., Engberts, J. B. F. N. (1993) Hydrophobe Effekte - Ansichten und Tatsachen, *Angew. Chem.* **105**, 1610-1648
13. Bunton, C. A., Savelli, G. (1986) Organic reactivity in aqueous micelles and similar assemblies, *Adv. Phys. Org. Chem.* **22**, 213-309
14. Meisel, D., Matheson, M. S. (1989) Photocatalysis in organized assemblies, in N. Serpone, E. Pelizzetti (eds.) *Photocatalysis*, Wiley & Sons Inc., New York, pp. 385-419 see also 11.
15. Lecomte, L., Sinou, D., Bakos, J., Toth, I. Heil, B. (1989) Chiral sulphonated phosphines II. Influence of water on the enantioselectivity in the reduction of dehydro-aminoacids, *J. Organomet. Chem.* **370**, 277-284
16. Oehme, G., Paetzold, E., Selke, R. (1992) Increase in activity and enantioselectivity in asymmetric hydrogenation reactions catalysed by chiral rhodium(I) complexes as a consequence of the action of amphiphiles, *J. Mol. Catal.* **71**, L1-L5
17. Achiwa, K. (1976) Asymmetric hydrogenation with new chiral functionalized bisphosphine-rhodium complexes, *J. Am. Chem. Soc.* **98**, 8265-8266
18. Kagan, H. B., Dang, T. P. (1972) Asymmetric catalytic reduction with transition metal complexes I. A catalytic system of rhodium(I) with (-)-2,3-O-isopropylidene-2,3-dihydroxy-1,4-bis(diphenylphosphino)butane, a new chiral diphosphine, *J. Am. Chem. Soc.* **94**, 6429-6433
19. Selke, R. (1989) Carbohydrate phosphinites as chiral ligands for asymmetric synthesis catalyzed by complexes V. New rhodium(I) chelates prepared by precipitation from the equilibrium between neutral and cationic complexes and hydrolysis of their bonded ligands, *J. Organomet. Chem.* **370**, 241-248
20. Grassert, I., Paetzold, E., Oehme, G. (1993) Influence of different types of amphiphiles on the rhodium(I) complex catalyzed asymmetric hydrogenation of (Z)-methyl-α-acetamidocinnamate in aqueous medium, *Tetrahedron* **49**, 6605-6612
21. Brij compounds (polyoxyethylene monoalkylethers) were purchased from Fluka, Tween compounds (polyoxyethylenesorbitan monoesters) were purchased from Sigma
22. Ueoka, R. Matsumoto, Y., Moss, R. A., Swarup, S., Sugii, A., Harada, K. Kikuchi, J., Murakami, Y. (1988) Membrane matrix for the hydrolysis of amino acid esters with marked enantioselectivity, *J. Am. Chem. Soc.* **110**, 1588-1595

23. Paleos, C. M. (1992) Polymerization of micelle-forming monomers, in C. M. Paleos (ed.) *Polymerization in Organized Media*, Gordon and Breach, Philadelphia, chapter 3

24 André, B., Boyer, B., Lamaty, G., Roque, J.-P. (1991) Efficient catalytic activity of polymerized micelle-forming surfactants, *Tetrahedron Lett.* **32**, 1881-1884

25. Anton, P., Köberle, P., Laschewsky, A. (1993) Recent developments in the field of micellar polymers, *Makromol. Chem.* **194**, 1-27

26. Flach, H. N., Grassert, I., Oehme, G. (in press) Polymeric surfactant systems in the asymmetric hydrogenation of amino acid precursors with a rhodium complex, *Macromol. Chem. Phys.*

27. Heller, B., Oehme, G., to be published

Bioorganometallic Chemistry: The Reactions of a (η^5-Pentamethylcyclopentadienyl)rhodium Aqua Complex with Nucleobases, Nucleosides, Nucleotides, and Oligonucleotides

Richard H. Fish*, David P. Smith, Hong Chen, Marcos F. Maestre, Marilyn M. Olmstead†, Moris S. Eisen≠, and Ariel Haskel≠

Lawrence Berkeley Laboratory, University of California, Berkeley, CA 94720
†Department of Chemistry, University of California, Davis, CA 95616
≠Department of Chemistry, Technion-Israel Institute of Technology, Haifa, 32000, Israel

Introduction

Bioorganometallic chemistry is a relatively new aspect of metal-carbon bond compound reactivity in that the reactions with biological substrates in aqueous solution was previously the purview of bioinorganic chemists.[1] Recently, there has been a dramatic new focus on the reactions of organometallic complexes with biologically relevant ligands and, interestingly, the majority performed in aqueous solution; the antithesis of our previous perception of metal-carbon bond reactivity.[2]

We have had a long standing interest in the reactions of (η^5-pentamethylcyclopentadienyl)rhodium ($[Cp^*Rh(S)_3]^{2+}$, S = solvent) complexes in anhydrous organic solvents with organonitrogen compounds for both synthetic and catalysis projects.[3] In a serendipitous manner, we started a bioorganometallic chemistry program that had the initial goal of anchoring, on one end, a single λ–DNA molecule of 50K base pairs to the glass/electrode surface of an epi-fluorescence microscopic stage in order to image this single DNA molecule with ethidium bromide. The manner in which we anchored the single λ–DNA was to selectively bind a Cp*Rh aqua complex to a single strand 24 base oligonucleotide (24 mer) that had part of its sequence complimentary to the terminal sequence of λ–DNA; i. e., Cp*Rh was only bound to a selective portion (dA_{12}) of the 24 mer and, simultaneously to the glass/electrode surface, but not directly to λ–DNA.[4]

Clearly, this latter binding mode(s) of the Cp*Rh aqua complex to a 24 mer was going to be extremely difficult to ascertain spectroscopically. Therefore, we decided to

I. T. Horváth and F. Joó (eds.), Aqueous Organometallic Chemistry and Catalysis, 259–276.

study the reactions of the less complicated nucleobases, nucleosides, nucleotides, and oligonucleotides in aqueous solution with the Cp*Rh aqua complex. In this chapter, we will demonstrate the rich and diverse structural chemistry uncovered via reaction of $[Cp^*Rh(H_2O)_2(OTf)_2]_x$ with 9-methyladenine, 1-methylcytosine, 9-ethylguanine, 9-methylhypoxanthine, 9-ethylhypoxanthine, adenosine, guanosine, 5'-adenosine, methyl-5'-adenosine, and 3'-adenosine monophosphates in aqueous solution at various pH values.[2a-d]

We will also discuss the structure and equilibrium of the starting Cp*Rh aqua complexes as a function of pH and will demonstrate the interesting use of the cyclic trimer structures of several of the Cp*Rh complexes with nucleobases, nucleosides, and nucleotides as molecular recognition hosts for aromatic amino acid guests. In addition, we will also briefly discuss the oligos, dA_{12}, a 12 mer, and the 24 mer, in their interaction with the Cp*Rh aqua complex by circular dichroism spectroscopy studies.

Cp*Rh Aqua Complexes as a Function of pH

It was apparent from all the above-mentioned reactions we ran with the DNA/RNA bases that we had to determine the structure of the Cp*Rh aqua complex(es) as a function of pH. Therefore, we have completed an indepth NMR/FAB/MS/potentiometric titration study to show that from pH 2-5 a monomeric Cp*Rh aqua complex was present. However, as the pH was raised to 5-7, a rapid equilibrium existed between this monomer and several [Cp*Rh(μ-OH)] dimers and, as well, at pH 10 only one dimer was present.[5] Scheme 1 summarizes the Cp*Rh aqua chemistry and, what we must keep in mind, is that at a pH range of 5-7, where we conducted our reactions with nucleobases, nucleosides, nucleotides, and oligonucleotides, the predominate Cp*Rh aqua complexes are the [Cp*Rh(μ-OH)] dimers.

Scheme 1

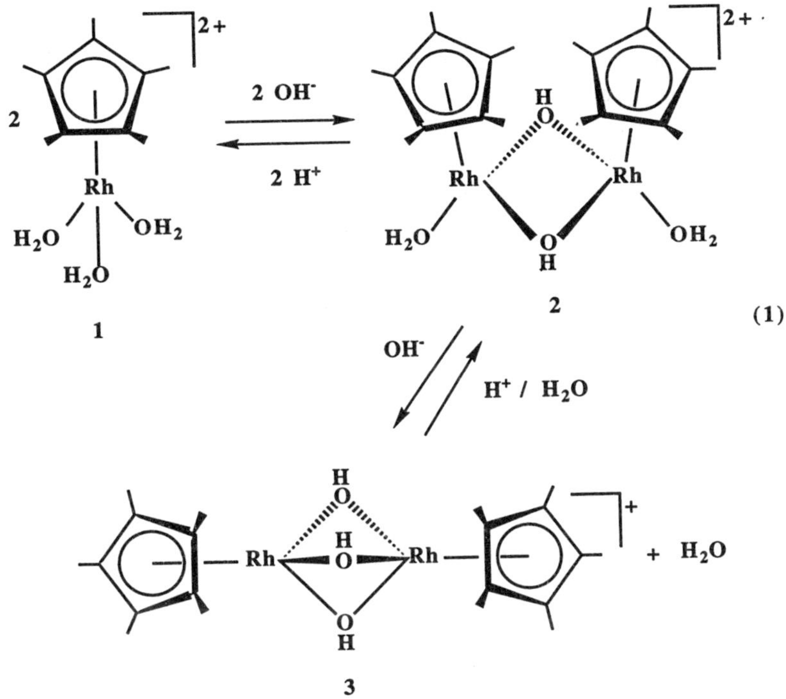

Reactions of the Cp*Rh Aqua Complexes with Nucleobases, Nucleosides, and Nucleotides

The Chart provides the name and structures of the DNA/RNA bases utilized in these studies. We initiated our perusal of reactivity of nucleobases with the Cp*Rh aqua complexes by studying 9-methyladenine (9-MA). Reaction of this nucleobase in D_2O at pD 7.2 provided, by 500 MHz ^1H NMR spectroscopy, evidence for formation of a Cp*Rh-9-MA complex with dramatic chemical shifts for H2 and H8 in comparison to free 9-MA at 8.83 and 7.67 ppm, respectively. By utilizing 9-MA-d that is selectively deuterated at H8, we were able to unequivocally assign the chemical shifts to each proton and show that H8 was shifted 0.75 ppm downfield from free 9-MA, while H2 was shifted 0.47 ppm upfield!

The product from this reaction was isolated and purified by recrystallization from methanol to yield an orange solid (26%) that was found by ^1H NMR, FAB/MS (m/z =1456.7; M-OTf), single-crystal X-ray crystallography, and elemental analysis to have the unusual and unprecedented structure of a cyclic trimer, **4**. Previously reported studies have verified a N6-N7, 5-membered ring chelate of 9-MA with a Cp_2Mo center and a μ-N1 and N7 bridging of 9-MA between metal centers (Ag^+, Co^{2+}), but no complex has been reported as yet incorporating both bonding features in one molecule as is shown in **4**. Clearly, the formation of the 12 membered ring cyclic trimer results from the ability of the Cp*Rh cation to act as an azaphile with the favorable geometry of N1 poised to form the third bond to a Cp*Rh of an adjacent [Cp*Rh(9-methyladenyl)](OTf) 5-membered ring NH6/N7 moiety.[2a]

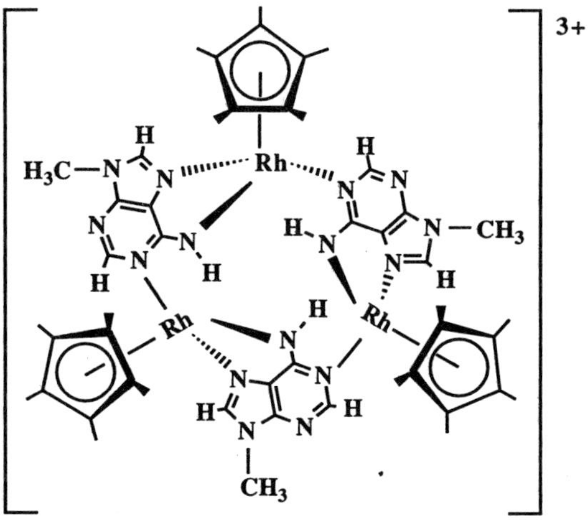

Top View of $[Cp^*Rh(9\text{-}MA)]_3^{3+}$, 4

Chart

Nucleobases

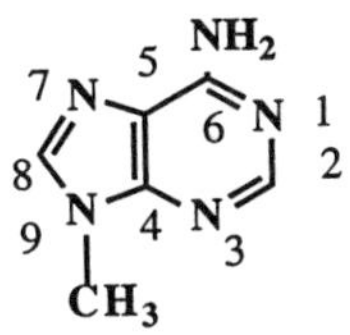

9-Methyladenine

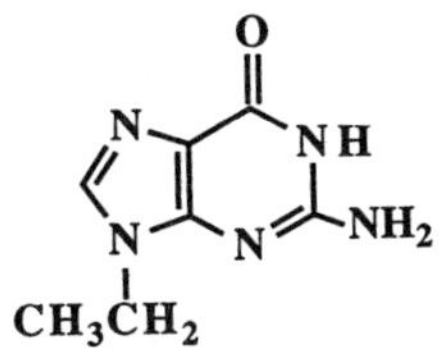

9-Ethylguanine

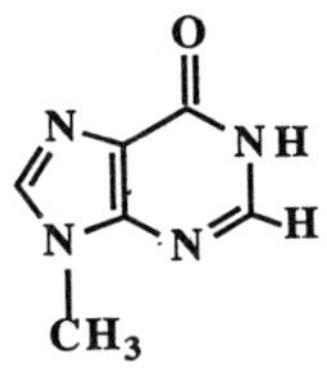

9-Methylhypoxanthine

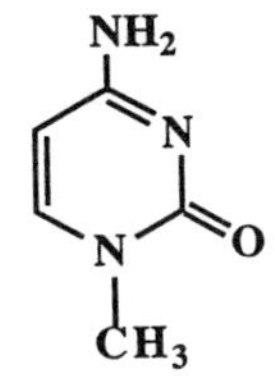

1-Methylcytosine

Nucleosides

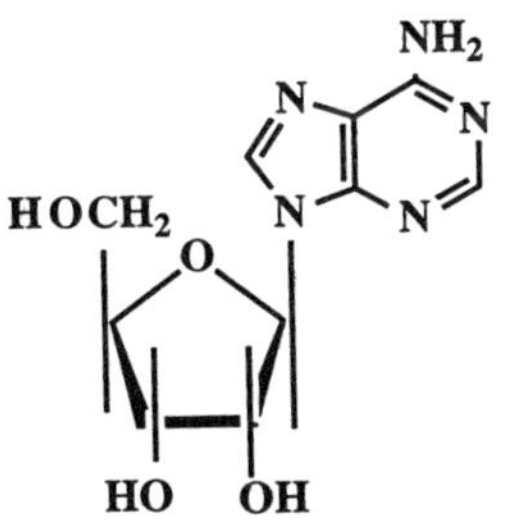

Adenosine

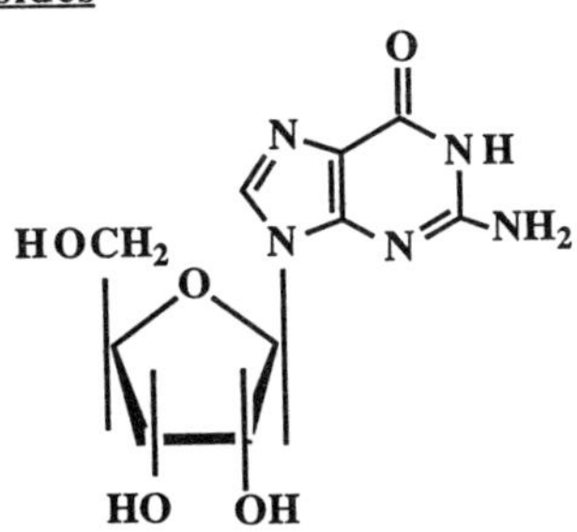

Guanosine

Nucleotides

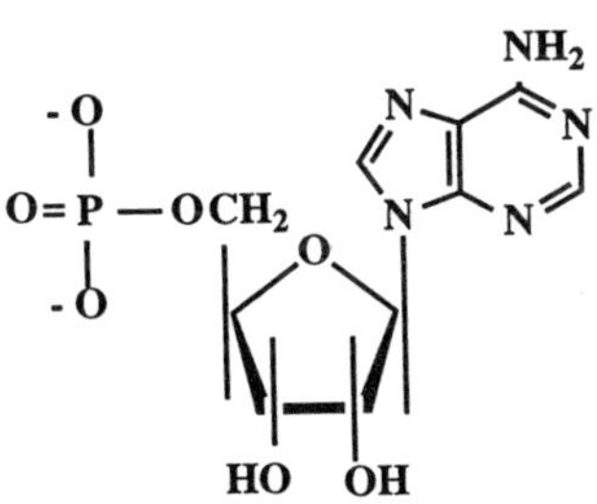

5'-AMP

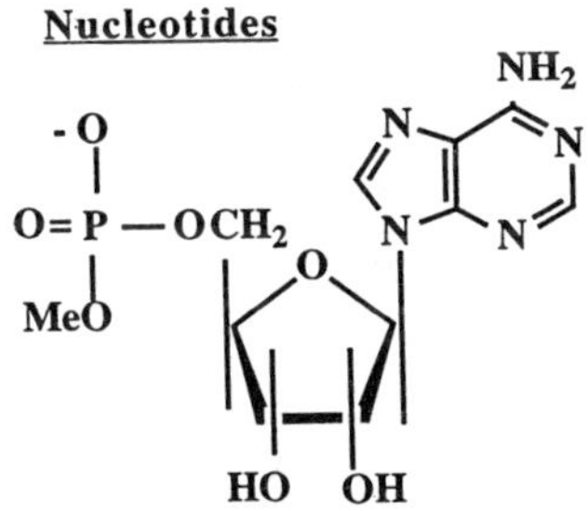

5'-Methyl-AMP

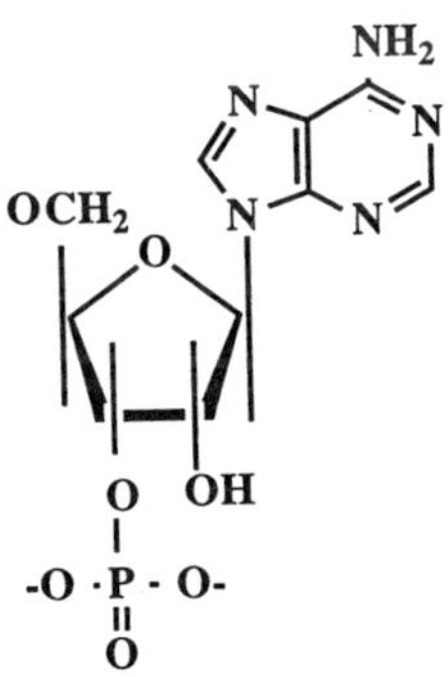

3'-AMP

Oligonucleotides

$d(A)_{12}$

dCpCpCpGpCpCpGpCpTpGpGpA

$d(A)_{12}$dCpCpCpGpCpCpGpCpTpGpGpA

Reaction of the Cp*Rh aqua complexes with 1-methylcytosine in H_2O at pH 5.4 provides by ^1H NMR, FAB/MS, elemental analysis, and single-crystal X-ray crystallography, a *trans*- μ-hydroxy dimer, with the formula *trans*-[Cp*Rh(1-methylcytosine)(μ-OH)]$_2$(OTf)$_2$, **5**. The 1-methylcytosine binds at N3 and forms intramolecular hydrogen bonds to the μ-OH with the NH_2 groups and the μ-OH with the C=O, but does not form a 4-membered chelate with N3/NH_2 in a similar fashion as observed with **4**.[2b]

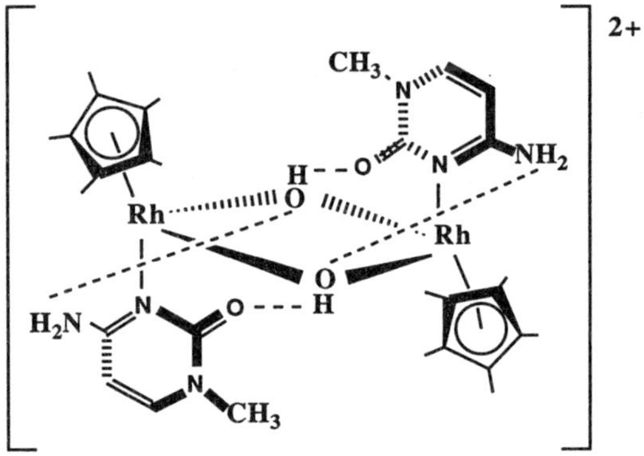

Hydrogen-Bonded View of 5

In another study with a nucleobase, the reaction of one equivalent of 9-ethylguanine (9-EG) with *in situ* generated [Cp*Rh(MeOH)$_3$](OTf)$_2$ in methanol produces [Cp*Rh(η^1(N7)-9-EG)(CH_3OH)$_2$](OTf)$_2$, **6**, whose ^1H NMR in MeOH-d$_4$ provided evidence of exclusive Cp*Rh-N7 bonding with an H8 downfield shift in comparison to free 9-EG of $\Delta = 0.7$ ppm. Similarly the 9-methylhypoxanthine (9-MH) complex, [Cp*Rh(η^1(N7)-9-MH)(CH_3OH)$_2$](OTf)$_2$, **7**, was also prepared and showed a ^1H NMR (DMSO-d$_6$) downfield shift for H8 in comparison to free 9-MH of $\Delta = 0.53$ ppm and H2 of $\Delta = 0.15$ ppm, which is also consistent with Cp*Rh-N7 binding.

The single-crystal X-ray structure of **7**, determined on crystals grown from methanol/Et_2O, verifies this N7 coordination mode.[2c] In addition, the X-ray structure of **7** also reveals an interesting pseudo-chelation of the rhodium center that involves a covalent Rh-N(7) bond and a hydrogen bond between the C=O6 of 9-MH and a coordinated MeOH ligand (1.83 Å). Hydrogen bonding also occurs between coordinated MeOH and one of the triflate oxygens (CF_3SO_2O--$HOCH_3$,1.86 Å), as well as the other triflate anion to NH1 (CF_3SO_2O--HN, 1.89 Å).

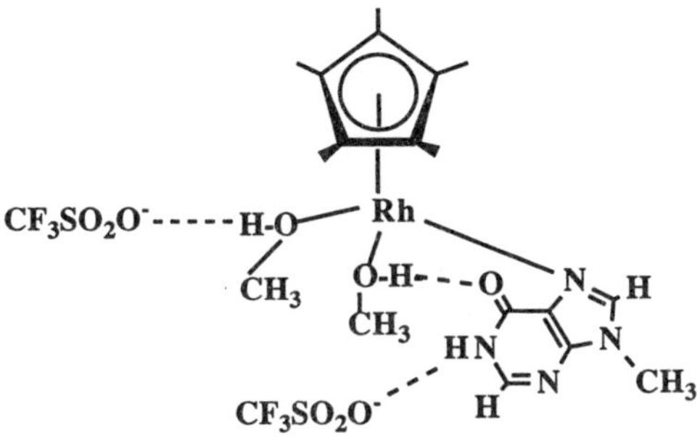

7

To further investigate the effect of an aqueous reaction media, as a function of pH, an ^{1}H NMR study was performed with **6** and **7**, since it would also allow us to determine the steric role, if any, of the NH_2 group at C2 (**6** only, H8) and the bonding mode of NH1 (**7** only, H8, H2). The pH profile of **7** was studied by ^{1}H NMR in D_2O. At pD 2.45 to 5.13, the downfield chemical shifts for both H8 (8.48 ppm, $\Delta = 0.44$ ppm) and H2 (8.35 ppm, $\Delta = 0.17$ ppm; selective H8 exchange in refluxing D_2O verified these assignments) compared to free 9-MH (H8, 8.04 ppm and H2, 8.18 ppm) are consistent with exclusive N7 binding. However, at pD 6.45, the Cp*Rh-9-MH complex provides the dramatic chemical shifts we have found to be diagnostic for cyclic trimer formation, especially for H8 (downfield shift) and H2 (upfield shift) at 8.60 (H8, $\Delta = 0.56$ ppm) and 7.78 ppm (H2, $\Delta = 0.40$ ppm) as well at 3.73 ppm (9-CH_3, $\Delta = 0.09$ ppm), and 1.84 ppm (Cp*) and strongly suggests the presence of the unusual and unprecedented Cp*Rh-guanine nucleobase derivative structure, [Cp*Rh-μ-η^1(N1):(η^2(O6, N7)-9-methylhypoxanthyl)]$_3^{3+}$, **8**, analogous to similar structures found for adenine derivatives, 9-methyladenine, adenosine, and methyl-5'- and 3'-adenosine monophosphates.[2a,c,d] We could not obtain suitable crystals of **8**; however, the 9-ethylhypoxanthine analogue of **8** did indeed provide, at pD 6.45, suitable crystals for X-ray study and this analysis was completed.[6] Two structural features of the ethyl analogue of **8** merit comment. First, the cationic portion has a triangular dome-like cavity, with three Cp* groups stretching out from the top of the dome, three ethyl groups pointing to the bottom, three hypoxanthine planes forming the surrounding shell, and three Rh atoms embedded on the top of the dome. The cation also possesses a C3 axis, which passes from the top of the dome to the bottom. Secondly, the C6-O6 bond distance of 1.296(24) Å falls between the single bond distance of 1.42 Å, found in an alcohol, and the double bond distance of 1.233(4) and 1.230(7) Å, which were observed in inosine. This result suggests that a significant amount of multiple bond character still exists in the C6-O6 bond. This argument is supported by the long bond distance of the C6-N1 bond (1.401(29) Å) in comparison to a typical C4-N3 double bond distance of 1.358(10) Å that was found in the reported *trans* -[(Cp*)Rh(η^1-(N3)-1-methylcytosine)(μ-OH)]$_2^{2+}$ complex.[2b]

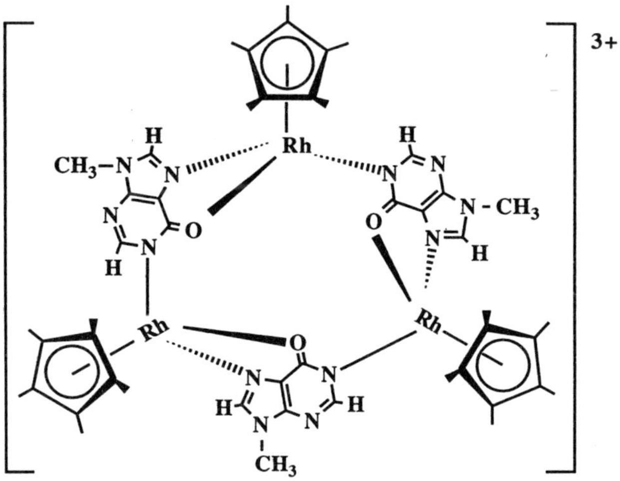

8

A similar pH profile with **6** *shows no discernable cyclic trimer formation.* This conclusion is based on no prominent H8 signals in the 8.5-8.8 region, no upfield chemical shifts for the methylene protons of the ethyl group, as well as no downfield shift for the Cp*Rh signal; these ^{1}H NMR shifts were diagnostic for other cyclic trimer structures such as **4**. This result suggests that the NH_2 group at C2 on the guanine nucleus, whose steric and electronic effects have not been previously well defined during the metal coordination process, *plays a significant steric role*, in this instance, in preventing cyclic trimer formation; steric rather than electronic due to NH1 pKa similarities for 9-EG and 9-MH.

Reaction of the Cp*Rh aqua complex with the nucleoside, adenosine (Ado), in D_2O at pD 7.1 also provided by 500 MHz ^{1}H NMR spectroscopy similar chemical shifts for H8 (8.92 ppm) and H2 (7.62) that were observed for complex **4** and, thus, we tentatively conclude that adenosine also forms a cyclic trimer complex. In contrast to cyclic trimer formation for Ado, it was found that the nucleoside, guanosine (Guo), reacted with the Cp*Rh aqua complexes at pH 5.4 to provide an isolated product that by elemental analysis and FAB-MS (m/z = 670.1, [Cp*Rh(Guo)(OTf)]; 556.1, [Cp*Rh(Guo)$(H_2O)_2$(OH)]), was a monomer with the formula [Cp*Rh(Guo)(H_2O)(OH)](OTf), **9**. The structure, **9**, was elucidated by 500 MHz ^{1}H NMR spectroscopy in DMSO-d_6 to show a substantial downfield shift for H8 at 8.93 ppm (Δ = 1.01 ppm), which is consistent with N7 binding to the Guo nucleus. The NH1 group was also shifted downfield by 0.53 ppm and this may be indicative of the 6-C=O group interacting with the Rh metal center as shown.[2a]

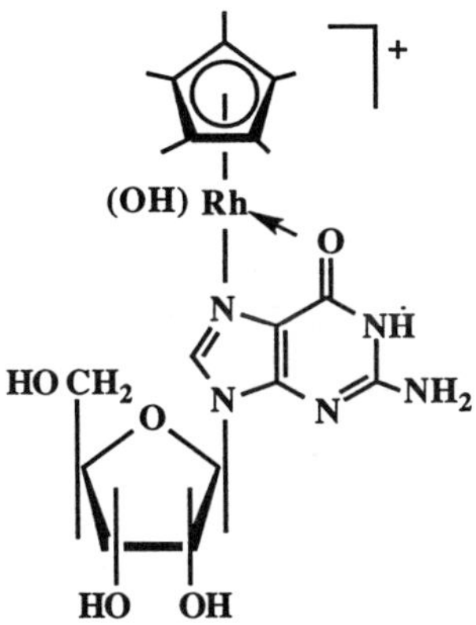

9

The reactions of adenosine 5'-monophosphate (5'-AMP) in water (pH 5-8) with a Cp*Rh aqua complex showed, by ^{1}H and ^{31}P NMR spectroscopy, that the 5'-$P(O)_2$=O group significantly inhibited cyclic trimer formation, [Cp*Rh(μ-η^1(N1): η^2(N6,N7)-5'-AMP)]$_3$, via competition with N1 and NH6 for the Cp*Rh site and, as well, provided other mononuclear and dinuclear Cp*Rh-O-P complexes. In contrast, both the phosphate methyl ester of 5'-AMP and 3'-AMP formed the cyclic trimer structures, [Cp*Rh(μ-η^1(N1): η^2(N6, N7)-methyl-5'-/3'-AMP)]$_3$. The consequence of steric effects, as demonstrated by the position and substitution of the phosphate group attached to the ribose, on the diastereoselectivity of cyclic trimer formation, as observed by ^{1}H and ^{31}P NMR spectroscopy and circular dichroism analysis over time, shows that the phosphate methyl ester of 5'-AMP provides a greater diastereoselectivity (6:1) after one week of equilibration compared to 5'-AMP (1.5:1) and 3'-AMP (1.2:1).

The role of the 5'-$P(O)_2$=O group inhibits cyclic trimer formation and this directional influence interferes with the condensation reaction between the exocyclic amine of 5'-AMP and the reactive Cp*Rh-hydroxy species. Thus, with the less nucleophilic phosphate, methyl-5'-AMP, or the positional isomer, 3'-AMP, minimal interference was observed for cyclic trimer formation. Therefore, Cp*Rh-OP interactions effectively compete with cyclic trimer formation only in the 5'-AMP example and represents a new role of the 5'-$P(O)_2$=O group in directing nucleotide reactions.[2d]

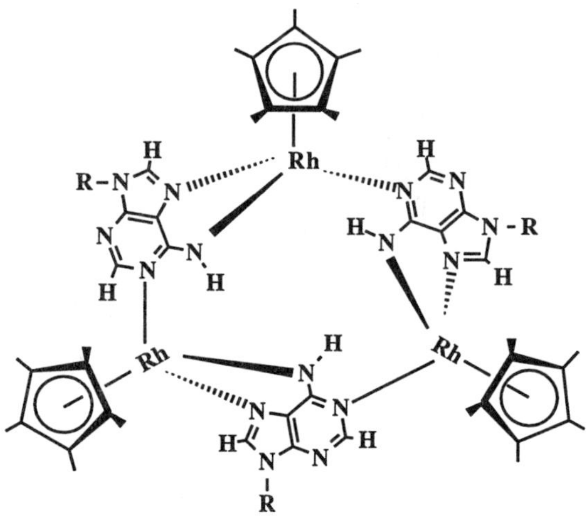

R =

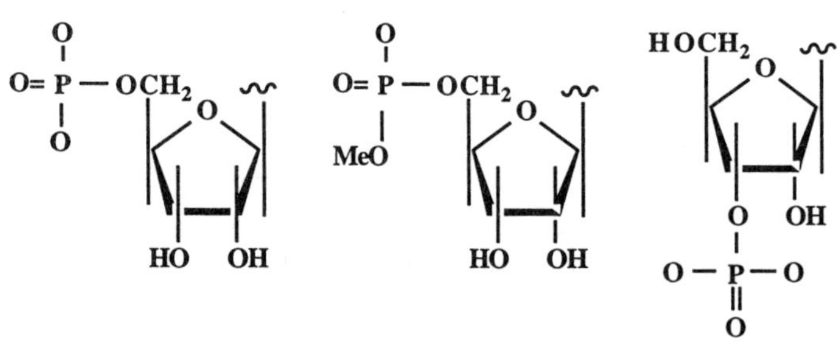

In similar, but unpublished experiments, the ^{1}H and ^{31}P NMR experiments were conducted with Cp*Rh aqua complex and nucleotides, 5'-guanosine monophosphate

(GMP), 5'-cytidine monophosphate (CMP), and 5'-thymidine monophosphate (TMP) at pD 7.2. The ^{1}H and ^{31}P NMR spectra show broadened and shifted signals at pD 7.2 for GMP, but only broadened ^{31}P NMR spectra for CMP and TMP and no shifts in the ^{1}H NMR spectra. The broadened and only slightly shifted ^{31}P NMR spectra may be consistent with a weak interaction of the P-O^{-} group with a Cp*Rh complexed H_2O molecule during a fluxional process. Alternatively, GMP shows a ^{1}H NMR spectrum with a downfield shift for H8 of 0.6 ppm, indicative of N7 bonding, as well as a broadened, slightly shifted ^{31}P NMR signal for a similar weak interaction of a complexed H_2O molecule with the P-O^{-} group as observed for AMP.[7]

In attempts to define nucleotide and nucleoside selectivity for the Cp*Rh aqua complex, we carried out several competitive experiments via ^{1}H NMR spectroscopy. Thus, competitive reactions of the Cp*Rh-AMP complex with free GMP (1:1 mmol) shows displacement of AMP and formation of Cp*Rh-GMP complex, while CMP or TMP do not alter the ^{1}H NMR spectrum of the Cp*Rh-AMP complex. Clearly, the Cp*Rh aqua complex prefers the following selectivity order: GMP > AMP >>> CMP ~ TMP.[7]

Moreover, the reaction of [Cp*Rh(Guo)(H_2O)(OH)](OTf), with one equivalent of Ado in D_2O at pD 7.3, provides the Ado cyclic trimer, complex **4**, and displacement of the Guo ligand; a result opposite to that found with AMP and GMP. Therefore, we tentatively conclude that the differences between AMP and Ado in reactivity with GMP and Guo, respectively, must involve the absence of the phosphate group and the favorable formation of the very stable Ado trimer.[7]

Reactions of the Cp*Rh Aqua Complexes with Olignucleotides

The above-mentioned studies with nucleobases, nucleosides, and nucleotides clearly were important to understand bonding modes of these DNA/RNA bases, as a function of pH, with the Cp*Rh aqua complexes. However, a perusal of the literature in the very limited area of bioorganometallic chemistry, focused on reactions with oligonucleotides, clearly revealed that no definitive studies, other than computer modelling to discern that these sterically demanding complexes could not possibly intercalate between two base pairs on a single DNA/RNA strand[2h] as does, for example, cis-[$Cl_2(NH_3)_2Pt$],[1] on the reactions of oligonucleotides with organometallic aqua complexes. From our biological application, described in the introduction section, we wanted to start our Cp*Rh-oligomer studies with the 24 mer, which we used to anchor a single λ-DNA molecule to the surface of an epi-fluorescence microscopic stage.[4]

Our unpublished circular dichroism studies on the titration of the Cp*Rh aqua complex with the 24 mer (Chart) and its components, the 12A (dA_{12}) and 12 mer, at pH

7.0 in a phosphate buffer allowed us to tentatively determine what portion of the 24 mer selectively interacted with the Cp*Rh aqua complex. The CD spectra of the Cp*Rh-24 mer titration is shown and the results with no added, 2, 4, and 8 mmoles of Cp*Rh/24 mer provided evidence that the Cp*Rh aqua complex appears to interact selectively with the 12A portion of the 24 mer. The CD Cotton Effect at 220 nm, the 12A region, is dramatically affected by the addition of the Cp*Rh aqua complex, while the 12 mer region of the CD spectrum at 280 nm does not appear to be affected to any significant extent. The dramatic decrease in the Cotton Effect at 220 nm (12 A region) is reminiscent of the classical melting experiments of DNA molecules conducted using CD analysis; i. e., there is a decrease in the CD Cotton Effect of the DNA bases that are being observed. In the melting process, the DNA bases, that interact by a π-π mechanism, start to unstack, which is exactly what we observe in our titration of the 24 mer with the Cp*Rh aqua complex.[7]

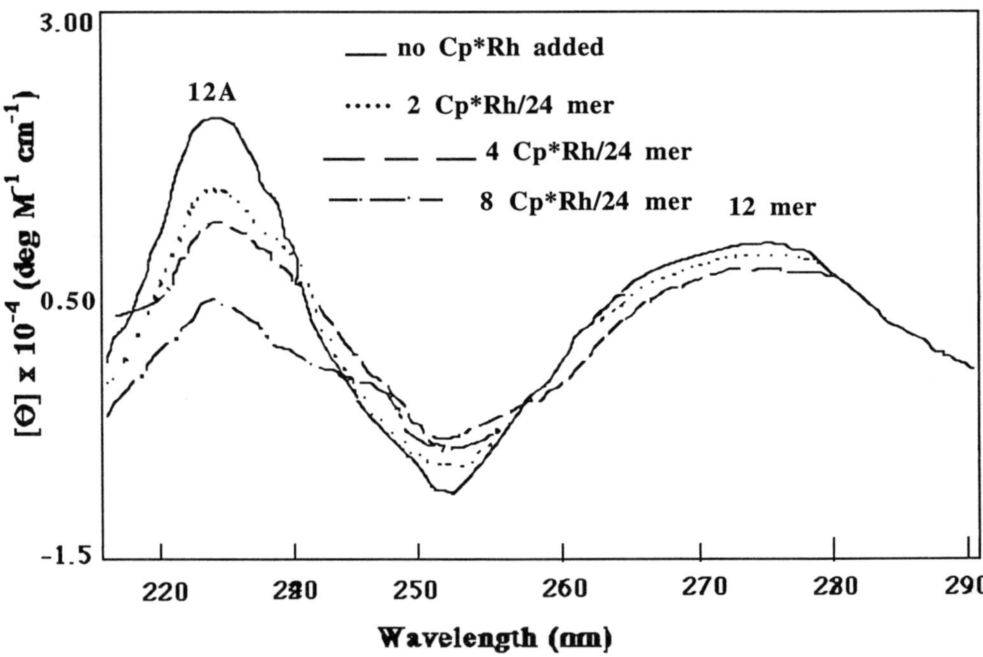

The more important question that we need to answer in the future is what type of interactions with oligomers and Cp*Rh are predominate, for example, hydrogen bonding, electrostatic, or terminal base N7 (adenine) binding, that causes the unstacking of larger oligomers. We are presently conducting NMR experiments with less complicated adenine oligomers, such as dApA, dApApA, and dApApApA, to ascertain the various bonding modes with the Cp*Rh aqua complex as was mentioned above.[8]

Molecular Recognition of Aromatic Amino Acids with the Cp*Rh-Nucleobase/Nucleoside/Nucleotide Cyclic Trimers

Finally, we wanted to see if the racemic mixtures and diastereomers of the various Cp*Rh-nucleobase/nucleoside/nucleotide cyclic trimers we have just described in this chapter could molecularly recognize any L- or D-amino acids in aqueous solution. The cyclic trimers have C3 symmetry and molecular models convey a supramolecular, bowl structure to this host with the possibility of hydrogen bonding or π-π interaction of the incoming amino acid guest molecules with the Cp*Rh complexes of DNA/RNA bases; if this molecular recognition phenomena occured, it would be the first reported example of an organometallic host, in aqueous solution, recognizing a biological guest molecule.[9]

We were quite surprised to find that indeed the cyclic trimers do molecularly recognize aromatic groups of amino acids, such as L and D-phenylalanine and tryptophan, but, interestingly, not L- or D-aliphatic amino acids such as alanine or histidine. Apparently, the aromatic groups interact by a classical π-π mechanism, a process that is well documented in organic host-guest molecule chemistry.[10]

The Figure depicts this interaction established by the use of ^{1}H NMR spectroscopy.

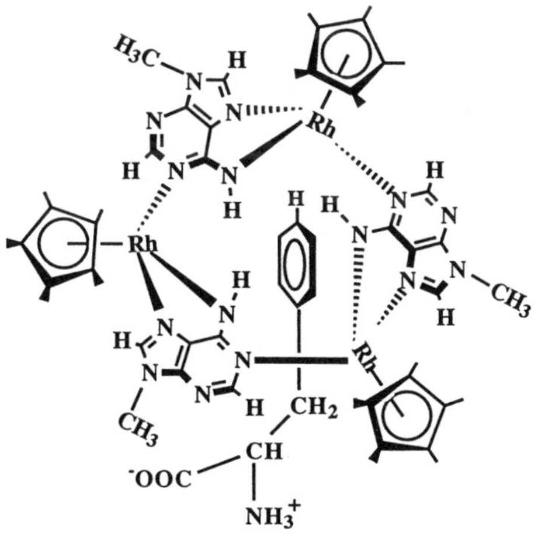

The aromatic hydrogens of the aromatic amino acids, designated below, affected by being in the influence of the π-system of the complexed nucleobases/nucleosides/nucleotides are all shifted upfield.

$$\textbf{Host} + \textbf{Guest} \xrightleftharpoons{K_a} \textbf{Host-Guest}$$

Host = [Cp*Rh(9-MA)]$_3$(OTf)$_3$, [Cp*Rh(Ado)]$_3$(OTf)$_3$, [Cp*Rh(Me-5'-AMP)]$_3$

Guest = L-Phenylalanine, L-Tryptophan, L-Histidine, L-Alanine

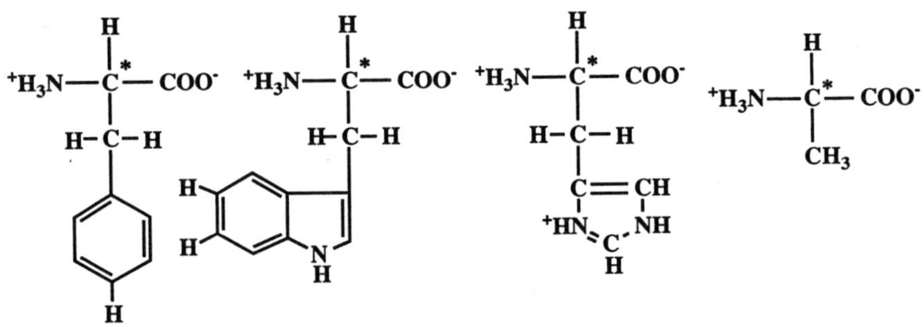

We used this data to determine the association constant (K_a) for the Host-Guest π-π interaction. It is noteworthy to mention that tryptophan, with its N donating atom, has the largest K_a value and that as you increase steric bulk at the 9-position from sugar to 5'-methyl phosphonated sugar, the Host-Guest interaction is dramatically decreased. This provides conclusive evidence that the π-π interaction is inside the Host supramolecular structure not outside. The estimated K_a values for the Host-Guest are as follows:

Host	Guest			
	L-Phe	L-Trp	L-His	L-Ala
Trimer (9-MA)	16	43	0	0
Trimer (Ado)	12	472	0	0
Trimer (Me-5'-AMP)	< 1	< 1	-	-

Conclusion

In this chapter, we have reviewed our results concerning the new area of bioorganometallic chemistry of DNA/RNA bases.[2a-d,4,6,7] Since all of this chemistry occurs in H_2O, the relevance to biological systems is paramount. This is especially borne out by the fact that our application of anchoring large DNA single molecules to glass and electrode surfaces as well as imaging this anchored DNA, to ultimately cut, amplify, and sequence the DNA segments, has been successfully demonstrated.[4] The structures of the Cp*Rh - DNA/RNA base complexes we identified by single-crystal X-ray and by NMR/FAB-MS analysis, that occur by reaction in H_2O, are unique and represent new vistas for aqueous organometallic chemistry. Finally, the ability of the Cp*Rh-DNA/RNA cyclic trimer complexes to molecularly recognize aromatic amino acids paves the way for utilization of the Cp*Rh aqua complexes as a new class of molecular recognition agents.[9]

Acknowledgment:

The studies at LBL were supported by NIH and Laboratory Directed Research and Development Funds and the Department of Energy under Contract No. DE-ACO3-76SF00098. We thank Dr. Mina J. Bissell, LBL Life Sciences Division Director, for generously supporting this project from its inception.

References

* To whom all correspondence should be addressed

1. Tullius, T.D. (1989) Metals and molecular biology, in Tullius, T.D.(Ed), *Metal-DNA Chemistry*; ACS Symposium Series (No. 402), American Chemical Society, Washington, D. C., pp. 1-24 and references therein.

2. (a) Smith, D.P., Baralt, E., Morales, B., Olmstead, M.M., Maestre, M.F., and Fish, R.H. (1992) Bioorganometallic chemistry. 1. Synthetic and structural studies in the reactions of a nucleobase and several nucleosides with a (η^5-pentamethylcyclopentadienyl)rhodium aqua complex, *J. Am. Chem. Soc.* **114**, 10647-10649. (b) Smith, D.P., Olmstead, M.M., Noll, B.C., Maestre, M.F., and Fish, R.H. (1993) Bioorganometallic chemistry. 2. Synthesis and structural studies of a nucleobase, 1-methylcytosine, with a (η^5-pentamethylcyclopentadienyl)rhodium aqua complex, *Organometallics.* **12**, 593-596. (c) Smith, D.P., Griffin, M.T., Maestre, M.F., and Fish, R.H. (1993) Bioorganometallic chemistry. 4. The bonding role of carbonyl and amino functionalities in the reactions of guanine nucleobase derivatives with (η^5-pentamethylcyclopentadienyl)rhodium aqua complexes in methanol and aqueous solutions, *Inorg. Chem.* **32**, 4677-4678. (d) Smith, D.P., Kohen, E., Maestre, M.F., and Fish, R.H. (1993) Bioorganometallic chemistry. 3. Role of the phosphate group during reactions of adenosine monophosphate derivatives with an (η^5-pentamethylcyclopentadienyl)rhodium aqua complex in the diastereoselective formation of cyclic trimers, $[Cp^*Rh(AMP)]_3$, *Inorg. Chem.* **32**, 4119-4122. (e) Toney, J.H, and Marks, T.J. (1985) Hydrolysis chemistry of the metallocene dichlorides $M(\eta^5\text{-}C_5H_5)_2Cl_2$, M = Ti, V, Zr. Aqueous kinetics, equilibria, and mechanistic implications for a new class of antitumor agents, *J. Am. Chem. Soc.* **107**, 947-953. (f) Toney, J.H., Brock, C.P., and Marks, T.J. (1986) Aqueous coordination chemistry of vanadocene dichloride, $V(\eta^5\text{-}C_5H_5)_2Cl_2$, with nucleotides and phosphoesters. Mechanistic implications for a new class of antitumor agents, *J. Am. Chem. Soc.* **108**, 7263-7274. (g) Kuo, L.Y., Kanatzidis, M.G., and Marks, T.J. (1987) Metallocene antitumor agents. Unusual $Mo(\eta^5\text{-}C_5H_5)_2Cl_2$ nucleotide/nucleobase aqueous coordination chemistry, *J. Am. Chem. Soc.* **109**, 7207-7209. (h) Kuo, L.Y., Kanatzidis, M.G., Sabat, M., Tipton, A.L., and Marks, T.J. (1991) Metallocene antitumor agents. Solution and solid-state molybdenocene coordination chemistry of DNA constituents, *J. Am. Chem. Soc.* **113**, 9027-9045. (i) Beauchamp, A.L., Cozak, D., and Mardhy, A. (1984) Synthesis and crystal structure of chlorobis (η^5-cyclopentadienyl)purinatotitanium(IV), a model compound for the interaction of the antitumor titanocene dichloride molecule with DNA bases, *Inorg. Chim. Acta*, **92**, 191-197. (j) Cozak, D., Mardhy, A., Olivier, M.J., and Beauchamp, A. L. (1986) N7/O6

chelation in a complex with an analogue of guanine. Preparation, spectroscopic study, and crystal structure of bis(η^5-cyclopentadienyl)(theophyllinato)titanium(III), *Inorg. Chem.*, **25**, 2600-2606. (k) Agnew, N.H., Appleton, T.G., Hall, J.R., Kilmister, G.F., and McMahon, I.J. (1979) A new mode of theophyllinate anion-metal bonding: N(7),N(9)-bridging in theophyllinate complexes of trimethylplatinum(IV), *J. Chem. Soc. Chem. Comm.* 324-326. (l) Ryabov, A.D., Menglet, D.L., and Levi, M.D. (1991) Direct binding of η^5-pentamethylcyclopentadienylrhodium(III) to nicotinamide cofactors: a step towards NAD^+/NADH recycling, *J. Organomet. Chem.* **421**, C16-C21. (m) Bauer, H., Sheldrick, G.M., Nagel, U., and Beck, W. (1985) Metallkomplexe mit biologisch wichtigen liganden, XL. Iridium(III)-komplexe mit 9-methylguanin, *Z. Naturforsch.* **40b**, 1237-1242. (n) Ryabov, A.D. (1991) The biochemical reactions of organometallics with enzymes and proteins, *Angew. Chem. Int. Ed. Engl.* **30**, 931-941. (o) Jaouen, G., Vessieres, A., and Butler, I.S. (1993) Bioorganometallic chemistry: A future direction for transition metal organometallic chemistry? *Acc. Chem. Res.* **26**, 361-369.

3. (a) Fish, R.H., Kim, H.-S., Babin, J.E., and Adams, R.D. (1988) Studies on the bonding of polynuclear heteroaromatic nitrogen ligands to (pentamethylcyclopentadienyl)rhodium dication: the role of nitrogen versus π-complexation on the regioselective hydrogenation of the nitrogen ring, *Organometallics* **7**, 2250-2252. (b) Fish, R.H., Kim, H.-S., and Fong, R.H. (1989) Bonding studies of nitrogen heterocyclic ligands to (η^5-cyclopentadienyl)ruthenium cation: a novel nitrogen to π rearrangement *Organometallics* **8**, 1375-1377. (c) Fish, R.H., Fong, R.H., Tran, A., and Baralt, E. (1991) Heteroaromatic nitrogen ligand studies with the (η^5-pentamethylcyclopentadienyl)rhodium cation: η^1(N) and $\eta^6(\pi)$ bonding modes and factors that influence a nitrogen to π rearrangement, *Organometallics* **10**, 1209-1212. (d) Fish, R.H., Baralt, E., and Kim, H.-S. (1991) Bonding of mono- and polynuclear heteroaromatic nitrogen ligands to the (η^5-pentamethylcyclopentadienyl)rhodium dication: structure-reactivity relationships in the formation of nitrogen (η^1) versus π (η^5,η^6) complexes and competition studies of the ligands for the rhodium metal center, *Organometallics* **10**, 1965-1971. (e) Baralt, E., Smith, S.J., Hurwitz, J., Horvath, I.T., and Fish, R.H. (1992) Homogeneous catalytic hydrogenation. 6. Synthetic and mechanistic aspects of the regioselective reductions of model coal nitrogen, sulfur, and oxygen heteroaromatic compounds using the (η^5-pentamethylcyclopentadienyl)rhodium tris(acetonitrile) dication complex as the catalyst precursor, *J. Am. Chem. Soc.* **114**, 5187-5196.

4. Zuccheri, G, Smith, D.P., Fish, R.H., and Maestre, M.F. (unpublished results).

5. Eisen, M.S., Haskel, A., Chen, H., Olmstead, M.M., Smith, D.P., Maestre, M.F., and Fish, R.H. Aqueous organometallic chemistry: Structure and dynamics in the formation of (η^5-pentamethylcyclopentadienyl)rhodium aqua complexes as a function of pH (submitted for publication).

6. Chen, H., Olmstead, M.M., Smith, D.P., Maestre, M.F., and Fish, R.H. Bioorganometallic chemistry. 6. The pH dependent synthesis and structural study of dimer and cyclic trimer complexes from the reactions of 9-methyl and 9-ethylhypoxanthine nucleobases with (η^5-pentamethylcyclopentadienyl)rhodium aqua complexes (submitted for publication).

7. Smith, D.P., Maestre, M.F., and Fish, R.H. (in preparation for publication).

8. Eisen, M.S., Maestre, M.F., and Fish, R. H. (unpublished results).

9. Chen, H., Maestre, M.F., and Fish, R.H. Bioorganometallic chemistry. 5. Molecular recognition of aromatic amino acid guests by Cp*Rh-nucleobase/nucleoside/nucleotide cyclic trimer hosts in aqueous solution (submitted for publication).

10. Lehn, J.-M. (1988) Supramolecular chemistry-scope and perspectives, molecules, supermolecules, and molecular devices, *Angew. Chem. Int. Ed. Engl.* **27**, 89-112 and references therein.

ON NATURE'S AQUEOUS ORGANOMETALLIC AND RADICAL CHEMISTRY WITH B_{12}-DERIVATIVES

BERNHARD KRÄUTLER
Institute of Organic Chemistry
University of Innsbruck, Austria

Nowadays, bio-organometallic chemistry is recognized to be of profound and widespread importance [1], in "simple" microorganisms, in particular [2]: In bacterial anaerobes, the newly discoverd pathway of autotrophic CO_2-fixation via acetyl coenzyme A is based largely on organometallic transformations [3].

The relevance of organometallic chemistry in Nature was discovered with the structure of coenzyme B_{12} (**3**, see Figure 1), an organometallic analog of vitamin B_{12} (**1**), and with the isolation from natural sources of methylcobalamin (**2**, see e.g. ref. [4]). The coenzyme **3** is recognized to be cocatalyst in about a dozen enzymatic reactions, thought to proceed via "bound" radicals [5]. Methylcorrins, such as **2**, serve in enzymatic methyl transfer reactions, crucial e.g. to bacterial C_1-metabolism, in the acetyl CoA pathway of CO_2-fixation [3].

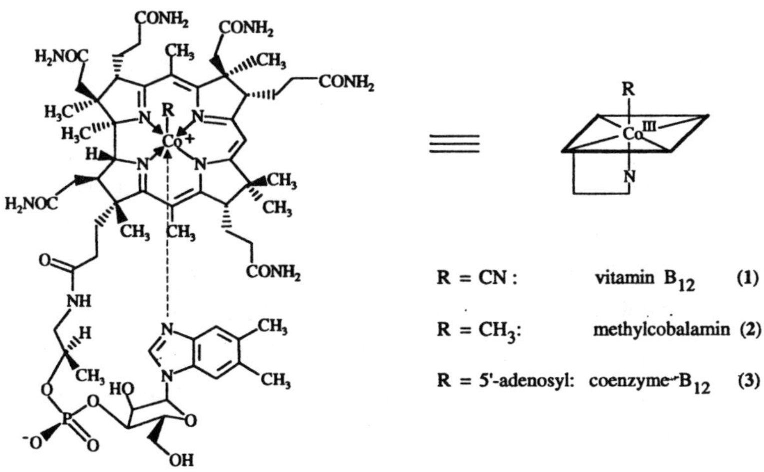

Figure 1: Structural formulae of some important B_{12}- derivatives

I. T. Horváth and F. Joó (eds.), Aqueous Organometallic Chemistry and Catalysis, 277–279.

The established B_{12}-dependent enzymatic reactions rely on the exceptional reactivities of the B_{12}-coenzymes (see Figure 2), characterized [6] by:

a. their ease of formation of organometallic B_{12}-forms, by alkylations involving highly nucleophilic reduced Co(I)-corrins or radicaloid Co(II)-forms;
b. the tendency of adenosylcorrins, such as **3**, to undergo homolytic cleavage of their organometallic bond, and of methylcorrins, such as **2**, to transfer their organometallically bound methyl group to nucleophiles;
c. the access under physiological conditions of the strongly reducing Co(I)-corrins and the high stability of Co(II)-corrins.

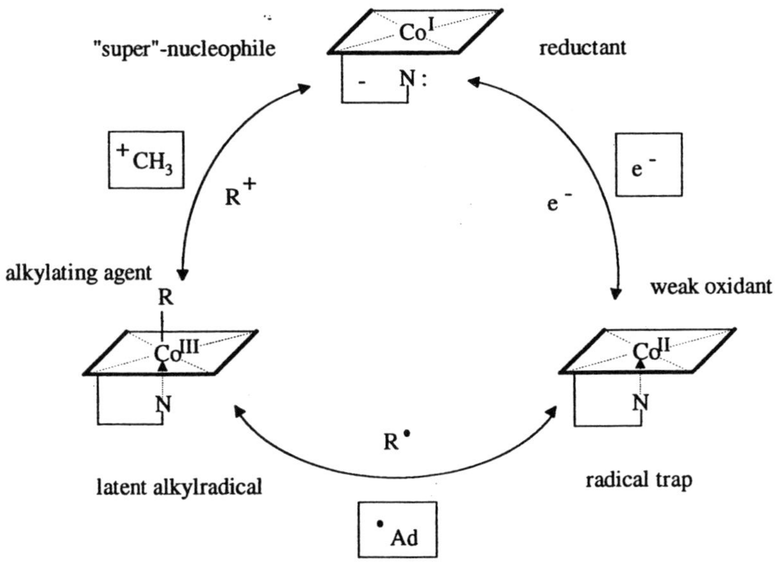

Figure 2: Biologically important reactivities of vitamin B_{12}-derivatives in organometallic and radical reactions [6].

Recently, work on the biosynthesis of novel ether lipids from the methanogen *Methanobacterium thermoautotrophicum* in the laboratory of D.Arigoni provided evidence for methylation reactions at alkyl functionalities, suggested to involve (C-C)-bond formation by abstraction of a B_{12}-bound methyl group by an alkyl radical [7]. In support of such a mode of (C-C)-bond formation, thermolysis or photolysis of organometallic alkyl-bridged B_{12}-dimers resulted in the formation of cycloalkanes (see Figure 3) [8,9].

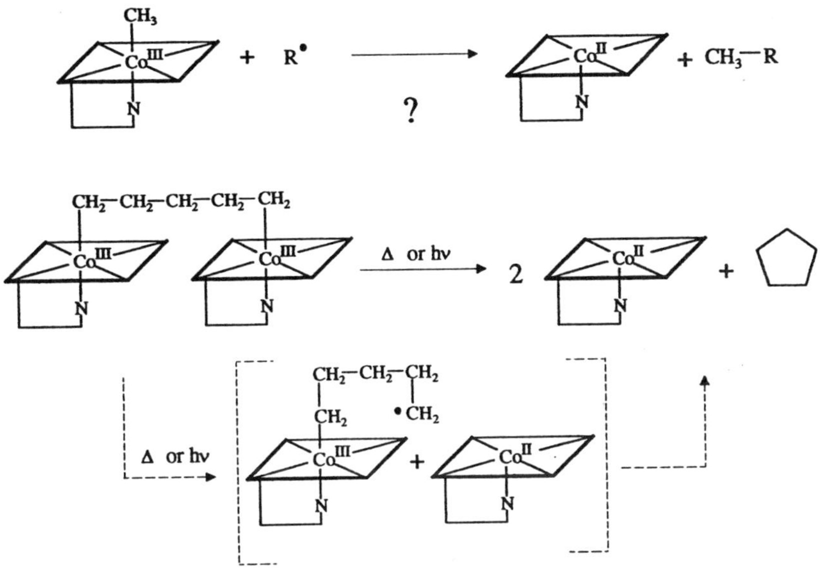

Figure 3: Reaction of a methylcorrinoid and an alkyl radical as hypothetical step in enzymatic (C-C)-bond forming reactions (top) [7] and intramolecular model reaction of an organometallic B_{12}-dimer [8,9] (bottom).

REFERENCES:

[1] Fausto da Silva,J.J.R and Williams,R.J.P. (1991) *The Biological Chemistry of the Elements*, Clarendon Press, Oxford.

[2] Thauer,R.K. (1988) *Nachr. Chem. Techn. Lab.* **36**, 993.

[3] Ragsdale, S.W. (1991) *CRC Crit. Rev. Biochem. Mol. Biol.* **26**, 261.

[4] Glusker,J.P. (1982) in D.Dolphin (ed), B_{12}, Wiley, New York, Vol. I, p 24

[5] Retey,J. (1991) *Angew. Chem.* **102**, 373; *Int. Ed. Engl.* **29**, 355.

[6] Kräutler,B. (1994) Cobalt: B_{12} Enzymes and Coenzymes, in R. B.King (ed) *Encyclopedia of Inorganic Compounds*, Vol. II, Wiley, New York.

[7] Arigoni, D. and Galliker, P.K., published in Galliker,P.K. (1990), Zur Biosynthese der Etherlipide aus Methanobacterium thermoautotrophicum, thesis, ETH-Zürich, No. 9119.

[8] Kräutler,B.,Derer,T.,Liu,P.,Mühlecker,W.,Puchberger,M.,Gruber,K. and Kratky,C. (1994), Angew. Chem., in press.

[9] Kräutler,B. (1991), Coord. Chem. Rev. **111**,215.

CATALYTIC HYDROGENATION OF BIOLOGICAL MEMBRANES: THE EFFECTS ON MEMBRANE PHYSICAL STATE AND FUNCTIONS

LÁSZLÓ VÍGH and FERENC JOÓ,
Institute of Biochemistry, Biological Research Center, Hungarian Academy of Sciences and Institute of Physical Chemistry, Kossuth Lajos University
H-6701 Szeged, POB 521. and H-4010 Debrecen, POB 7., Hungary.

About ten years ago our research group began a systematic study to understand more fully the diverse biological effects of fatty acid unsaturation. After assessing the particular advantages and disadvantages of each method to modify, and thereby define the exact role of membrane lipid unsaturation, we decided to apply the technique of homogeneous catalytic hydrogenation for answering many of specific questions raised in this subject. A vast number of soluble transition metal complexes have also been tested, but though they were able to catalyze the desired reaction, they did not fill the strict requirements for use in biomembrane hydrogenation (biocompatibility, high specific activity under physiological conditions, selectivity, stability, the possibility of removal after completion of reaction).

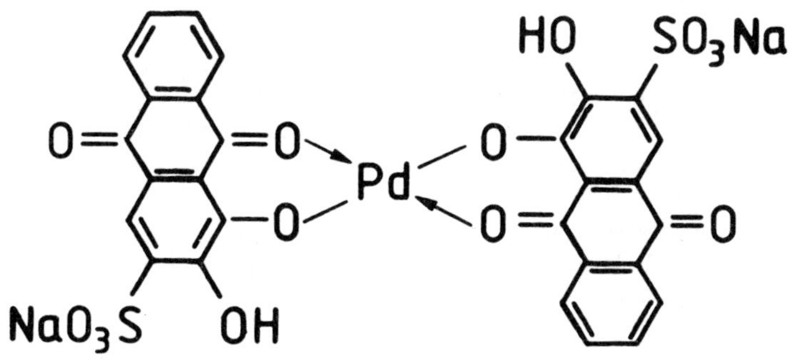

Figure 1. Palladium (II) complex of the alizarin red, (sodium 1,2-dihydroxy-9,10-anthraquinone-3-sulfonate) the superb catalyst for biomembrane hydrogenations. (Data from Joó et al., 1991)

By now it is well established, that one catalyst, palladium-di-(sodium alizarine monosulphonate), $Pd(QS)_2$ is far superior to any other complexes tested for

I. T. Horváth and F. Joó (eds.), Aqueous Organometallic Chemistry and Catalysis, 281–287.

membrane studies. This catalyst is water-soluble yet amphipathic enough to hydrogenate very efficiently and still selectively the cis-double bonds of lipid acyl chains in biomembranes either in vivo or in vitro (Figure 1.). Our major findings by using this catalyst are outlined as follows [1].

Comparison of the rates of hydrogenation, amongst various phospholipid classes a slight preference of the activated complex for phospatidylcholines was revealed and there was a preference also towards polyunsaturated molecular species. The catalyst had a better accessibility to substrates adapted to bilayer phase.

This complex was successfully exploited also to reveal the relationship between lipid saturation and lipid-protein interaction in mitochondria. The time courses of hydrogenation of individual complex lipids and their molecular species suggested that cardiolipin, being in tight association with some integral proteins is less accessible to the catalyst. The existence of the protein-immobilized membrane compartment has also been confirmed by ESR technique. To use Pd-complex as a structural probe to detect microdomains would become a promising tool in membrane research [2].

The effect of hydrogenation on the physical state, structure and function of pea chloroplast has been investigated in a series of experiments [3,4]. The effect of

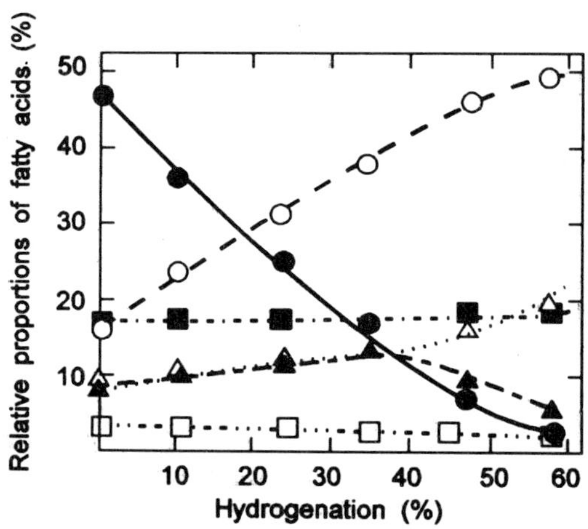

Figure 2. Relative proportions of various fatty acids during hydrogenation of thylakoid membranes plotted as function of percentage hydrogenation. 18:3 (l); 18:2 (▲); 18:1 (m); 18:0 (D); 16:1 (o); 16:0 (n). (Data from Vígh et al., 1985)

hydrogenation on the fatty acid pattern of pea chloroplast is presented in Figure 2.. Microdensitometry studies revealed, that the realignment of the hydrocarbon chains resulted in a thickening of the membranes (Figure 3.). Low temperature polarized fluorescence investigations indicated that neither the orientational pattern of chlorophyll-a molecules nor energy-transfer processes between photosystems were affected by hydrogenation. Hydrogenation did, however, inhibit whole chain electron

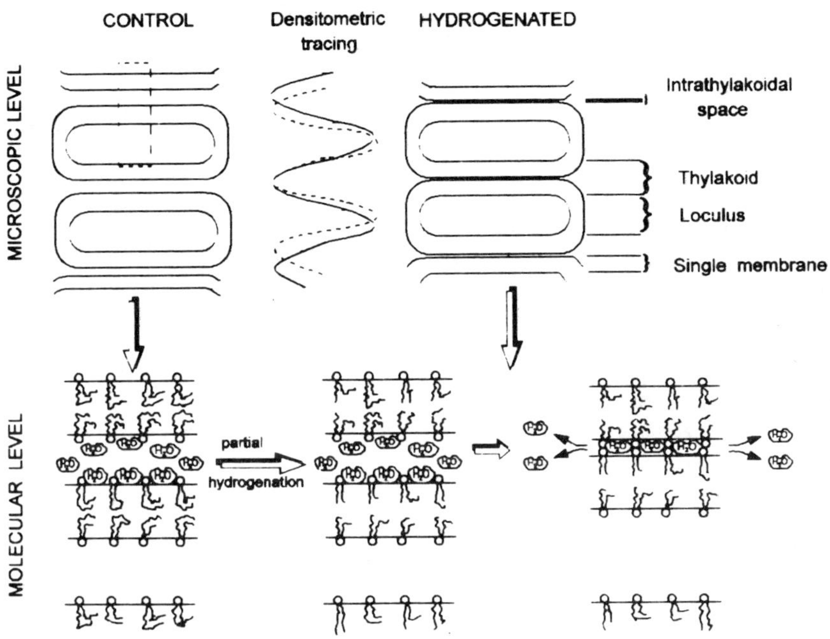

Figure 3. Proposed mechanism for structural changes in chloroplast membranes by hydrogenation. (Data from Horváth et al., 1986)

transport before inhibition of either PSII or PSI activity was observed (Figure 4/A). No change in the reduction rate of flash-oxidized cytochrome b6/f was detected leading to the conclusion that the rate of plastoquinol diffusion is unaffected by the reduced fluidity of thylakoid. Thereafter, a more detailed study of photosystems revealed, that there is an optimal level of unsaturation that is reguired to maintain an efficient electron transfer between QA and the plastoquinone pool. One main point of interest was to prove, that the lipid molecular order of thylakoids was not a linear function of the overall degree of unsaturation (Figure 4/B). According to our freeze-fracture EM studies, gel phase appeared above 30% hydrogenation [3,4]. Our above findings had a great impact to our knowledge in several respect. They underlined the fundamental importance of specific polyunsaturated fatty acyl chains in photosynthesis, whereas the contention, that fluidity *per se* is a key factor in governing photosynthetic reactions (e.g. during low temperature stress) had no experimental support.

The catalytic hydrogenation by $Pd(QS)_2$ has also been used to examine the mechanism of retailoring membrane lipids in the process of cold adaptation in *Tetrahymena* cilia membranes. Lipid saturation was associated with a dramatic decrease in endogenous PLA activity even at a slight hydrogenation, having not significant effect on the membrane fluidity. The way of this response, therefore, was not in full accordance with the proposed mechanism predicted to operate during thermal adaptation. Our recent investigations on mitochondrial membranes further corroborated, that PLA activity correlates with loss of polyunsaturated fatty acyls, rather than with change of the bulk membrane fluidity [5].

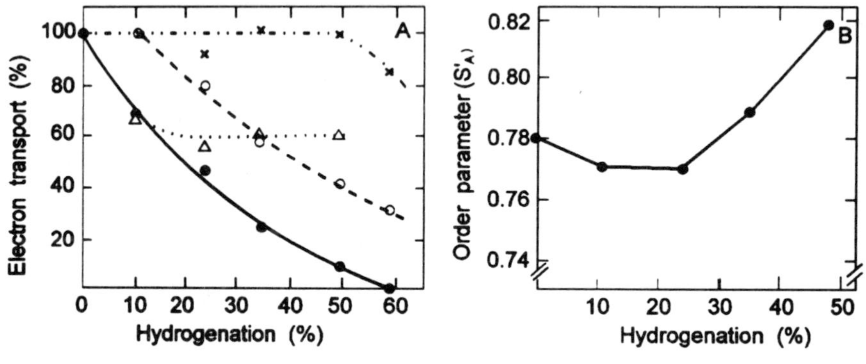

Figure 4. (A): Effect of hydrogenation on the activity of the different parts of the photosynthetic electron-transport chain. Whole electron transport measured from $H_2O \rightarrow$ methyl viologen (● — ●); PS II electron transport from $H_2O \rightarrow$ pBQ (O - - - O); PS I activity measured from $DQH_2 \rightarrow$ methyl viologen (Δ ····· Δ); or from DCIP → methyl viologen (X ----- X)

(B): Changes in the order parameter (S'_A) of fatty acid chains as a function of hydrogenation measured with 12-SASL.

(Data from Horváth et al., 1986 and Vígh et al., 1985, respectively)

A direct support for the hypothesis that the phase behavior of cytoplasmic membranes determines chilling susceptibility of cyanobacteria has been gained by *in vivo* hydrogenation on *Anacystis nidulans* cells. When the membranes of 28°C-grown cells were subjected to a specific hydrogenation they showed a similar susceptibility to chilling injury as 38°C-grown cells [6,7].Chilling sensitivity of cells was judged by measuring K^+ release, representing leakiness of the cytoplasmic membrane and by determining the rate of O_2 evolution, an index of the integrity of the thylakoid membrane (Figure 5.).

Valuable information on the mechanism of hydrogenation and restoration of lipid unsaturation have been obtained from studies on the unicellular green alga, *Dunaliella salina* (a model of higher plant cells) [8]. Stabilization of thylakoid membranes to heat stress has been shown in hydrogenated chloroplast. Hydrogenation caused an increased thermal stability of pigment protein complexes of the PS-II light harvesting apparatus and of PS-I region as well [9,10].

Biological organisms are capable of perceiving environmental changes caused by a variety of agents. Whereas sensors of light (rhodopsins in animals or phytochromes in plants) are well-characterized, no knowledge is available concerning the thermal sensors. The only clue, hitherto known about the nature of the thermal perception is that in general, temperature decrease would result in a reduction of

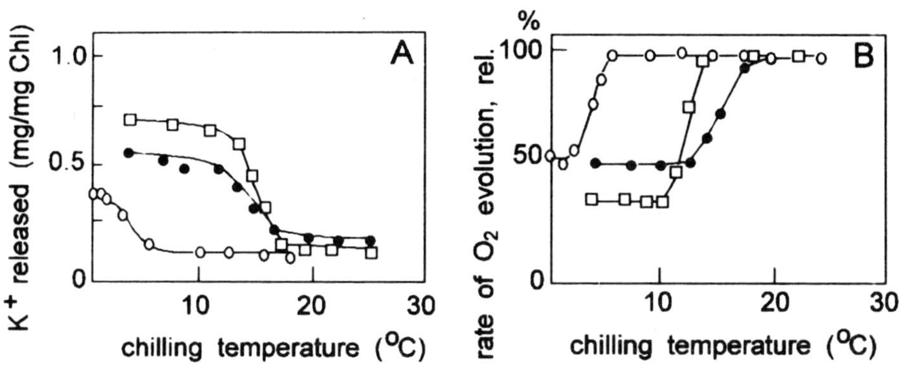

Figure 5. The effect of chilling temperature of *Anacystis nidulans* on (A) K^+ release (B) photosynthetic oxygen evolution. Cells grown at 28°C, (m);at 38°C, (o); at 28°C and hydrogenated in the presence of water-soluble palladium alizarin catalyst (l)
(Data from Vígh et al., 1983)

molecular motion in membranes, which is attained mainly via specific desaturation of fatty acids of membrane lipids. Recently we reported, that the increased transcription of the *desA* gene, coding for delta-12 desaturase, can be achieved not only by a decrease in temperature but also by direct in vivo catalytic hydrogenation [11,12]. The saturation induced rigidification and activation of gene expression simultaneously, was conduced on cyanobacteria. This was the first demonstration for the involvement of membrane physical state in the mechanism of thermal perception [13].. The brief hydrogenation induced expression of the *desA* in cyanobacterial cells is shown by Figure 6.

Another new approach of exploiting membrane selective saturation in order to follow the "fate" of specific membrane domains would be the method of catalytic deuteration introduced also quite recently by our team [14,15]. We demonstrated, that catalytic deuteration of lipids within model- and biomembranes is a simple and rapid method. FTIR spectroscopy of deuterated samples offered a selective information about deuterated lipid molecules by monitoring simultaneously the unreacted species as well. Thus, deuterating alkyl chains with double bonds at specific positions the saturated molecules can be used as internal "membrane structural probes". It was stated, that in case of some systems the live cell is able to keep the fluidity of its plasma membrane at a much higher value than that of was previously predicted by using isolated plasma membrane for measuring membrane physical state [16[. By applying reaction conditions enabling selective saturation of surface membranes [14], the above technique could directly be applied to live cells to measure the alterations in the conformational order and thermal characteristics of plasma membrane without its isolation. Besides to link these changes to immediate cellular responses (like elevetad expression of *desA* gene, changing thrmal stability, etc.), the kinetics of the "recovery"

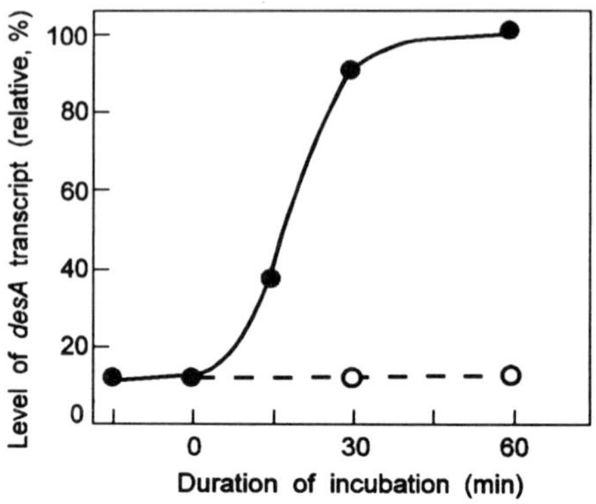

Figure 6. Changes in the level of the *desA* transcript after hydrogenation. After the cells had been hydrogenated at 36°C for 4 min, they were incubated at 36°C in the light under aerobic conditions for the designated periods of time. The level of the *desA* transcript was determined by Northern blot analysis and is expressed in units relative to the level at 60 min after hydrogenation. (●) Cells without hydrogenation, (○) cells hydrogenated for 4 min.
(Data from Vígh et al., 1993)

of manipulated membrane which might be expected due to retroconversion of lipid unsaturation in internal pools and a subsequent membrane flow could also be followed. The understanding the homeostatic controlling mechanisms by testing how the cells retain the optimal plasma membrane physical state, including bulk "fluidity", lipid-protein interactions [2] as well as restoration of preexisting lipid(fluidity) asymmetry within membrane leaflets would be the most interesting aspects of such kind of studies [15].

References

1. Joo, F., Balogh, N., Horvath, L.I., Filep, Gy., Horvath, I., and Vigh, L. (1991) Complex hydrogenation/oxidation reactions of the water-soluble hydrogenation catalyst, palladium-di(sodium alizarine-monosulfonate) and the know-how of homogeneous hydrogenation of lipids in isolated biomembranes and living cells. *Anal. Biochem.* **194**, 34-40.
2. Schlame, M., Horvath, L.I., and Vigh, L. (1990) Relation between lipid saturation and lipid-protein interaction in liver mitochondria modified by catalitycal hydrogenation with reference to cardiolipin molecular species. *Biochem. J.* **265**(1),: 79-85.
3. Vigh, L., Joo, F., Droppa, M., Horvath, L.I., and Horvath, G. (1985) Modulation of chloroplast membrane lipids by homogenous catalytic hydrogenation. *Eur. J. Biochem.* **147**,: 477-481.
4. Horvath, I., Vigh, L., Pali, T., and Thompson, G.A. (1989) Effect of catalytic hydrogenation of Tetrahymena ciliary phospholipide fatty acids on ciliary phospholipase A activity. *Biochim. Biophys. Acta* **1002**, 409-412.
5. Schlame, M., Horvath, I., Török, Zs., Horvath, L.I., and Vigh, L. (1990) Intramembraneous hydrogenation of mitochondrial lipids reduces the substrate availability but not the enzyme activity of endogeneous phospholipase A. The role of polyunsaturated phospholipid species. *Biochim. Biophys. Acta* **1045**, 1-8.

6. Vigh, L. and Joo, F. (1983) Modulation of membrane fluidity by catalytic hydrogenation affects the chilling susceptibility of the blue-green alga, Anacystis nidulans. *FEBS Letters* **162**, 423-427.
7. Vigh, L., Gombos, Z., and Joo, F. (1985) Selective modification of cytoplasmic membrane fluidity by catalytic hydrogenation provides evidence on its primary role in chilling susceptibility of the blue-green alga, *Anacystis nidulans*. *FEBS Letters* **191**, 200-204.
8. Vigh, L., Horvath, I., and Thompson, G.A. (1988) Recovery of *Dunaliella salina* cells following hydrogenation of lipids in specific membranes by a homogeneous palladium catalyst.*Biochim. Biophys. Acta* **937**, 42-50.
9. Thomas, P.G., Dominy, P.J., Vigh, L., Mansourian, A.R., Quinn, P.J., and Williams, W.P. (1986) Increased thermal stability of pigment-protein complexes of pea thylakoids following catalytic hydrogenation of membrane lipids. *Biochim. Biophys. Acta* **849**, 131-140.
10. Vigh, L., Gombos, Z., Horvath, I., and Joo,F. (1989) Saturation of membrane lipids by hydrogenation induces thermal stability in chloroplast inhibiting the heat-dependent stimulation of photosystem I-mediated electron transport. *Biochim. Biophys. Acta* **979**, 361-364.
11. Los, D., Horvath, I., Vigh, L., Murata, N. (1993) Low temperature dependent expression pf the desaturase desA in Synechocystis PCC 6803. *FEBS Letters* **318**,: 57-60.
12. Vigh, L., Los, D., Horvath, I., and Murata, N. (1993) The primary signal in the biological perception of temperature: Pd-catalyzed hydrogenation of membrane lipids stimulated the expression of the desA gene in Synechosystis PCC 6803. *Proc. Natl. Acad. Sci. USA* **90**, 9090-9094.
13. Maresca,B. and Cossins,A.R. (1993) Fatty feedback and fluidity. *Nature* **365**, 606-607.
14. Török, Zs., Szalontai, B., Joo,F., Wistrom, C.A., and Vigh, L. (1993)Homogeneous catalytic deuteration of fatty acyl chains as a tool to detect lipid phase transition in specific membrane domains: a fourier transform infrared spectroscopy study. *Biochem. Biophys. Res. Com.* **192**, 518-524.
15. Quinn, P.J., Joo, F., and Vigh, L. (1989) The role of unsaturated lipids in membrane structure and stability. *Prog. Biophys. Mol. Biol.* **53**,: 71-103.
16. Cameron,D.G., Martin,A., Moffatt,D.J. and Mantsch,H.H. (1985) Infrared spectroscopic study of the gel to liquid - crystal phase transition in live *Acheoplasma laidlawii* cells. *Biochemistry* **24**, 4355-4359.

AQUEOUS PHOSPHOLIPID DISPERSIONS AS VECTORS FOR SELECTIVE HOMOGENEOUS CATALYSIS

P.J. QUINN
Division of Life Sciences
King's College London, Campden Hill, London, W8 7AH, U.K.

1. Introduction

Catalysts of biochemical reactions are almost invariably specific for a particular reaction. The specificity is often such that enzymes are able to distinguish between very closely related molecules including optically-active enantiomers etc. This high degree of specificity is achieved by a "lock and key" principle in which the catalytic site on the enzyme represents a complementary region of the polypeptide structure where the substrate is recognised and binds with high affinity. The binding process often causes electron delocalisations about the reactive covalent bond of the substrate. These effects can be exacerbated by the close apposition of certain amino acid side chains or cofactors of the reaction. The formation of covalent intermediates between enzymes and substrates in biochemical reactions is also a common catalytic strategy.

Although enzymes have been categorised for a remarkably wide range of reactions, many chemical reactions of industrial importance are unsuited to conditions governing biochemical catalysis. This is because the conditions required to achieve reasonable rates of reaction are incompatible with retention of the native structure of proteins. Our task, therefore, is to design conditions under which chemical reactions are performed so as to maximise the specificity of reaction between different substrates for a particular reaction.

There are two ways to approach this problem. The first is to prepare catalysts that recognise and interact, with a high degree of specificity, only certain substrates from a

I. T. Horváth and F. Joó (eds.), Aqueous Organometallic Chemistry and Catalysis, 289–304.

mixture of related compounds. The second is to present the catalyst to different substrates so that only one substrate interacts preferentially with the catalyst. The fundamental difference between these two strategies is that the former relies on modifying the chemistry of the catalyst to tailor its action towards particular substrates and the latter attempts to provide the physical equivalent of the "lock and key" principle exploited by enzyme catalysed reactions. The subject of this review is to consider how methods of presentation of substrate to catalysts can be exploited to achieve reaction specificity and, in particular, how the ordered structures formed by phospholipids in water can act as reaction modulators for homogeneous catalytic hydrogenation and hydroformylation reactions.

2. Structures formed by Phospholipids in Aqueous Systems

Phospholipids of biological origin are weak amphipaths because of the attachment of long-chain fatty acids to carbon atoms 1 and 2 of the glycerol-3-phosphate group that represents the, so called, backbone of the molecule. The residue attached by a phosphodiester bond to the glycerol designates the class of phospholipid and, in the experiments that will be described in this chapter, the residue is choline. The critical micelle concentration of these phosphatidylcholines, or lecithins as they are often referred to, is in the nM range of concentration and consequently there is a strong tendency for the molecules to aggregate and to phase-separate from water. The phase behaviour of these phospholipid-water systems has been reviewed elsewhere [1].

The most obvious property of these molecules is that they form insoluble monomolecular films at an air-water interface. If the fatty acid chains are saturated the monolayers are typically condensed and the molecules form a close packed configuration. The introduction of one or more *cis* unsaturated bonds into the fatty acid chains produces monolayers of the liquid-expanded type because the conformation of the hydrocarbon chains does not permit a close packing arrangement that is possibe in their fully saturated counterparts.

When phosphatidycholines are hydrated they exhibit complex lyotropic mesomorphism. In addition, they also show thermotropic behaviour and undergo structural transitions at characteristic temperatures. This means that the phase structure of the phospholipid is a function of both water content and temperature. The penetration of water into an anhydrous phospholipid lattice only occurs when the temperature is sufficient to "melt" the hydrocarbon chains. In this case there is a simultaneous dissociation of the ionic lattice by the penetration of water and a dissordering of the hydrocarbon chain region. The temperature of the transition depends on the nature of the hydrocarbon chains and is higher for saturated chains as these interact extensivly through van der Waals cohesive forces. Since these forces tend to be additive, the cohesive force increases with increasing chain length. As with monomolecular films, the introduction of unsaturated residues dramatically decreases the temperature at

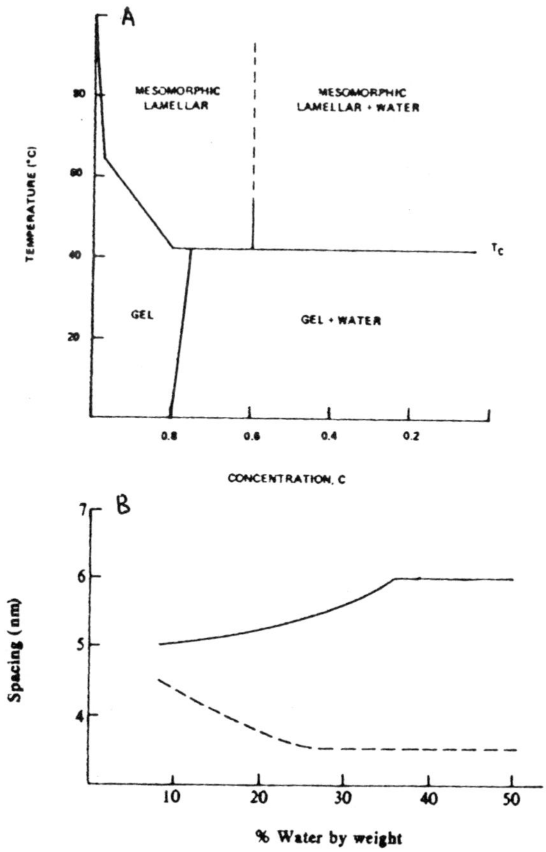

Figure 1. A. Phase diagram of 1,2-dipalmitoylphosphatidylcholine-water system. The T_c line delineates the mesomorphic lamellar phase from the gel and both phases are stable in excess water. B. Structural parameters of the lamellar liquid-crystalline phase. See ref [2] for details.

which the chains become disordered. The transition temperature also depends on the phospholipid class, the amount of water present and the inclusion of any solutes dissolved in the water.

The mesomorphic behaviour is best described by reference to a phase diagram. A simple phase diagram of lecithin-water is illustrated in Fig 1A for phospholipid with palmitoyl residues attached to the glycerol. This shows that the addition of water progressivly reduces the temperature of transition of the hydrocarbon chains from a gel to a liquid-crystalline configuration. Eventually this temperature reaches a limiting value, T_c, in excess water. Increasing water content also causes the transition to become more cooperative as the phospholipid molecules orient at the aqueous domain boundary. The value of T_c is the minimum temperature required for water to penetrate between the layers of lipid molecules. Above the T_c line the lecithin-water system exists in a mesomorphic lamellar phase, in which the hydrocarbon chains are in a liquid-like disordered state. The composition of the system at maximum hydration is about 40% water; addition of more than 50% water gives rise to a two-phase system consisting lamellar phase at maximum hydration dispersed in excess water.

The structural parameters associated with the penetration of water into the lecithin lattice are presented in Fig 1B. This shows that the X-ray Bragg spacing of the mesophase increases with uptake of water reaching a maximum limiting value at full hydration. The thickness of the phospholipid bilayer, by contrast, decreases to a limiting value in excess water.

If more vigorous methods are employed to disperse the phospholipid in excess water, such as the use of ultrasonic irradiation, the multilamellar liposome structures break down and eventually vesicles are formed of a diameter of 250-300nm. These vesicles consist of a single bilayer of phospholipid molecules. The formation of such vesicles is greatly assisted by the inclusion of a small proportion of amphipaths in the phospholipid which carry a net positive or negative charge.

3. Catalysis of Lecithin Structures

Lecithins which have unsaturated hydrocarbon chains can be hydrogenated readily in organic solvents in the presence of both heterogeneous and homogeneous catalysts. Since the unsaturated phospholipids are soluble in organic solvent they are able to orient at the catalytic surface of Adam's catalyst or form reactive complexes in solution with homogeneous catalysts like Wilkinson's catalyst.

If the solvent is made progressivly more polar the rate of hydrogenation in the presence of Wilkinson's catalyst decreases. This is illustrated in Fig. 2 which shows the initial rate of hydrogenation of soya lecithin suspended in solvent mixtures consisting of tetrahydrofuran-water. It can

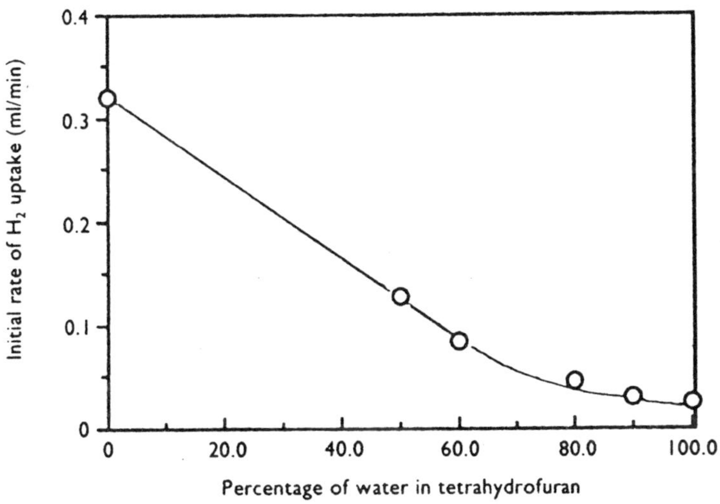

Figure 2. Initial rate of hydrogenation of soya lecithin in the presence of Wilkinson's catalyst in reactions performed in solvent mixtures of tetrahydrofuran and water. Data from ref [3].

be seen that the initial rate of hydrogenation decreases as the proportion of water in the system increases and reaches a limiting value when the phospholipid assumes a bilayer configuration. Despite differences in initial reaction rates, virtually complete hydrogenation is eventually achieved in all solvent mixtures. This is not the case with Adam's catalyst. Bilayer structures of lecithin in water are completely unreactive when incubated under hydrogenating conditions in the presence of heterogeneous catalyst. The apparent reason for this is that the substrate is unable to partition into the aqueous phase and orient at the catalytic surface of the metal. Wilkinson's catalyst, by contrast, partitions into the bilayer structure where it is able to form complexes with the unsaturated fatty acyl residues of the phospholipid even when it is arranged in a bilayer structure.

Water soluble derivatives of homogeneous catalysts were first synthesized by Joo and Beck [4] and the strategy was to replace the triphenyl phosphine residues with sulphonated triphenylphosphine. The water-soluble derivative of Wilkinson's catalyst [5] and other transition metals [6] appear to have similar properties to those of their non-sulphonated triphenylphosphine counterparts. The presence of the charged sulphonyl group renders the catalyst soluble in neutral aqueous solutions [7] and the solubility can be modulated by salt concentration or changes in pH. The solubility of metal complexes of phosphines in water can be increased by using multisulphonated triphenyphosphine [8,9]. These types of catalysts do not pentrate readily into lipid substrates but partition can be influenced by the use of complexes with phosphine-like ligands [10] or attachment of amphiphilic, long chain aliphatic ternary phosphines to the metal [11].

Hydrogenation of unsaturated phospholipids dispersed in aqueous systems using a water-soluble homogeneous catalyst was first reported by Madden and Quinn [12]. The catalyst was a sulphonated derivative of Wilkinson's catalyst which did not appear to alter the structure of substrate bilayers at least with respect to their permeability barrier properties [13]. Fig. 3 shows the relative rates of hydrogenation of multilamellar liposomes of soya lecithin and single bilayer

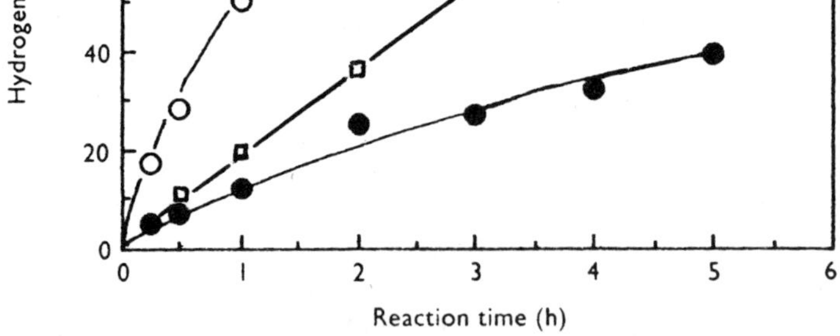

Figure 3. Time-course of hydrogenation of soya lecithin dispersed in water as multibilayer liposomes (●) or single bilayer vesicles (○) incubated under hydrogenating conditions in the presence of water-soluble derivative of Wilkinson's catalyst. The progress of the reaction is also shown for multibilayer liposomes dispersed in an aqueous solution of the catalyst (□).

dispersions of the phospholipid incubated in the presence of the sulphonated derivative of Wilkinson's catalyst. The single bilayer vesicles were completely hydrogenated after three hours reaction whereas an incubation period of about 20 hours was required for complete reaction of the substrate presented in multilamellar dispersions. The reason for this difference is that accessibility of the substrate to the catalyst is hindered in multilamellar liposomes of the substrate by the need for the catalyst to penetrate successive bilayers of the structure. This is consistent with results obtained when dispersions of the substrate are performed in aqueous solutions of the catalyst; hydrogenation in this case proceeds to completion, also in three hours, but with reaction kinetics indicating much lower catalyst concentrations accessible to all the substrate [12]. Comparisons of the rate of hydrogenation of unsaturated molecular species of lecithins, which typically form smectic mesophases, with unsaturated phosphatidylethanolamines in hexagonal-II structure in the presence of water-soluble catalyst [14] have been reported. The catalyst was found to be more active against substrate in bilayer compared with hexagonal-II conformation. Furthermore, there was a distinct preference for polyunsaturated molecular species as opposed to monounsaturated molecular species of lecithins.

The reactivity of the water-soluble paladium catalyst, paladium di(sodium)alizarine monosulphonate [$Pd(QS)_2Na_2$], has been examined in multilamellar dispersions of unsaturated phospholipids [14]. With substrates of lecithin containing dioleoyl chains there is a transient appearance of *trans w-9* derivatives of the substrate. This indicates that the reaction proceeds via a *cis-trans* isomerisation followed by reduction of the *trans* double bond. Hydrogenation of di-18:*cis*2 and di-18:*cis*3 derivatives of lecithins show highly complex patterns of partially saturated molecular species without convincing evidence of bond migrations.

4. Orientation of Amphipathic Substrates in Phospholipid Bilayers

Because of their highly ordered arrangement and orientation with respect to the aqueous phase, phospholipids have potential to influence the presentation of amphipathic substrates to water-soluble homogeneous catalysts. Thus, the location of the reactive bond of the substrate could be located at precise distances from the aqueous interface, the orientation within the phospholipid structure can be determined and the extent of penetration of the catalyst into the structure can be modulated. The last parameter can be influenced by either inclusion of salts or adjusting the pH of the aqueous medium or altering the electrostatic potential at the surface of the phospholipid structure. The manipulation of any of these factors, or combinations of them, can potentially serve to increase specificity of catalytic reactions. In the following disscussion ways in which substrates may interact and orient in phospholipid dispersions will be considered and finally examples of how specificity of catalysis can be achieved will be exemplified.

5. Phase Separation of Substrate Lipid

When unreactive, fully-saturated phospholipid is codispersed with weakly polar or non-polar substrates in aqueous systems, phase separation occurs. In this situation, the phospholipid acts as a detergent, essentially creating an oil-in-water emulsion. In terms of reaction rate such systems offer considerable advantages over reactions that take place at simple oil-water interfaces since the exposure of substrate molecules to water-soluble catalyst increases enormously by creation of such emulsions.

Phase separated systems of this type are created by triglycerides when codispersed with phospholipids. The only polar groups on triacyl glycerols are the three carbonyl oxygens and these are complemented by three long-chain hydrocarbon substituents attached to the glycerol backbone. The affinity of the carbonyl oxygens for water is not sufficient to intercalate the molecules between the phospholipids so that they orient within the phospholipid bilayer domain. When present in relativly low proportions with respect to the phospholipid, the triglycerides occupy a position in the centre of the bilayer structure. As the proportion of triglyceride in the phospholipid increases, the triglyceride accumulates in lenses and eventually into droplets that are surrounded by a monolayer of phospholipid.

Catalytic reactivity in such structures depends on the extent to which water-soluble catalysts are able to extend into the phospholipid monolayer and the tendency of the phase-separated substrate to orient about the penetrating catalyst. This type of reaction is illustrated in Fig. 4 in which cod-liver triglycerides were hydrogenated in the presence of $RhCl(dpm)_3$ catalyst [15]. The reaction mixture consists of an

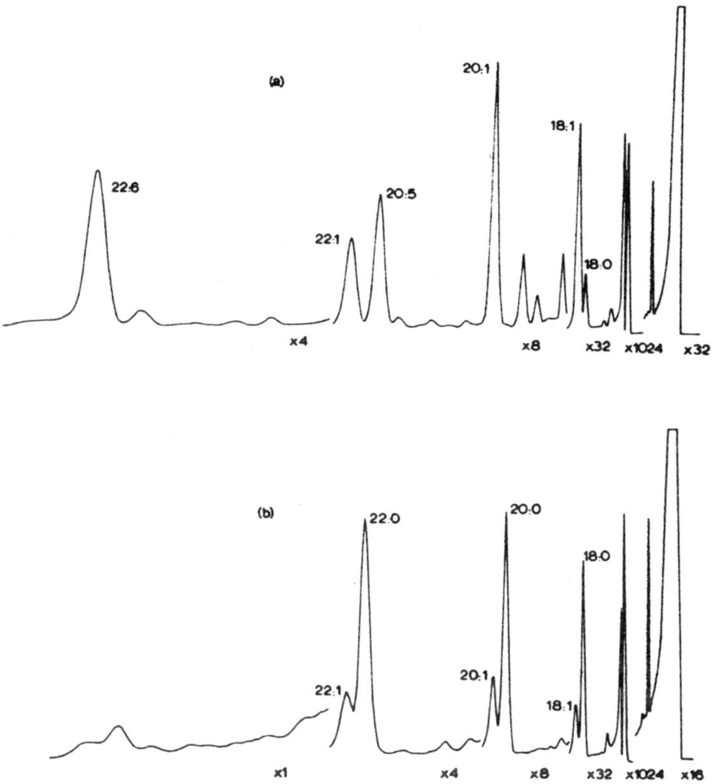

Figure 4. Gas chromatographic analysis of cod-liver triglycerides before (A) and after (B) incubation under hydrogenating conditions for 15 hours in the presence of $RhCl(dpm)_3$. The triglycerides were codispersed in neutral aqueous buffer with an equimolar amount of saturated lecithin. Data from ref [15].

approximatly equimolar amount of saturated lecithin and unsaturated cod-liver triglyceride dispersed in buffer at neutral pH. The fatty acid composition of the substrate before and after 15 hours incubation under hydrogenating conditions in the presence of the catalyst is shown in the figure. Clearly, almost complete saturation of the substrate has been achieved. The reaction rate could be increased markedly by increasing the concentration of NaCl in the aqueous reaction medium which presumably aids the penetration of the catalyst into the phospholipid monolayer separating the substrate from the aqueous phase. There was no evidence that the catalyst partitions completely into the hydrocarbon domain of the structure and the reaction is likely to take place in the interfacial region.

6. Intercalation of Substrates into Phospholipid Bilayers

If the amphiphilic balance within substrate molecules is favourable the polar affinity may serve to orient them at the aqueous interface between the phospholipid molecules of the bilayer structure. Two types of organisation can take place. The simplest type is where there is no specific interactions between substrate and phospholipids and the two components become randomly distributed in the plane of the lipid bilayer. A second type of arrangement is where interaction between phospholipid and substrate leads to the formation of stoichiometric complexes which coexist with uncomplexed phospholipid in certain regions of the binary phase diagram. Complexes of this type have been described for phospholipids and free fatty acids [16] and for phospholipids and diglycerides [17,18].

The phase separation of stoichiometric complexes of

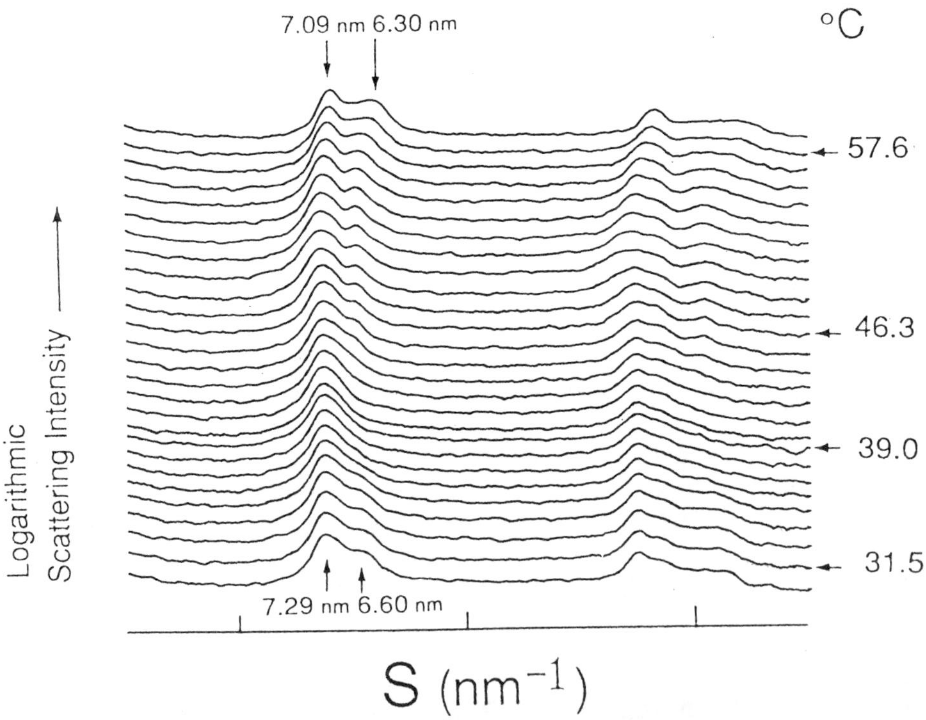

Figure 5. Successive X-ray diffraction patterns of lamellar repeat spacings recorded during heating of a mixed dispersion of dipalmitoyl derivatives of lecithin and diglyceride (5:1 mole ratio). The diffraction peak at longer spacing is from a stoichiometric 1:1 complex and the peak at shorter spacing is pure lecithin. Data from ref [19].

phospholipid and diglyceride is sufficiently extensive to produce an ordered lattice producing a detectible lamellar repeat spacing from X-ray examination. This is illustrated in Fig. 5 which shows a series of low-angle X-ray diffraction patterns recorded during heating of an aqueous dispersion of phospholipid and diglyceride. Two lamellar repeat spacings can be seen, one which changes as the temperature increases in a manner identical to the thermotropic phase transitions of the pure phospholipid and the other corresponding to a 1:1 stoichiometric complex of phospholipid and diglyceride. The data is consistent with an immiscibility of the complex in both fluid and gel phases. The formation of such complexes may be expected to result in very precise orientation of the diacyl glycerol (and also the phospholipid) with respect to the aqueous phase and represent a highly ordered presentation of the potential substrate to the catalyst complex.

7. Homogeneous Catalysis of Substrates in Phospholipid Bilayers

Possibly one of the most convincing demonstrations of the use of phospholipid bilayers as vectors for selective homogeneous catalysis is the reaction of olefins codispersed with saturated phospholipid in the presence of sulphonated Wilkinson's catalyst [20]. Hydrogenation or hydroformylation of 9-decene-1-ol and 10-undecene-1-ol in organic solvent showed no selectivity in reaction in the presence of the catalyst. This is illustrated in the case of hydroformylation of the two substrates dissolved in organic solvent and incubated under hydroformylating conditions in the presence of Wilkinson's catalyst (Fig. 6). There is clearly no preferential reaction of either substrate. Hydroformylation, using the water-soluble derivative of the catalyst, in a heterogeneous system consisting of substrate codispersed with saturated lecithin in water shows a marked preference for the C-10 alcohol which is hydroformylated at a significantly faster rate than the C-11 alcohol throughout the reaction period. The role of the orientation of the substrate in the phospholipid bilayer in acheiving preferential reactivity was confirmed by perturbing the phospholipid bilayer. This was achieved by addition of an equimolar volume of tetrahydrofuran to the reaction mixture. This procedure serves to "loosen" the bilayer structure without destroying the overall arrangement of the molecules. The rates of reaction of the two substrates were found to be identical under these

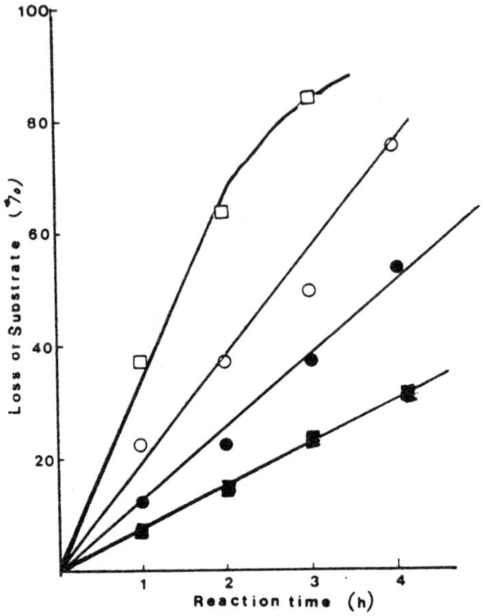

Figure 6. Percent hydroformylation of C-10, ○ and C-11, ● terminal enols codispersed with saturated lecithin in water and reacted with water-soluble derivative of Wilkinson's catalyst. Hydroformylation of both substrates when dispersed with phospholipid in water:tetrahydrofuran (1:1, by vol.) —□— . Hydroformylation of both substrates in the presence of Wilkinson's catalyst in a solvent of benzene, —■— , is also shown. Data from ref [20].

conditions as shown in Fig. 6.

Similar selectivity in reaction between the C-10 and C-11 alcohols were also observed when the reaction mixture was incubated under hydrogenating conditions (Fig. 7). Again it can be seen that there is a marked preference for hydrogenation of the C-10 substrate compared with the C-11 alcohol. Disturbance of the bilayer structure, as in the case of hydroformylation, by addition of an equal volume of tetrahydrofuran causes an increase in the overall reaction rate and loss of selectivity between the two substrates.

In both hydroformylation and hydrogenation reactions described above, the reaction temperature plays a crucial role in selectivity of the reaction. The temperatures at which the reactions illustrated in Figs. 6 and 7 were performed was lower than the gel to liquid-crystalline phase transition of the phospholipid vector. This means that the hydrocarbon chains of the phospholipid were in a close-packed gel phase configuration. If the reaction were performed at temperatures above the phase transition of the phospholipid the selectivity of the reaction between the C-10 and the C-11 alcohols was

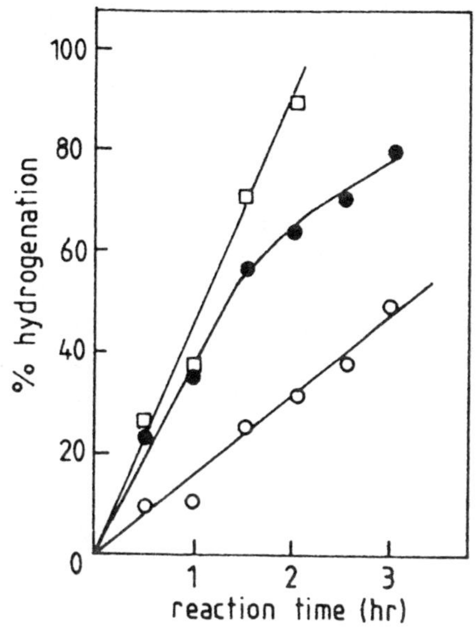

Figure 7. Percent hydrogenation of C-10, ● and C-11, ○ terminal olefins codispersed with saturated lecithin and reacted in the presence of water-soluble derivative of Wilkinson's catalyst. Reaction performed in aqueous medium, ——, and a medium consisting of water:tetrahydrofuran (1:1, by vol.) —□— . Data from ref [20]

again lost. This suggests that the highly ordered gel phase provides the appropriate orientation of the substrate with respect to the catalyst that is responsible for selectivity. When this is lost as the hydrocarbon chains of the phospholipid become disordered in the liquid-crystalline state the relative orientations of the two substrates become equivalent. This requirement for mutual orientation of substrate in gel phase phospholipid and catalyst in the aqueous phase to achieve preferential reaction is consistent with results obtained for reactions performed under identical conditions but using Wilkinson's catalyst instead of the water-soluble derivative. In this case the catalyst partitions into the phospholipid bilayer and is able to orient about the C-10 and C-11 substrates with equal facility. No preferential reaction was observed irrespective of whether the phospholipid was in a gel or liquid-crystal phase.

The simplest intepretation of these observations is that the phospholipid bilayer serves to orient the substrate molecules with the polar hydroxyl group anchored at the aqueous interface and the alkene chain extending into the hydrophobic interior of the bilayer structure. The catalyst, on the other hand, interacts with the surface of the bilayer with the sulphonate groups restricting its location to the interfacial region. The catalytic surface would be able to penetrate into the bilayer interior where the reactive bonds of the substrate molecules are located. Preferential reaction of the C-10 alken-1-ol could be explained by the fact that the susceptible bond is positioned slightly closer to the bilayer surface compared to the C-11 substrate.

Another consistent feature observed in these reactions is that both hydroformylation and hydrogenation reactions performed on substrates oriented in phospholipid bilayers were invariably faster than when performed in completely homogeneous reaction systems at equivalent substrate and catalyst concentrations. This may be explained by the fact that both substrates and catalyst partition from water and associate with the phospholipid bilayer where they become concentrated and oriented in a manner favourable for reaction to take place. In this respect, the phospholipid bilayer serves in a way not unlike enzymes which bind substrates (and cofactors) with high affinity thereby concentrating them within the system and bringing about mutual orientations that are conducive to reaction.

8. Conclusions

Specificity of homogeneous catalysis can be achieved by performing reactions in heterogeneous systems. In such systems, the substrate and catalyst may be bought together in a way that achieves selectivity of reaction. Synthetic phospholipid analogues of biological phospholipids form highly ordered structures in aqueous systems. Codispersion of substrate molecules induces a highly ordered presentation of the substrates with precise orientation and location with respect to the lipid-water interface. Preferential reaction in the presence of water-soluble homogeneous catalysts can be achieved in such systems depending on the distance of the reactive bond from the aqueous-phospholipid interface. Further refinements in terms of physical presentation of substrates and catalysts may result from formation of specific complexes between phospholipids and substrates. Further studies are required to explore the potential of such systems in achieving selective reactions in the presence of water-soluble homogeneous catalysts.

Acknowledgements: The studies described were originally inspired by a fruitful collaboration with Dennis Chapman and later by the skill of Ferenc Joo and Laszlo Vigh in synthesis and application, respectivly, of homogeneous catalysts.

9. References

1. Quinn, P.J. (1989) Membrane lipid phase behaviour and lipid-protein interactions, *Subcell. Biochem.* **8**, 25-95.
2. Quinn, P.J. and Chapman, D. (1980) The dynamics of membrane structure *Crit. Rev. Biochem.* **8**, 1-117.
3. Chapman, D. and Quinn, P.J. (1976) A method for the modulation of membrane fluidity: homogeneous catalytic hydrogenation of phospholipids and phospholipid-water model biomembranes *Proc. Nat. Acad. Sci., U.S.A.* 73, 3971-3975.
4. Joo, F. and Beck, M.T. (1973) Redox reactions of phosphine complexes in aqueous solution. New possibilities in homogeneous solution catalysis *Magy. Kem. Foly.* **79**, 189-191.
5. Borowski, A.F., Cole-Hamilton, D.J. and Wilkinson, G. (1978) Water-soluble transition metal phosphine complexes and their use in two-phase catalytic reactions with olefins *Nouv. J. Chim.* 2, 137-144.
6. Oishi, S. (1987) A water-soluble Wilkinson's complex as homogeneous catalyst for the photochemical reduction of water *J. Mol. Catal.* 39, 225-232.
7. Salvesen, B. and Bjerrum, J. (1962) Studies on metal ion coordination in solution *Acta Chem. Scand.* **16**, 735-741.
8. Fontal, B., Orlewski, J., Santini, C.G. and Basset, J.M. (1986) Synthesis of a new family of water-soluble clusters: $Ru_3(CO)_{12-x}L_x$ (x=1-3), $Os_3(CO)_{10}L_2$ and $r_4(CO)_9L_3$ (L=trisodium salt(hydrated) of tris(m-sulphonatophenyl)phosphine) *Inorg. Chem.* 25, 4320-4322.
9. Kuntz, E.G. (1987) Homogeneous catalysis in water *Chemtech.* **17**, 570-575.
10. Wilson, M.E., Nuzzo, R.G. and Whitesides, G.M. (1978) Bis(2-diphenylphosphinoethyl)amine - flexible synthesis of chelatine diphosphines *J. Amer. Chem. Soc.* **100**, 2269-2270.
11. Farin, F., van Gaal, H.L.M., Bonting, S.L. and Daemen, F.J.M. (1982) Studies on the catalytic hydrogenation of biomembranes. Hydrogenation of phosphatidylcholine liposomes by two water-soluble rhodium-phosphine catalysts *Biochim. Biophys. Acta* 711, 336-344.
12. Madden, T.D. and Quinn, P.J. (1978) Modulation of model biomembrane fluidity by hydrogenation with an amphiphilic catalyst *Biochem. Soc. Trans.* **6**, 1345-1347.

13. Vigo, C., Goni, F.M., Quinn, P.J. and Chapman, D. (1978) The modulation of membrane fluidity by hydrogenation processes II Homogeneous catalysis and model biomembranes *Biochim. Biophys. Acta* **508**, 1-14.

14. Brown, G.M., Brunschwig, B.S., Creutz, C., Endicott, J.F. and Sutin, N. (1975) Homogeneous catalysis of the photo-reduction of water by visible-light. Mediation by a tris(2,2'-bipyridine)ruthenium(II)-cobalt(II) macrocycle complex *J. Amer. Chem. Soc.* **101**, 1298-1300.

15. Madden, T.D., Peel, W.E., Quinn, P.J. and Chapman, D. (1980) The modulation of membrane fluidity by hydrogenation processes IV Homogeneous catalysis of liposomes using a water-soluble catalyst *J. Biochem. Biophys. Meth.* **2**, 19-27.

16. Koynova, R.D., Tenchov, B.G., Quinn, P.J. and Laggner, P. (1988) Structure and phase behaviour of hydrated mixtures of L-dipalmitoylphosphatidylcholine and palmitic acid. Correlation between structural rearrangements, specific volume changes and endothermic events *Chem. Phys. Lipids* **48**, 205-214.

17. Heimburg, T., Wurz, U. and Marsh, D. (1992) Binary phase diagram of hydrated dimyristoylglycerol-dimyristoylphosphatidylcholine mixtures *Biophys. J.* **63**, 1369-1378.

18. Lopez-Garcia, F., Villalain, J., Gomez-Fernandez, J.C. and Quinn, P.J. (1994) The phase behaviour of mixed aqueous dispersions of dipalmitoyl derivatives of phosphatidylcholine and diacylglycerol *Biophys. J.* **66**, 1991-2004.

19. Quinn, P.J., Takahashi, H. and Hatta, I. (1994) Characterisation of complexes formed in fully hydrated dispersions of dipalmitoyl derivatives of phosphatidylcholine and diacylglycerol *submitted*

20. Quinn, P.J. and Taylor, C.E. (1981) Selective homogeneous catalysis in heterogeneous reaction systems: phospholipid bilayers in water *J. Mol. Catal.* 13, 389-396.

MODIFICATION OF BIOMEMBRANES BY CATALYTIC HYDROGENATION

L. NÁDASDI, É. PAPP and F. JOÓ
Institute of Physical Chemistry, Lajos Kossuth University
Debrecen 10, P.O. Box 7, H-4010 Hungary

1. Introduction

The integrity and function of biomembranes critically depends on their fluidity which in turn is largely, though not entirely, determined by the fatty acid composition of polar lipid constituents of the membrane. Heterogeneous catalytic hydrogenation has been widely used for modification the fatty acid composition of fats and oils. However only with the introduction of homogeneous catalysis did hydrogenation become a rather general, usefool tool of membrane biochemistry. Homogeneously catalyzed hydrogenation was applied for the study of the most widely differring organisms, from algae to human peripheric lymphocytes, from plant protoplasts to rat liver mitochondria and gave information on the most diverse membrane associated processes in isolated organelles or in living whole cells [1,2].

In case of each and every organism the conditions of hydrogenation must be carefully chosen as to provide acceptable hydrogenation rate without being detrimental for the cells. A thorough knowledge of the hydrogenation mechanism is therefore essential. Since unsaturated lipids are found together with other reducible substrates (carotenoids, quinones) chemoselectivity towards isolated olefins is also required.

The unsaturated fatty acyl residues can be found in highly organized and spatially well separated compartments of the cell. In most cases it is highly desirable to achieve topologically selective hydrogenation *i.e.* discrimination between such compartments.

For the above reasons we studied the mechanism of hydrogenations catalyzed by the water soluble $RhCl(PTA)_3$ complex (PTA=1,3,5-triaza-7-phosphaadamantane) and the use of $Pd(QS)_2$ (QS=alizarin red) supported on anion exchange materials.

I. T. Hórváth and F. Joó (eds.), Aqueous Organometallic Chemistry and Catalysis, 305–311.

2. Hydrogenations with $RhCl(PTA)_3$ in aqueous solutions

Recently it has been found that $RhCl(PTA)_3$ is an active catalyst for the selective reduction of unsaturated to saturated aldehydes using an aqueous HCOONa solution as hydrogen source [3]. Hydrogenation of several other substrates is also catalyzed by this complex under mild conditions (Table 1).

TABLE 1. Hydrogenation of unsaturated fatty acids catalyzed by $RhCl(PTA)_3$

Substrate	Rate (turnover/hour)
Maleic acid	69
Fumaric acid	342
Crotonic acid	331
Itaconic acid	453

$[Rh]=4.9.10^{-4}$ M, [substrate]=0.045 M, T=37°C, $P_{total}=1$ bar, pH=4.70

An interesting feature of these hydrogenations is in that fumaric acid is reduced approximately five times faster than maleic acid, contrary to the behaviour of Wilkinson's catalyst, which, in organic solvents, is considerably more active for hydrogenation of *cis* than for *trans*-olefins. In this respect $RhCl(PTA)_3$ resembles $RhCl(TPPMS)_3$ (TPPMS=*mono*sulfonated triphenylphosphine) [4].

Detailed kinetic investigations show that the rate of crotonic acid hydrogenation is first order in the catalyst concentration, levels off with crotonic acid concentration and varies in a slighlty non-linear way with the partial pressure of dihydrogen. Excess of PTA decreases the rate considerably, however the reaction is not completely inhibited even at phosphine concentrations.

All these findings could be interpreted in terms of a classical dihydridic mechanism of hydrogenations catalyzed by Wilkinson type rhodium phosphine complexes (CA=crotonic acid, BA=butyric acid):

$$RhCl(PTA)_3 + H_2 \rightleftharpoons (H)_2RhCl(PTA)_2 + PTA \qquad (1)$$
$$(H)_2RhCl(PTA)_2 + CA \rightleftharpoons (HCA)(H)RhCl(PTA)_2 \qquad (2)$$

$$(HCA)(H)RhCl(PTA)_2 + H_2 \rightleftharpoons (H)_2RhCl(PTA)_2 + BA \qquad (3)$$

Although this simplified scheme does not show all possible elementary steps, it is obvious, that the overall rate of such a reaction sequence should be independent of pH since no proton is involved in any of the consecutive reactions.

However, the hydrogenation of crotonic acid is very strongly influenced by the pH. At pH 4.7 a sharp maximum rate is observed not only with crotonic acid but in the hydrogenation of the non-ionizable substrate: allyl alcohol, too (Table 2).

TABLE 2. Effect of pH on the hydrogenation rate (turnover/hour) of crotonic acid and allyl alcohol catalyzed by $RhCl(PTA)_3$ [5]

pH	Crotonic acid	Allyl alcohol
4.1	242.6	111.5
4.6	275.4	137.7
4.7	289.0	172.3
4.9	260.3	140,9
5.2	252.8	138.5

$[Rh]=4.9.10^{-4}M$, $[CA]=0.017$ M, $[AA]=0.045$ M, $T=37°C$, $P_{total}=1$ bar

It has already been established by pH-static measurements that in aqueous solutions the reaction of $RhCl(PTA)_3$ and H_2 results in proton liberation [3]:

$$RhCl(PTA)_3 + H_2 \rightleftharpoons HRh(PTA)_3 + Cl^- + H^+ \qquad (4)$$

Considering this proton production and the pH-dependence of the catalyzed hydrogenations we regard it highly probable that hydrogenations proceed partly, if not entirely, on a monohydridic pathway. A further support of this suggestion comes from the fact that when hydrogenations are carried out with H_2 in D_2O the products are extensively deuterated. In case of α-acetamido-cinnamic acid this *monodeuteration* is complete and 100% selective towards the α-carbon atom. Deuteration of the fatty acyl chains in membrane lipids coupled with FTIR detection is an emerging technique of determining gel to liquid crystalline phase transitions of biomembranes [6] and $RhCl(PTA)_3$ holds promise for such applications.

Both the dihydridic and monohydridic mechanisms give rise to the reversible formation of an intermediate rhodium-alkyl species [*e.g.* reaction (2)] giving a chance of olefin isomerization. Indeed, when hydrogenations were carried out using 5 v/v% H_2 in Ar, and the reactions were stopped at low conversions 14-17% fumaric acid was detected by HPLC (Table 3).

Table 3. Isomerization of maleic to fumaric acid catalyzed by $RhCl(PTA)_3$

Time (min)	Fumaric acid (%)
5	16.2
10	17.4
20	14.2

$[Rh]=4.9.10^{-4}$ M, $[MA]_0=0.045$ M, T=37°C, pH=5.30, 5%(v/v) H_2 in Ar, $P_{total}=1$ bar

Finally, $RhCl(PTA)_3$ was succesfully applied for the hydrogenation of dioleoylphosphatidylcholine (DOPC) liposomes (Table 4). Under mild conditions, at the optimum pH of hydrogenations with this catalyst the Δ^9-double bond was hydrogenated with biologically acceptable rate (see also [7]).

TABLE 4. Hydrogenation of dioleoylphosphatidylcholine with $RhCl(PTA)_3$

Time (min)	Fatty acids (%)	
	18:0	18:1
0	0	100
2	49	51
5	81	19

Conditions: 2.8 mg DOPC in 20 ml Na-phosphate buffer (pH=4.7), 5 mg $RhCl(PTA)_3$, T=37°C, $P_{total}=1$ bar

It should be mentioned that under these conditions the reaction is closer to being stoichiometric than catalytic. This is the usual case in biomembrane hydrogenations where the substrate to catalyst ratio is generally very low [1].

3. Membrane hydrogenations with supported $Pd(QS)_2$

A major step towards wider utilization of hydrogenation as a technique of biochemistry was the introduction of the palladium complex of sulfonated alizarin, $Pd(QS)_2$ [8].

The most possible mechanism of hydrogenations catalyzed by $[Pd(QS)_2]$ involves the formation of a palladium(II) monohydride complex:

$$[Pd(QS)_2] + H_2 \rightleftharpoons [HPd(QS)_2]^- + H^+ \qquad (5)$$

which rearranges to give the Pd(II) complex of the corresponding anthrasemiquinone radical anion. As a consequence olefin hydrogenations are fast and their rate is practically independent of temperature (characteristic for free radical reactions). Moreover, highly selective monodeuteration can be achieved using D_2O as solvent:

$$[HPd(QS)_2]^- + C{=}C + D^+ \rightarrow CH\text{-}CD + Pd(QS)_2 \qquad (6)$$

As can be assumed from reaction (6), the rate of hydrogenation is strongly influenced by the pH of the medium; the catalyst can be used conveniently between pH 5-9. Hydrogenations by $Pd(QS)_2$ are usually accompanied by extensive isomerization [9].

$Pd(QS)_2$ binds strongly *via* the sulfonate group on commercially available anion exchangers (Dowex, Sephadex, Molselect, *etc.*). These anionites usually carry protonated amino (*e.g.* diethylammonium-ethyl, DEAE) or quaternary ammonium functional groups, which may not be innocent in a liposome or towards the constituents of biomembranes. Therefore covering the residual exchange groups by strongly binding ions (*e.g.* SO_4^{2-}) is recommended. Catalyst samples were prepared by equilibrating an acid/base treated ion exchanger with a solution of the Pd-alizarin red complex previously reacted with H_2 with O_2 (form *B* [8]). The heterogenized $Pd(QS)_2$ catalyst undergoes the same color changes on the surface of the ion exchanger than in homogeneous solutions, therefore the same reaction mechanism and selectivities can be expected as for the dissolved palladium complex.

$Pd(QS)_2$ on Sephadex-QAE (QAE = quaternary ammonium-ethyl) was applied for the hydrogenation of soybean lecithin in aqueous dispersions. At 37°C it actively removed 18:3 (linolenic acid) and most of 18:2 (linoleic acid) in 45 min under 1 bar hydrogen pressure with the concomitant increase in the proportion of 18:1 (oleic acid) and 18:0 (palmitic acid) (Table 5).

Table 5. Hydrogenation of soybean lecithin with $Pd(QS)_2$ heterogenized on QAE-Sephadex anion exchanger

Time (min)	Fatty acids (%)		
	18:0	18:1	18:2
0	10.2	27.2	62.6
15	37.3	44.8	17.9
45	40.3	54.5	5.2
60	49.3	50.7	-

Conditions: 1 mg lipid in 5 ml Na-phosphate buffer (pH=7), 3 mg $Pd(QS)_2$ on 100 mg QAE-Sephadex, T=37°C, P_{total}=1 bar

Following the reaction the catalyst can be easily collected by filtration or centrifugation. However, the lipid adheres extensively on the surface of the supported catalyst. Using pentadecanoic acid (15:0) as internal standard it was shown that up to 95% of the substrate was bound onto the beads when the latter was removed from the hydrogenation mixture by centrifugation. Experiments are in progress to modify the adhesion properties of the supported catalyst.

4. Conclusions

Hydrogenation of lipids either in form of aqueous dispersions (liposomes) or as constituents of biomembranes (including those of living cells) can be achieved under biologically acceptable conditions. Both $RhCl(PTA)_3$ and $Pd(QS)_2$ act *via* a monohydridic mechanism and offer possibility for deuteration of membrane lipids in D_2O. $Pd(QS)_2$ heterogenized on ion exchange materials is useful in heterogeneous hydrogenation of liposomes under ambient conditions.

5. Acknowledgement

We are grateful for financial support provided by the Hungarian National Research Foundation (OTKA Grant 1699) as well as by the Ministry of Culture and Education (MKM 209/92). Thanks are due to Albright and Wilson Ltd. for a gift of tetrahydroxyphosphonium chloride and to Johnson Matthey Ltd for a loan of $(NH_4)_2PdCl_4$. Funds to buy an HPLC facility was provided by the TEMPUS scheme of EU (2237/90). L.Nádasdi thanks the Universitas

Foundation (Kereskedelmi Bank Rt) for a stipend to support his work.

6. References

1. Quinn, P.J., Joó, F. and Vígh, L. (1989) The role of unsaturated lipids in membrane structure and stability, *Prog. Biophys. molec. Biol.*, **53**, 71-103
2. Maresca, B. and Cossins, A.R. (1993) Fatty feedback and fluidity, *Nature*, **365**, 606-607
3. Darensbourg, D.J., Stafford, N.W., Joó, F. and Reibenspies, J.H. (1995) Water-soluble organometallic compounds. 5. The regio-selective catalytic hydrogenation of unsaturated aldehydes to saturated aldehydes in an aqueous two-phase solvent system using 1,3,5-triaza-7-phosphaadamantane complexes of rhodium, *J. Organometal. Chem.*, accepted for publication
4. Joó, F., Csiba, P. and Bényei, A. (1993) Effect of water on the mechanism of hydrogenations catalyzed by rhodium phosphine complexes, *J.C.S. Chem. Commun.*, 1602-1604
5. For a more detailed discussion of pH-effects on the rate of aqueous hydrogenations see: Joó, F., Nádasdi, L., Bényei, A., Csiba, P. and Kathó, Á. (1995) Aqueous organometallic chemistry. Effects of a reactive solvent, *This Volume*
6. Török, Zs., Szalontai, B., Joó, F., Wistrom, C. A. and Vígh, L. (1993) Homogeneous catalytic deuteration of fatty acyl chains as a tool to detect lipid phase transitions in specific membrane domains: a Fourier transform infrared spectroscopic study, *Biochem. Biophys. Res. Commun.*, **192**, 518-524
7. Vígh, L. (1995) Catalytic hydrogenation of biological membranes: the effects on membrane physical state and stability, *This Volume*
8. Joó, F., Balogh, N., Horváth, L. I., Filep, Gy., Horváth, I. and Vígh, L. (1991) Complex hydrogenation/oxidation reactions of the water-soluble hydrogenation catalyst palladium di(sodium alizarinmonosulfonate) and details of homogeneous hydrogenation of lipids in isolated biomembranes and living cells, *Anal. Biochem.*, **194**, 34-40
9. Vígh, L., Horváth, I., Joó, F. and Thompson, Jr., G.A. (1987) The hydrogenation of phospholipid-bound fatty acids by a homogeneous, water-soluble palladium catalyst, *Biochim. Biophys. Acta*, **921**, 167-174

SUBJECT INDEX

S

T

U

V

W

AUTHOR INDEX